MW00715538

Survival

Dinosaurs survived for about 165 million years.

Homo sapiens, about 300,000 so far.

Can we do better than the dinosaurs?

Survival

Evolutionary Rules for Intelligent Species Survival

SAMUEL LAYNE

MAIJAI PRESS

PO Box 754220, Forest Hills, NY 11375
www.maijaipress.com
info@maijaipress.com

First published in hardcover, softcover and ebook in 2020

For information about special discounts for bulk purchases,
please contact info@Maijaipress.com.

Publisher's Cataloging-in-Publication Data:
Names: Layne Samuel, author.
Title: Survival: evolutionary rules for intelligent species survival/ Samuel Layne -
Includes bibliographical references and index.
Identifiers: LCCN 2019905817 | ISBN 978-1-7337555-1-1 (Hardcover) |
ISBN 978-1-7337555-0-4 (Paperback) | ISBN 978-1-7337555-2-8 (ebook)

Subjects: LCSH: Human evolution. | Human beings—Origin. | Human beings—Effect of climate on. | Climatic changes. | Mass extinctions. | Evolution (Biology). | Extinction (Biology) | Nature—Effect of human beings on. | Environmental degradation. | Technology and civilization—History. | Human ecology. | Social evolution. | Human beings—Forecasting. | Human beings—Migrations. | Sustainability. | BISAC SCIENCE / Global Warming & Climate Change | SCIENCE / Life Sciences / Evolution
Classification: LCC GF71. L37 2019 | DDC 599.938--dc23

Illustrations & images by Sherry Wang
Book cover design by The Frontispiece
Interior design by Firewire Creative

This book has been composed in Granjon LT Std. and ITC Avante Garde Gothic
∞ Printed on acid-free paper.

10 9 8 7 6 5 4 3 2 1
First Edition
Survival 01 ms x 01 04 2020 05 14 na 02 c
Printed in the United States of America

To David

Survival
Unique and
impossible to forget,
You brillant
were
yet gentle, very
private
yet charming,
intelligent,
mischievous
and a very
Kind
person.
May your
be as Spirit
and Free
unrestrained as
You LONGED FOR
Life to be.
We will
Love always
and
Remember
You.
Like the
World
you
were a
beautiful
troubled Mystery
of Light
darkness
and
Love.

ACKNOWLEDGEMENTS

THIS BOOK WOULD NOT BE POSSIBLE without the guidance and help of the publishing industry professionals with whom I collaborated.

I value the time a number of people have spent reading different parts of the draft manuscript at various stages, and appreciate their wise suggestions about style and content, but especially for catching some glaring misunderstandings and errors of fact.

The appearance and message of the book have been enhanced immeasurably by use of the NASA *Earthrise* and other photos, and I wish to thank NASA for permitting its use.

Finally, I must thank my wife for designing and illustrating numerous tables and diagrams, as well as, her patience and untiring efforts working with me through all production stages to transform the manuscript into this book.

TABLE OF CONTENTS

INTRODUCTION

The Struggle for Life and Survival

Can Homo sapiens survive or is extinction inevitable?

xv

PART ONE

EVOLUTION VERSUS THE UNIVERSE AND THE STRUGGLE FOR LIFE AND SURVIVAL

CHAPTER 1

Earth

The Planet Evolution Terraformed for Life and Survival

5

CHAPTER 2

Does the Universe Come with Evolutionary Species Survival Mechanisms?

Are There Evolutionary Rules for Intelligent Species Survival?

17

CHAPTER 3

Is Species Survival the Default Evolutionary Outcome?

33

CHAPTER 4

An Introduction to Intelligent Species Survival Patterns

51

PART TWO

THE MAKING OF AN INTELLIGENT SPECIES

FROM SLIME TO APE-MAN TO SPACEMAN

CHAPTER 5

Species-Driven Evolution
A Grand Evolutionary Experiment

Did Evolution Deal Intelligent Species a Hand in Its Own Evolution?

67

CHAPTER 6

The Role of Climate Change in the Making of an Intelligent Species

Evolution's Tool to Prod Intelligent Species to Greater Maturity and Survival or Extinction

91

PART THREE

EVOLUTIONARY RULES FOR INTELLIGENT SPECIES SURVIVAL

CHAPTER 7

Habitats

The Conditions Necessary for the Evolution and Sustainment of Life

133

CHAPTER 8

Sustainability

Habitat Survival Drives Species Survival Drives Habitat Survival

183

CHAPTER 9

Adaptation

Must be Survival-driven to Preserve Habitats and Biodiversity

213

CHAPTER 10

Innovation

From Ape-man to Spaceman in Three Million Years—to . . . ?

253

CHAPTER 11

Migration

A Built-in Evolutionary Survival Response

To Survive, an Intelligent Species
Must be Free and Able to Migrate

309

PART FOUR

SURVIVAL

THE REASON INTELLIGENT SPECIES EVOLVED

AN EVOLUTIONARY GAME OF LIFE

CHAPTER 12

Survival

The Reason Intelligent Species Evolved

The Default Outcome in the Evolutionary Game of Life

339

CHAPTER 13

Evolutionary Maturity

The Only Path to Intelligent Species Survival

367

CHAPTER 14

Extinction

A Human Universe without Humans

405

PART FIVE

GOING BACK TO THE PAST TO SAVE THE FUTURE

FROM CIVILIZATION TO ECOLIZATION

CHAPTER 15

Freedom to Choose

Survival vs. Extinction

Has Homo Sapiens Lost the Ability to Choose?

431

CHAPTER 16

From Civilization to Ecolization

Going Back to the Past to Save the Future

481

GLOSSARY

515

NOTES & REFERENCES

531

LIST OF PHOTOS & CHARTS

545

LIST OF TABLES

549

INDEX

551

Introduction

THE STRUGGLE FOR LIFE AND SURVIVAL

Can Homo sapiens survive or is extinction inevitable?

AFTER MORE THAN 4 BILLION YEARS AND MULTIPLE mass extinctions, intelligent species finally evolved on Earth. What prompted this? Was it inevitable? There are many opinions, but the only explanation that should really matter is the one that can be found in the history of evolution's struggle to maintain life in a universe ensnared in a perpetual cycle of birth, life, and death.

From an evolutionary perspective, it seems as if the entire reason for the existence and evolutionary journey of an intelligent species is to mature past the point of self-extinction and become capable of surviving the universe's existential threats. Put simply, to keep life alive. This book explores the possibility that intelligent species emergence was an evolutionary response to the universe's endless propensity for destroying life, and, the roles of climate change and Homo sapiens, in that response.

Homo sapiens evolved in a universe in the grip of an endless cycle of birth, life, and death. Notwithstanding, evolution found a way to perpetuate life by evolving species capable of reproducing themselves. Evolution has managed to get life going on Planet Earth and has kept it going ever since, repeatedly jump-starting life in the face of numerous mass extinctions. Yet, after all of this effort to evolve and sustain life, species survival seems to remain at the mercies of the universe. Even after surviving for millions of years, species like the dinosaurs could do nothing to avoid extinction at the hands of this Universe.

That colossal meteorite impact that put an end to more than half of the Earth's species, including all non-avian dinosaurs, was felt across the evolutionary species universe and may have changed the course of evolution forever. It certainly got evolution's attention. It's possible that the evolution of an intelligent species that can choose its own fate—survival or extinction—might have been evolution's response to the inability of otherwise-successful species' to save themselves when confronted with such random galactic, solar, or planetary scale existential threats.

It is unsurprising that it wasn't long, in geologic time, after the elimination of the dinosaurs—widely believed to have made the rise of mammals possible—that intelligent species, in the genus *Homo*, including hominins and eventually Homo sapiens, entered the picture. About 6 million years before Homo sapiens showed up, evolution began grooming and winnowing a clutch of emerging hominin species. Most went extinct, but some survived and eventually gave rise to what we now know as Homo sapiens. One evolutionary question might have been whether one of these intelligent species would survive past the point of self-extinction to evolutionary maturity, be able to migrate from planet to planet and star to star and, eventually, perhaps to other galaxies if necessary to survive.

No matter what one might believe about why we are here, then, intelligent species evolution might be nothing more or nothing less than evolution's response to the universe's existential threats to the survival of life. Evolution, it seems, wanted to find a way for life to survive, whatever the universe might throw at it, and the evolution of mature intelligent species might well have been the answer—so far.

So, more than 6 million years ago, evolution set out to evolve a new kind of animal species that would be free to choose its own evolutionary destiny. A species that would in time evolve the capability to transcend and largely replace the instincts that control the behavior of all other animal species. A species with an emergent intelligence that would enable it to evaluate potential outcomes and make choices to shape its own evolutionary outcome in ways no prior species could. A species that could make choices so far-reaching that it could potentially, perhaps inadvertently, bring about its own extinction and the extinction of numerous other species, or instead make choices to mature beyond the point of self-extinction, to evolutionary maturity.

No prior species could do anything other than follow the integral biological routines endlessly executing in its brain; none could reprogram itself or do anything beyond its natural instincts. But with this fixed programming also came an advantage.

Such species never had to ask, "Why am I here?" or wonder about their purpose in life; they lacked the intellectual tools to do so, and that purpose was hardwired inside them. Every area of their lives had been preprogrammed. Nor were such species going to come up with the means to drive themselves or other species to extinction: they would never develop the ability to change Earth's atmosphere, precipitate global warming, poison the food chain, or hunt or fish other species to extinction. They could never enslave others of their own species nor wage global warfare against them. Such species had no need to grow to evolutionary maturity as they evolved; they were as fully matured as they would ever be.

Homo sapiens, though, would need to reach evolutionary maturity just to survive and avoid self-extinction. The attributes that enabled an intelligent species to transcend the limitations of prior species—intelligence and the freedom to choose its own evolutionary destiny—were simultaneously its greatest strengths and its greatest weaknesses.

With intelligence and the freedom to choose came the awesome responsibility to choose to grow to full evolutionary maturity—a challenge no prior species had ever had to face. Like human infants, intelligent species require guidance and time to grow into full evolutionary maturity, or, like immature adult humans, they will self-destruct and take themselves out and potentially all other species with them, and that, it would seem, is the survival struggle twenty-first-century Homo sapiens now confront.

Whether we are aware of it or not, our species, Homo sapiens, is engaged in an evolutionary battle for its very survival, and the enemy it faces is within. Homo sapiens' freedom to choose its own evolutionary outcome, and the intelligence that comes with it, can be used to either reach that level of evolutionary maturity that other species begin with, or to imagine, innovate, and create the means of its own extinction. The evolutionary pattern at work here is uncomplicated, and innovation has little if anything to do with it. Notwithstanding its ability to adapt and innovate, an intelligent species must reach evolutionary maturity, or it will go extinct. No maturity, no survival. It's that simple.

The greatest threat to Homo sapiens' survival is not a meteorite impact, not a death star, not WMDs (weapons of mass destruction), and perhaps not even AI (artificial intelligence). Though possible, most of these are unlikely in the short to intermediate term, and others, such as WMDs and AI, are but symptoms of the real threat. The real threat, the clear and present danger, is Homo sapiens itself.

But alas! The search for evolutionary maturity and survival is not exactly a topic that preoccupies Homo sapiens today, and therein lies the problem. This may be due to a lack of a true understanding of Homo sapiens' reason for being and potential role in the greater evolutionary scheme for life in the universe. It may also be due to not understanding the dire consequences of getting the relationship between its survival and its maturity wrong.

Homo sapiens appears to have conjured up its own reasons for being and missed those landmarks that might have led it in a direction more consistent with survival and evolutionary maturity. Today humans have arrived at a point where a handful of men are literally capable of extinguishing our species.

Fortunately, it does not appear that evolution has left this intelligent species to fend for itself. It may have provided numerous survival rules and patterns, assembled here into a survival manual, a Species Survival Maturity Model, to show the way, a kind of yellow brick road that runs past the point of self-extinction and on to evolutionary maturity and survival. In the following chapters I attempt to delineate that yellow brick road and identify the landmarks along the way, making it easier to follow in the hope that our species may yet notice and change direction before it's too late.

This is a book about survival. It asserts that in a universe ensnared in endless repeating cycles of birth, life, and death, evolution appears to be firmly on the side of life and intelligent species survival, and, consequently, extinction is not inevitable.

It explores the backstory—why, from an evolutionary perspective, an intelligent species like Homo sapiens may have evolved in the first place and what it needs to do to continue to survive in such a universe. It explains the yin and yang relationship between intelligent species' survival and its attainment of evolutionary maturity.

It depicts the struggle to mature past the point of self-extinction to reach evolutionary maturity in terms of an evolutionary survival game of life that may well have begun more than 5 million years ago with the evolution of the genus *Homo*, possibly Earth's first intelligent species and forebears of Homo sapiens.

It identifies what appear to be survival rules and patterns that evolution may have strewn like Hansel and Gretel's bread crumbs, instructions for intelligent species to follow in order to play to win and thereby survive, and it organizes these into a Species Survival Maturity Model.

It shows the potential link between reaching evolutionary maturity and developing the abilities necessary to survive existential threats such as those that took out the dinosaurs and achieving interplanetary migration.

It concludes with a peek at where Homo sapiens' evolution could be headed (extraterrestrial mature intelligent species evolution) should it manage to mature past the point of self-extinction. It also looks at the potential opportunities Homo sapiens might well have to collaborate with evolution in the spread of mature intelligent species life across the galaxy.

And, finally, it considers what could be in store should Homo sapiens fail to mature. It explores which of these two outcomes Homo sapiens is likely to choose and why: maturing past the point of self-extinction to survive, or, failing to mature and going extinct. These are the only possible outcomes in this evolutionary game of life. Choose well.

PART ONE

Evolution versus the Universe and the Struggle for Life and Survival

LIFE AND SURVIVAL VERSUS DEATH AND EXTINCTION

WHEN IT COMES DOWN TO THE QUESTION of species extinction versus species survival, evolution appears to have always sided with species survival, endowing species with the ability to survive and persist through reproduction and to preserve their survival through migration.

Viewed from this perspective, it begins to look like an interplay, perhaps even "co-opetition," between the universal forces of birth, life, and death on one hand and evolution on the other, with life and survival as the prize. When this interplay and delicate dance is traced back through Earth's history, it begins to seem unbelievable that life somehow survived being handed off, back and forth, from one of these galactic forces to the other.

It is even more amazing that order was lurking deep inside the incomprehensible chaos of the universe and the early days of the Earth. Despite seeming like a purposeless process, evolution comes across as a type of order that brought life and consciousness into the world, and through evolution, life managed to find a way. Paraphrasing the words of Nobel Prize-winning Austrian physicist Erwin Schrödinger, "Evolution and thereby life sucked order from a universe of disorder."

There is an untold story here of the heroism and persistence of the evolutionary processes, a seemingly mindless, repeating universal pattern that drives order and life in the midst of a universe of seeming chaos, from microbial to complex animal to intelligent species life. Across galactic time scales, possibly across solar systems, galaxies, and universes, through innumerable extinction events on Earth, evolution enabled and sustained life and survival over death and extinction.

From this perspective, one can no longer think of evolution as a detached and disinterested process; instead, it becomes a set of repeating universal processes that create a repeating universal pattern, and in the midst of a universe seemingly consumed by chaotic processes enables order in the form of life and survival over death and extinction.

Evolution versus the Universe and the Struggle for Life and Survival

---

CHAPTER 1

Earth

The Planet Evolution Terraformed for Life and Survival

CHAPTER 2

Does the Universe Come with Evolutionary Species Survival Mechanisms?

Are There Evolutionary Rules for Intelligent Species Survival?

CHAPTER 3

Is Species Survival the Default Evolutionary Outcome?

CHAPTER 4

An Introduction to Intelligent Species Survival Patterns

Fig. 1-1 *Earthrise* photo from Apollo 8

Apollo 8 Astronauts became the first to look back at their home planet and see the entire world in one glimpse. The view they shared had an everlasting impact.

(Source: http://www.nasa.gov/topics/history/features/apollo_8.html)

1.

Earth

THE PLANET EVOLUTION TERRAFORMED FOR LIFE AND SURVIVAL

CAN YOU RECALL HOW IT FELT THE FIRST TIME YOU saw the photo *Earthrise*, taken by astronaut William Anders during the Apollo 8 mission in 1968? According to NASA, "They also became the first to look back at their home planet and see the entire world in one glimpse."

Like seeing oneself in a mirror for the first time, after 300,000 plus years or so evolving to our present state, Homo sapiens had finally discovered Earth. The image of our blue planet far beneath those astronauts has had an everlasting impact on our collective consciousness.

For the first time, we were yanked away from the cares of our individual lives to collectively gaze in awe and wonder at the Earth from the perspective of the moon, each of us slowly realizing that we stood somewhere on that rotating surface along with everyone else, as it receded into insignificance, becoming like a "pale blue dot" barely visible when seen from beyond Pluto, 6 billion kilometers out, as photographed by Voyager 1.

Fig. 1-2 The pale blue dot
(https://en.wikipedia.org/wiki/Pale_Blue_Dot)

We are trapped here, adrift like polar bears driven by erratic currents on a chunk of ice, endlessly rotating around the sun, the lone star in Earth's solar system, one among 300 billion stars in the Milky Way galaxy, itself just one of 200 billion galaxies. And yet, for most of the 300,000 years or so we have been evolving, we hadn't realized how lonely and precarious, and how precious, our existence has been. Judging from where we are heading, we still haven't.

But while it was for us the very first time we had actually seen the Earth floating in the darkness of space, we might not have been the only ones looking. Over the last 4 billion plus years, evolution has been keeping watch over the Earth like a hen over her chicks, getting it ready to evolve and sustain complex animal and intelligent species life. And while we may not have realized how lonely, precarious, and precious our existence has been, evolution would no doubt had known all along.

Fig. 1-3 Polar bears on iceberg
https://www.livescience.com/15129-gallery-polar-bears-arctic-ocean.html

It is amazing how different that pale blue dot seems when viewed from this perspective. Perhaps we aren't alone; perhaps we never were.

During the course of human evolution, we have learned the value of living in communities and the benefits of having neighbors who can lend a helping hand in times of trouble. But where are Earth's neighbors? Do we even have neighbors? Where and how do we even begin to look?

Fig. 1-4 Our neighborhood: The Solar System
Are they neighbors? Is there any other intelligent life in our solar system?

The distances to planets in our solar system are mindboggling. This is a whole new challenge confronting our species. Within the span of 30,000 years, our ancestors travelled from humanity's ancestral home in Africa across continents and oceans until they had migrated to every corner of the planet. But until now, we had never ventured away from its surface to visit another planet.

We have been to the moon, but in astronomical terms, the moon is in our own backyard, and we found no trace of life there, let alone intelligent life. The distance from Earth to the moon on average is about 238,855 miles, or 384,400 kilometers. In contrast, the distance to the sun is about 92,900,000 miles, or 149,600,000 kilometers, or one astronomical unit (AU). For good reason, except for NASA's Parker Solar Probe to study the sun, we have no plans to go there anytime soon, even though we could use some of the fusion energy that powers it.

How about Mars? There's lots of talk, plenty of interest, and even some plans for going to Mars. That's a distance of about 0.52 AU, which at approximately 78,340,000 kilometers, or 48,678,219 miles, is no road trip either.

According to NASA, a manned vessel would take roughly six months to travel to Mars and another six months to travel back. In addition, astronauts would have to stay from eighteen to twenty months on Mars before the planets realigned for a return trip. In all, the mission would take roughly two and a half years. In other words, if we have any neighbors, don't expect to meet them in the near future.

So, though the search continues, we have found no intelligent life in our solar system so far.

For perspective, it's important to remember that the technologies needed to search for intelligent life are less than 200 years old. It was only in 1844 that Samuel Morse sent his first telegraph message, from Washington, D.C., to Baltimore, Maryland; by 1866, just twenty-two years later, a telegraph line had been laid across the Atlantic. Similarly, television was first successfully demonstrated in 1927.

We are so immersed in technology today that it's easy to forget that in less than a single lifetime, we have gone from the Wright Brothers' four brief flights in 1903 to landing men on the moon in 1969. As Brian Cox says in *Human Universe*, Homo sapiens has gone from "ape-man to spaceman" in less than 250,000 years.

It was not until the middle of the twentieth century that we had the computers and radio telescopes that allowed us to begin looking for extraterrestrial intelligent life. Now, after more than sixty years of searching for any sign that there are other intelligent species out there, the SETI (Search for Extraterrestrial Intelligence) program has come up with nothing, not just in our solar system, but also across several star clusters in our galaxy. Nada. But SETI continues to look and listen.

Can this conspicuous absence of other emerging technological species in our solar system—and as far as we know, our galaxy—be purely accidental? And not only have we yet to find another emergent intelligent species elsewhere in our galaxy, there are no other hominin species left on Earth either (no Homo erectus, no Neanderthals, or even Homo floresiensis, the "hobbits" who hung around until about 18,000 years ago). Are these two things purely random, or could they be somehow related? If the evolutionary goal is the survival of at least one emerging intelligent species, then a not-so-subtle hint might be the trouble Homo sapiens have had getting along with each other, let alone other intelligent species.

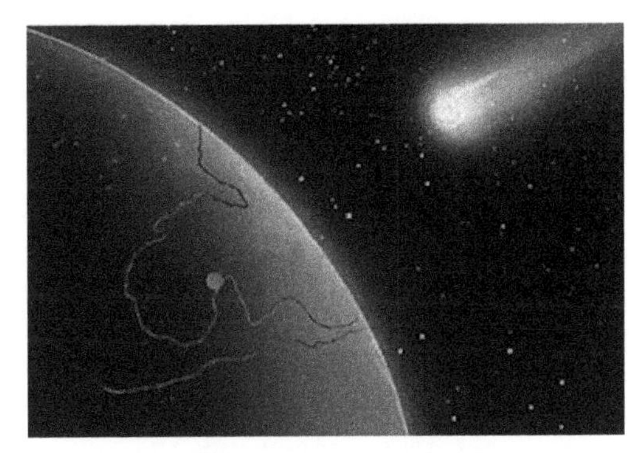

Fig. 1-5 The asteroid approaches Earth
Conceptual images by Mary Parrish, Smithsonian National Museum of Natural History (NMNH).

And is the location of Earth, as a place on which to evolve complex animal and intelligent species life, also accidental? From anyone's perspective, finding a quiet corner of the galaxy to evolve such life forms—as far as possible from the potential galactic mayhem of black holes, gamma-ray bursts, and other agents of death that, however inadvertently, can quickly and unceremoniously snuff out life—seems an eminently reasonable evolutionary choice, even if random.

Imagine what it would have been like for early humans. Imagine, say, you lived somewhere in Canada and suspected other humans might live somewhere in Australia. Imagine, too, a world in which you had no means of transporting yourself from Canada to Australia, nor any means of contacting that neighbor. In fact, if a message somehow got through, there would be no way of knowing it had been received, let alone understood. Finally, consider the odds of communicating if it took fifty years for a message to get to Australia, another fifty for a reply to be received, when your life expectancy was just fifty years!

As an analogy of our current plight, this isn't overstated, and it is definitely not funny. Should something really bad happen in Canada—no, not like the Alberta Tar Sands bonfire, as bad as that was. I mean really bad, bad like what happened during the Mesozoic Era, about 65 million years ago, that left Mexico's Chicxulub impact crater buried underneath the Yucatán Peninsula. Yes, that big, bad asteroid, roughly fifteen kilometers (nine miles) wide that hurtled into Chicxulub Mexico and took out more than 50% of the species on the planet, including the dinosaurs that had been roaming the Earth for about 165 million years.

If something like that were to happen again, and it could (we don't have the technology to deflect a potential killer rock with Earth in its cross hairs), we couldn't expect, say, Vulcans to come to our rescue. If they exist, they would not have found us as yet as we still have not developed warp drives. In fact, we would be like those polar bears floating on a lone chunk of sea ice; no one would be coming to look for survivors, as far as we know. Like the dinosaurs, that would be it for us.

Millions of years later, some subsequent species, perhaps humanoid, perhaps not, might well be heating their homes by burning our fossilized remains, just as we have done with the fossilized remains of other species, dinosaurs among them.

But if not the Vulcans, certainly evolution might have been looking when the dinosaurs were taken out. In fact, that event, as this book suggests, might have been the reason we are here.

So, there might not be neighbors in our solar system to come to our aid, but might there be intelligent, technological species elsewhere in the Milky Way Galaxy, our galaxy? Many scientists think the answer has to be a resounding yes, based on the number of Earth-like planets orbiting their stars in the habitable zone, and knowledge of how intelligent life came into being on Earth. The habitable zone, sometimes referred to as the "Goldilocks zone," is the orbital distance from a star that allows liquid water to exist on a planet's surface.

Scientists now estimate that there are about 300 billion stars in the Milky Way Galaxy; extrapolating from the number of planets detected by the Kepler Space Telescope (Kepler's limited field of view is one-quarter of 1% of the visible galaxy), there might be as many as 100 billion planets orbiting those stars, of which about 50 billion are believed to be rocky planets, like Earth.

In November 2013, based on Kepler space-mission data, astronomers reported that there could be as many as 40 billion Earth-sized planets orbiting their stars in habitable zones in the Milky Way, 11 billion of which may be orbiting sun-like stars.

The following table (courtesy of NASA) shows the number of Earth-like planets discovered through May 2016.

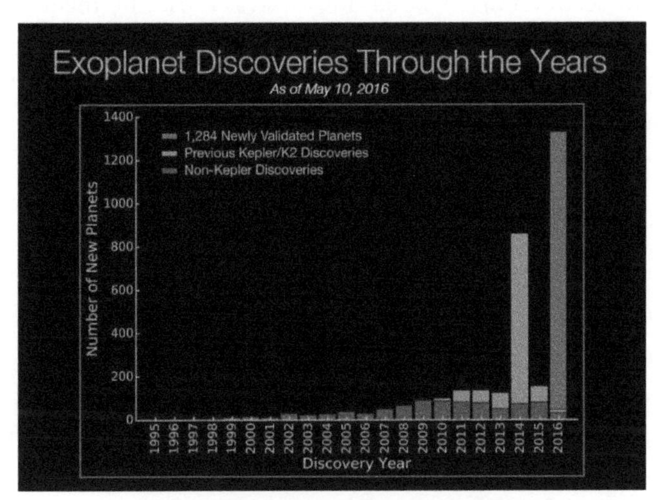

Fig. 1-6 Exoplanets discoveries through the years
(Courtesy of NASA)

Given such a large number of Earth-like planets orbiting sun-like stars in the habitable zone, scientists have concluded that life may have started on some of these planets as well. But what are the conditions that must have existed on Earth for life to evolve?

While the galaxy is thought to be nearly as old as the universe, some 13.8 billion years old, our solar system is thought to have formed only about 4.6 billion years ago. Single-celled life is believed to have sprung into existence on Earth as soon as the conditions made it possible some 3.5 billion years ago. Homo sapiens, the species, has existed for 300,000 years or so, and has had technology with the ability to contact other intelligent species for little more than the last 100 years. Yet to find other intelligent species they must first exist. How long should it take for an intelligent technological species to evolve?

It essentially took about 4 billion years for intelligent life to evolve on Earth. Since many of these stars with Earth-like planets orbiting in the habitable zone are billions of years older than the Earth, then intelligent, technological civilizations could have arisen billions of years ago on at least a few of these. But if that is so, one must ask, as Enrico Fermi did, "Where are they?" (the so-called Fermi Paradox). Why is there no answer to the SETI signals, and why has nobody tried to make contact, as the Vulcans did in the *Star Trek* movie *First Contact*? Where have all these intelligent, technological, communicating civilizations gone?

Some scientists believe that while life may have arisen on many of these exoplanets, intelligent, technological life with the ability to seek out and communicate with others might occur so rarely and last so briefly as to preclude opportunities for contact between intelligent species. Scientists call this the Great Filter theory. It's a notion that's often invoked to explain why we've never been contacted or visited by extraterrestrials—although the UFO community might disagree with that conclusion.

Some astrobiologists theorize that *all* advanced technological civilizations hit an insurmountable developmental wall and self-destruct, a topic this book explores in some detail. But if so, why would an intelligent extraterrestrial species go extinct, and how could that be avoided? What was it about their evolutionary journey that led to their demise? Were they destroyed by, or in spite of, their own technological advancements? Did they make it that far along the evolutionary journey only to be wiped out by a planetary or interstellar catastrophe, as the dinosaurs were even after 165 million to 175 million years of roaming the Earth?

Compared to the dinosaurs, our species, Homo sapiens, has only been around for about a quarter of a million years. Notwithstanding, there are those who argue that Homo sapiens might be the only extant intelligent species, not only in our solar system, but in the entire Milky Way Galaxy. They base this view on how incredibly lucky Homo sapiens seems to have been to evolve at all, but that's a conclusion that is no doubt reached from a purely anthropocentric perspective, and it's one with which evolution might beg to differ.

The sobering question we should all be asking ourselves is: Lucky or not, will humans, who began only 200 years ago to develop the kind of technology that enables a search for other advanced alien species, continue to be lucky and get past this developmental wall, this survival wall, or go extinct?

But does survival of an intelligent species come down to just a series of lucky breaks? Is there really some event that needs to break our way, such as dodging the next dino-destroying-size rock that might be headed in our direction? Are there planetary, solar, and galactic speed bumps that work against getting over that so-called insurmountable developmental wall? Or is there some sort of Species Survival Maturity Model (SSMM), as yet unknown, a path to survival if followed, with which Homo sapiens needs to align its evolutionary journey?

The question this book explores is: Can we—*will* we—survive? While scientists have not yet been able to prove that intelligent, technological, communicating alien species exist on other Earth-like planets, or have already come and gone, we know for certain that the dinosaurs came, hung around on the Earth for 165 million years to 175 million years, and went. Can we do better than the dinosaurs? Is there even a way to know?

Given our current trajectory, many people would agree that it's more likely than not that we'll eventually do ourselves in. Indeed, it's difficult not to be pessimistic when considering Homo sapiens' prospects. For those who think Homo sapiens can survive, comfort can no doubt be found in the words of the late Stephen Hawking, one of the greatest minds, if not the greatest, to come along in a lifetime.

He said: "It is clear that we are just an advanced breed of primates on a minor planet orbiting around a very average star, in the outer suburb of one among a hundred billion galaxies. BUT, ever since the dawn of civilization people have craved for an understanding of the underlying order of the world. There ought to be something very special about the boundary conditions of the universe. And what can be more special than that there is no boundary? And there should be no boundary to human endeavor. We are all different. However bad life may seem, there is always something you can do, and succeed at. While there is life, there is hope."

So, what is it we need to do or change as an emerging intelligent species to scale that insurmountable developmental barrier and survive? If there is such a thing as a Species Survival Maturity Model that can help our species survive, we owe it to ourselves to find it.

Evolution began the incubation of complex animal and intelligent species life here on Earth, a planet in an obscure arm of the Milky Way Galaxy, far from any neighbors, far from its center, out of the way of most of the galactic mayhem, perhaps to give life a better chance at survival. It was going to take evolution most of 4 billion years to prepare Earth for the emergence of possibly its first line of intelligent species in the genus *Homo* 6 million years ago, whence came the line of hominins and, 300,000 years ago, Homo sapiens.

It would seem beyond strange if after going to such lengths, evolution didn't leave intelligent species, if not a user manual, at least a few bread crumbs, like in the Hansel and Gretel fairy tale, to guide the way to survival. Homo sapiens' quest, it would seem, would be to find and follow them all the way to survival. The purpose of this book is to identify them, if they exist.

Going forward we will attempt to identify the evolutionary laws or patterns that affect intelligent species survival, as well as those that hasten their extinction. By reviewing early Homo sapiens' survival history, we hope to see how these patterns and laws facilitated or impeded migration out of Africa. Further on, we will attempt to gather step by step the components, then go on to assemble what may well be an intelligent species survival roadmap—a Species Survival Maturity Model.

2.

Does the Universe Come with Evolutionary Species Survival Mechanisms?

ARE THERE EVOLUTIONARY RULES FOR INTELLIGENT SPECIES SURVIVAL?

THE UNIVERSE OFFERS A VAST RANGE OF LAWS, patterns, and models that enable an understanding of the most massive of multiverses, galaxies, and solar systems and the most minute of cells, atoms, and subatomic and quantum particles. Wouldn't it be surprising if such a universe did not also include laws, patterns, and models that enable an understanding of the processes that drive intelligent species' survival and extinction? Wouldn't it also be surprising if such a universe did not show just as clearly what models, laws, and patterns an intelligent species must follow to escape extinction and ensure its survival?

The universe is understandable because it follows a model. Homo sapiens evolved in a universe that science has shown is governed by a set of laws. In his book *The Big Picture,* cosmologist Sean Carroll says, "A law of physics is a pattern that nature obeys without exception." Given the state of the universe at any specific time, these laws specify how the universe would develop from that moment forward, or its history up to that point in time. Biological processes are also governed by universal laws, the laws of physics, chemistry, and quantum-particle laws, and therefore might be just as deterministic as the orbits of the planets at some levels, and as probabilistic as the laws governing quantum phenomenon at others.

Indeed, according to quantum physics, says Steven Hawking in his book *The Grand Design*, "Nature does not dictate the outcome of any process or experiment, even in the simplest situations. Rather, it allows a number of different eventualities, each with a certain likelihood of being realized. It is, to paraphrase Einstein, as if God throws the dice before deciding the result of every physical process." Hawking continues: "The laws of nature determine the probabilities of various futures and pasts rather than determining the future and the past with certainty."

It is precisely because of the orderliness and predictability that result from these patterns and laws that Homo sapiens can send space probes and satellites to survey the other planets of our solar system or rendezvous with comets. It is because of them that we know with certainty that the next solar eclipse would have occurred on July 2, 2019, and that Halley's comet will again appear in Earth's vicinity in 2061.

Similarly, because of such patterns and laws, biologists are now able to conceive of an ecosystem as a society of organisms interacting with one another through a food chain; doctors can see the body as a collection of organs communicating with one another through the nervous and endocrine systems; and microbiologists can picture life in cells as a society of macromolecules bound together by a complex system of communications regulating both their synthesis and their activity. In the words of Sean Carroll, in his book *The Serengeti Rules*, "Every cell contains a society of molecules, every organ a society of cells, every body a society of organs, and every ecosystem a society of organisms, and understanding the interactions within each of these societies has been the primary aims of molecular biology, physiology, and ecology."

Given that there are rules, patterns, and models that govern the universe and regulate the cells of the body and number and type of animals in an ecosystem, what, then, are the rules, patterns, and models that must also exist to guide an intelligent species to survival? What do nature's probabilities say about intelligent species' survival? Is there such a thing as a Species Survival Maturity Model that would enable an intelligent species such as Homo sapiens to survive if followed? Are there planetary, or even galaxy-wide, patterns that work to accelerate survival, or serve as speed bumps that unintentionally slow or block survival?

If a bacterium, barely visible in a microscope, without a nervous or endocrine system, has been shaped by evolution to "know" how to make the right enzyme for whatever sugar it encounters, wouldn't evolution have endowed an emergent intelligent species with the ability to make the right choices to survive? Yes and no.

Yes, because the evolutionary rules and patterns are more than likely there and will lead an intelligent species to survival if discovered and followed. No, because unlike bacteria that evolved hardwired programming to internalize those rules and patterns, and thus know exactly how to make the right enzyme in response to a specific chemical, an emergent intelligent species has to first discover those rules and patterns (what we now call chemistry and microbiology). It must then go on to write its own mental programs on how to apply them in order to survive.

The existence of these universal natural laws, patterns, and models, which we now know predate intelligent species evolution, is by itself no guarantee they'll be discovered and used to aid intelligent species survival. Just to be able to understand and do what nonintelligent but preprogrammed species do routinely must be anything but easy, as it took evolution 4-billion-plus years to get Earth's first intelligent species from "slime" to ape-man to spaceman, modern Homo sapiens.

A cursory review of the history of science and early intelligent species' discovery and application of these laws documents the emergence of a nascent intelligent species that evolved from an instinct-driven, tree-climbing ape on all fours to a bipedal, erect, intelligent, self-aware spaceman. The journey took several million years, involved the evolution and extinction of multiple hominin species, and the rise and fall of innumerable prior Homo sapiens civilizations. By the look of things, Homo sapiens might still have a long way to go.

For almost the entire 300,000-year existence of our species, Homo sapiens discovered relatively little about the laws of the universe and nature, let alone the evolutionary patterns and laws that can potentially guide intelligent species survival. Paraphrasing Sean Carroll, we gathered fruits, nuts, and plants; hunted and fished for the animals that were available; and like the wildebeest and zebras on the Serengeti, we moved on when the resources ran low. Even after the advent of farming and "civilization," and the development of cities, Homo sapiens remained vulnerable to the whims of the weather, climate change (and still is), and to famine and epidemics. It was only in the last 100 years that knowledge of these universal and natural laws accelerated. During those 300,000 years Homo sapiens was controlled mostly by biology, but in the last 100 or so years, has begun to be controlled more by its intelligence and has taken control, not only of biology, but of most areas that impinge upon Homo sapiens existence. Well, almost.

First came knowledge of the universe and physical laws that regulate how the universe works. Then followed knowledge and awareness of the laws of nature that enabled the move away from hunting and gathering to agriculture, which in turn led to the formation of cities and towns. As knowledge in these and related areas grew, so did our ability to take control of our destiny. This increase in knowledge turbocharged the development of Homo sapiens' civilizations. But even as our knowledge and abilities grew in these areas, knowledge of the evolutionary laws and patterns that can guide an intelligent species to survival appears to have gotten lost along the way.

Modern Homo sapiens has emerged with astonishing capabilities, but the price paid for them is lack of growth in the one area that will matter most for its survival: evolutionary maturity, the ability of an intelligent species to mature beyond the point of self-extinction. A bacterium knows how to make the right enzyme for whatever sugar is available, but it also knows when to stop. Homo sapiens has learned to apply many of the laws of nature, but unlike that simple bacterium, has yet to learn when to stop. What's more, Homo sapiens has yet to discover and implement the evolutionary laws and patterns that will ensure an intelligent species' survival. And, in the absence of this knowledge, can our species survive, or go extinct like the dinosaurs?

Starting in this chapter and throughout this book, we will go back to the beginning and look at various events in the life story of our forebears over geologic time. We will consider the emergence of what might have been Earth's first intelligent species, and pick up the trail to find out how our species evolved and the methods evolution seems to have employed to guide intelligent species to maturity. Along the way, we'll try to identify any survival patterns, rules, and models that emerge, attempting to keep a wary eye on evolution and try to decipher, if possible, what might have led to the evolution of intelligent species.

Homo sapiens was neither the original nor the only species of hominins that came out of Africa. In fact, Homo sapiens, the last of the hominins, is just the most recent in a long line of hominin species that evolved, thrived, and then went extinct. There's no guarantee we won't go extinct, too. Extinction is a normal part of life in the universe, and despite the high regard in which we currently hold ourselves, we, too, might one day be replaced by a superior version of our own species as Homo erectus was by us. Perhaps we will be supplanted by "Homo machina," with a biology that is just the right mix of man and machine for interplanetary and interstellar travel—but more about this later.

From the fossil record, it appears some species of hominins survived longer than others; on average, they lasted between 900,000 and 1 million years before going extinct. Two million years ago in Africa, several species of human-like creatures roamed the landscape. Our own species appeared around 300,000 years ago, at a time when several others existed.

But as recently as 30,000 years ago, there were only four other hominin species around: the Neanderthals in Europe and western Asia; the Denisovans in Asia (about whom not much is known); the "hobbits," from the Indonesian island of Flores; and more recently discovered, *Homo luzonensis*—found in a cave on Luzon, the largest island in the Philippines—who lived some 50,000 to 80,000 years ago when there were multiple archaic humans.

Today, only we remain, which raises the question: "How did Homo sapiens manage to survive when all others, even those closest to our species, have gone extinct?"

The answer might be quite different if it could be asked from evolution's perspective, a perspective that we, as a species, might do well to ponder. The following is an incomplete timeline of known species of hominins going back nearly 6 million years.

The hobbits, Homo floresiensis, could have survived until as recently as 18,000 years ago, and according to geological evidence, they may have been wiped out by a large volcanic eruption. Living on one small island, as they did, makes a species much more vulnerable to extinction when disaster strikes, and that might also turn out to be true of Homo sapiens, who now live on a single planet, this island Earth. (It certainly turned out to be true for the dinosaurs.)

The first discovery of Neanderthals was in the mid-nineteenth century, at a site in the Neander valley of Germany; "Neanderthal" means "Neander valley" in German. Their earliest ancestors, called archaic Homo sapiens, evolved, like all hominins did, in Africa, and migrated outwards into Europe and Asia. Neanderthals migrated to Eurasia long before humans did and lived as far north and west as Britain, through part of the Middle East, and east to Uzbekistan. Estimates put the peak Neanderthal population at around 70,000, though some scientists believe the number was drastically lower, with as few as 3,500 females.

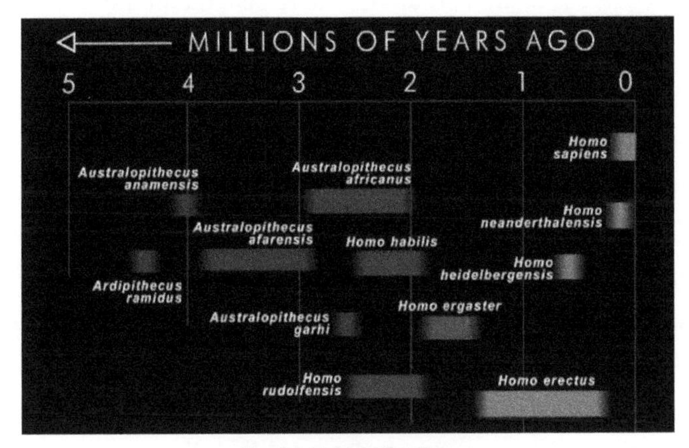

Fig. 2-1 Hominin Timeline
(https://www.exploratorium.edu/evidence/lowbandwidth/INT_hominin)

They lived as hunter-gatherers and scavengers until about 30,000 years ago, when they disappeared from the fossil record. For the last 10,000 years of their existence, Neanderthals shared Europe with Homo sapiens, and apparently led fairly similar lifestyles.

No one knows exactly why Neanderthals went extinct and humans survived. Some scholars theorize that gradual or dramatic climate change led to their demise. They had lived during the Ice Age, often taking shelter from the ice, snow, and otherwise unpleasant weather in Eurasia's plentiful limestone caves. They struggled more with a changing climate than Homo sapiens did, and there were very unfavorable climatic conditions at the time of their extinction. Others blame dietary deficiencies. Some theorize that humans killed the Neanderthals. Until recently, the hypothesis that Neanderthals didn't go extinct but simply interbred with humans until they were absorbed into our species was popular, and studies reveal that in European and Asian populations, between 1% and 2.5% of human DNA comes from Neanderthal DNA.

Why did Homo sapiens survive while other hominin species went extinct? What might have been the hurdles Homo sapiens successfully negotiated that enabled its survival? Perhaps the more general question is: What are the evolutionary factors that select for survival when lacking them leads to extinction?

When viewed from this perspective, it also becomes useful to ask which of these survival factors are due to "nature" or "nurture." What we might think of as "nature" are those that are innate to a species—the size of their brains and how smart, innovative, and naturally adaptable they are. And what we might think of as "nurture" are external factors that would include planet-wide disasters, such as an asteroid impact (remember the dinosaurs), glaciation (ice ages), and supermassive volcanic eruptions. Also, extreme environmental and climatic conditions, such as killer storms, floods, and droughts, might have made staying in familiar, ancestral areas untenable due to increasingly unsustainable or unavailable habitats and increasingly inadequate and inaccessible food and water sources. And of course, there's also predation, as our ancestors were as often prey as they were predator.

A lot has been made of the larger brain size, and thereby the smarts and inventiveness of Homo sapiens, in contrast to some of the earlier hominins. But Neanderthal brains were as large as Homo sapiens', even thought to be larger, yet they went extinct. Neanderthals, and possibly some earlier hominin species, are believed to have developed language, as well as art, as demonstrated by cave paintings in parts of Europe long before humans are thought to have arrived.

So, could the naturally occurring innate capabilities (nature) have been so significant that they can account for Homo sapiens survival and Neanderthal extinction by themselves? If so, what can explain the survival of seven species of great apes (two species of chimpanzees, two species of gorillas, and three species of orangutans), which few would consider innately more capable, or "smarter," than the Neanderthals or other hominin species?

It seems hard to conclude, then, that Homo sapiens' innate abilities, however superior and which must have conferred some necessary evolutionary advantages, could have been sufficient to ensure survival. Consider the dinosaurs again—their extinction appears to have had nothing to do with how big their brains were, or any other innate abilities. So, given the environment and changes in climate that were occurring at the time, it seems more likely that external factors played a much greater role when it came to the survival of Homo sapiens.

What was the Earth like when Homo sapiens were leaving Africa? This becomes apparent when one examines the prevailing environmental and climate conditions thought to have accompanied early human migration out of Africa and its impact on Homo sapiens evolution.

It is believed that all later species of the genus *Homo* evolved during the Pleistocene Epoch (2,600,000–11,700 years ago). This was generally a time of more extreme world cooling and recurrent glaciations (ice ages). During the coldest periods, global temperatures dropped by about 9°F (5°C), and long-lasting ice sheets spread out from the poles and high mountains. Between the four or more major glaciations of the Pleistocene, there were interglacial warming periods, when temperatures were similar to what they are now. Both the glacial periods and interglacial periods lasted tens of thousands of years. Some scientists believe it's likely the Earth is still in an interglacial period that began 10,000 to 12,000 years ago.

The continents of the Northern Hemisphere were more affected by glaciations than those in the Southern Hemisphere, which generally remained mostly tropical and subtropical during the ice ages, though more humid. The coldest regions of the world became arctic deserts. However, during the Pleistocene, the great deserts of North Africa and western North America were mostly grasslands, with large permanent lakes and abundant large game animals. Sea levels during the coldest periods were as much as 450 feet (137 meters) lower than today, due to a substantial volume of the world's water being locked on the continents in 1–2-mile-thick (1.5–3 kilometers) glacial sheets covering thousands of square miles. As a consequence, vast areas that are now shallow sea and ocean bottoms were exposed for thousands of years. Twice during the last ice age, lowered sea levels resulted in Siberia being connected to Alaska by a 1,200–1,300-mile-wide (1,900–2,100 kilometer) corridor. Asian hunters are believed to have used this route to migrate into the Western Hemisphere to become the first Native Americans.

Ever since our ancestors branched off the primate evolutionary tree millions of years ago, there have been drastic swings between moist and dry periods, as well as long-lived glacial freezes and thaws. It is highly likely that early humans had to survive an ice age. What is generally thought of as the last ice age, 12,000–110,000 years ago, came near to being a Homo sapiens extinction event.

According to the genetic and paleontological record, earlier hominin species (Homo erectus, Neanderthals) migrated from Africa many thousands of years earlier, but Homo sapiens started to leave Africa only between 60,000 and 70,000 years ago. What set this in motion is uncertain, but it is highly likely that it had a lot to do with major climatic shifts, including, but not limited to, the eruption of a super volcano, Mount Toba, in Sumatra, also around 70,000 years ago. That may have led to a "nuclear winter," a sudden cooling of the Earth's climate driven by the onset of one of the worst parts of the last ice age that resulted in a 1,000-year ice age.

These changes in environment and climate are evident to scientists in layers of sediment, each layer indicating changes in vegetation, moisture, the animals that were around, and other survival challenges faced by our predecessors. There is some evidence that the ice came and went in regular cycles, driven by changes in Earth's orbit.

The cool temperatures of the Quaternary Period, the most recent glacial period, are believed to have left their mark on our species. Cold snaps would have made life extraordinarily difficult for our African ancestors, and, indeed, the genetic evidence points to a sharp reduction in population size around that time. In fact, the human population likely dropped to about 2,000 individuals. As a species, Homo sapiens appears to have been holding on by a thread.

"Shortly after Homo sapiens first evolved, the harsh climate conditions nearly extinguished our species," said Professor Curtis Marean of the Institute of Human Origins at Arizona State University. Professor Marean discovered ancient human artifacts in the isolated caves around an area known as Pinnacle Point, in South Africa. According to Marean, the caves contain archaeological remains going back at least 164,000 years.

"Recent finds suggest the small population that gave rise to all humans alive today just happened to survive by exploiting a unique combination of resources along the southern coast of Africa," said Marean. The strip of land on Africa's southern coast, around 240 miles east of Cape Town, may have been the only place that remained habitable during the devastating Ice Age, some scientists think. The sudden change in temperature wiped out many species elsewhere around 195,000 years ago. Some scientists even believe that the human population may have fallen to just a few hundred individuals.

Homo sapiens evolution, then, was very likely strongly affected by the dramatic climate swings of the Pleistocene. These changes are believed to have presented powerful new natural selection pressures. Many animal species were driven to extinction by the advancing and retreating ice ages.

Homo sapiens may well have survived, it would appear, not because of our big brains and capacity for innovation, but primarily because a handful of individuals were lucky enough to find a not-so-cold patch in Southern Africa. They may have been able to survive because of rich vegetation available in the area. Although many other explanations for their extinction have been suggested, Neanderthals, with whom our species shared the planet until just before the last glacial maximum, 20,000 years ago, may have struggled to survive as the rising and falling ice ate away at their habitat, and, unlike Homo sapiens, failed to find sustainable habitats.

Variability in Earth's early environment, then, may have been the biggest challenge to Homo sapiens evolution and survival. These repeated, dramatic shifts in the environment challenged many species, and may have actually selected for the features that ensured survival and have come to typify Homo sapiens, like the ability to alter our immediate surroundings in face of a growing threat.

The evolutionary survival selection factors mentioned previously may have included the ability to respond to those changes. These environmental events would have put immense pressure on early humans. As noted above, it seems that early Homo sapiens were just barely able to survive these changes.

Although early humans had the advantage of being in the Southern Hemisphere, in Africa, during some of the Ice Age, which afforded a somewhat less harsh habitat in which to survive, there are those who think they were able to survive these extreme conditions only through cooperating with one another. This may have led to the formation of close family groups or tribes and the development of some of the modern Homo sapiens behaviors we are familiar with today.

Humans would have had to develop new cultural technology to deal with cold environments and changing food sources, especially during the last quarter million years. One of the greatest problems in the cold regions would have been the relative scarcity of plant foods to eat during the winters. In response, Homo sapiens would have had to become more proficient at hunting, especially the large animals that provided more calories. This in turn would have required the development of more sophisticated hunting skills as well as better weapons and tools. These changes in subsistence patterns (turning to different sources of food and different ways to obtain them, i.e., scavenging and hunting) had become essential for Homo sapiens survival.

Hence, some believe that *change* itself may have been an evolutionary survival selection mechanism, and still is. A growing number of scientists think that major climate shifts may have also forged some of the defining traits of humanity. A few large evolutionary leaps in particular, such as bigger brains and complex tool use, seem to coincide with significant climate change.

There are others who believe that the climate change we're experiencing now, and the changes coming in the next century, may well be the catalyst for the next change in Homo sapiens subsistence patterns and will engender another burst of survival-focused innovation and cooperation. Still others think this respite from the ice is likely to prove short-lived, at least in geological terms, as interglacials (during which all hominin species evolved) are estimated to have lasted no more than 10% of the Pleistocene, which lasted from 2,600,000 until 11,700 years ago. This means that for the remaining 90% of the period the Earth was in the glacial part of its climate cycle.

Research by Eelco Rohling of the University of Southampton in England suggests we are now 2,000–2,500 years overdue for another ice age, and the reason it has not arrived yet is the impact Homo sapiens has had on the global climate. Specifically, human activities like deforestation and the burning of fossil fuels have resulted in an atmospheric "greenhouse effect" that has prolonged relatively warm interglacial conditions.

Homo sapiens' effects on the climate notwithstanding, the cycle will continue. This hothouse period we have helped bring about will someday come to an end, and the ice sheets will descend again.

Unlike nonintelligent species that are preprogrammed and a priori evolutionarily mature, a nascent intelligent species must discover for itself how the world works, then use that knowledge to program its own mind to avoid extinction and survive. It's probably as unlikely that Homo sapiens could have developed radio, television, and most electronics without first discovering Maxwell's equations and the theory of electromagnetism as it is for a nascent intelligent species to avoid extinction and survive if it fails to uncover and follow potential evolutionary species survival rules.

So, does a universe filled with rules, laws, patterns, and models that prescribe how most things work also come with evolutionary species survival mechanisms? What are the planetary, solar system, and galactic speed bumps and patterns that unintentionally work to accelerate, slow, or block intelligent species survival? Can one begin to identify some essential survival "must-haves" from this brief recounting of our species' evolution and emergence from Africa?

Put simply, will Homo sapiens survive, or go extinct like the dinosaurs, and has evolution given us a way to find out? If the summary below hints at the beginnings of patterns, evolution might have already done so.

The following are survival-impacting factors one might glean so far:

Luck, location, and timing of independent planetary events.

Climate change driven by galactic, solar, and planetary events.

Habitat and habitat sustainability or erosion driven by climate change.

Ability to adapt in the face of change.

Innate smarts and ability to innovate to survive when faced with change.

Freedom to choose and the freedom to migrate to survive when all else fails.

In succeeding chapters we will discover how these and others come together to form a Species Survival Maturity Model.

3.

Is Species Survival the Default Evolutionary Outcome?

IF EXTINCTION MEANS THE END OF A SPECIES, THEN survival must mean the continuation of a species. But is extinction inevitable? Probably not, and here is why this might be so.

As our knowledge of the universe increases, we are beginning to realize that nothing in the universe lasts forever. Not planets, not stars, and apparently not even galaxies. The tale of the universe is a tale of birth, life, and death. Planets form, survive, even for billions of years (Earth was formed some 4.5 billion years ago), then get absorbed into their stars as the stars age and expand into the orbits of their encircling planets, on the way to their own extinction.

To survive in such a universe, a species will need a home world located, hopefully, in a relatively quiet part of the galaxy, millions of light-years from a black hole or supernova. The planet should not be in the path of gamma ray bursts, nor should it cross the path of any rogue planet or wandering meteorite or asteroid such as struck Earth around 65 million years ago. The species will need time to evolve, obviously, but it must also have enough time to develop and integrate a common civilization and cultural survival narrative. And it will need to prioritize and develop the technologies, innovations, and capabilities to migrate from planet to planet, from one solar system to another, possibly even from one galaxy to another, if it is to escape this natural cycle of birth, life, and death that characterizes the component bodies of the universe.

It is not unreasonable to imagine that such a species will need to become a multiplanet species if it is to avoid extinction, even as Homo sapiens needed to become a multicontinent species in order to survive. Just as we were once seafarers, leaping from continent to continent, crossing oceans and scaling mountains, and braving the near extinction of the last ice age, adapting and innovating and creating the technologies needed to survive each phase along the way, we must now become spacefarers, migrating from one planet to another, from one solar system to another.

It may well turn out that our species will look back at its initial survival-driven migration across the planet and realize it was but a practice run, a survival pattern that we will have to keep repeating, not just across the planet, but eventually across the galaxy.

What's often missing from many accounts of early human history is the role played by evolution and the planet itself—its climate, environment, and the consequent sustainability of the local habitat. Similarly, given the way things are shaping up in the early twenty-first century, it's possible that the planet, its climate, environment, and habitat sustainability—this time not just of local habitat but of the entire Earth—will become again the accelerant, if not the major factor, in yet another species-wide survival-driven migration. This time, however, it might require Homo sapiens to become a multi-planet species.

The big question is how long does an emerging intelligent species have before it must be ready to pack up and migrate to another life-sustaining moon or planet, or even to another solar system?

Sometimes, to better see what might lie ahead, it helps to look at where we've been. It's safe to assume these early humans weren't seriously considering the need to abandon their familiar habitat in the Great Rift Valley of Ethiopia in order to survive any more than leaving Earth for another planet preoccupies us today. More than likely, they probably couldn't even imagine what life away from the valley might be like, nor is it likely they had much time to prepare.

And now, 70,000 years later, it is beginning to look like we, their descendants, are no more prepared to migrate to another planet in order to survive than they were to leave Africa. The obvious pattern here seems to suggest we might find ourselves in a similar position, having to leave in a real hurry, relatively speaking, ready or not. Survival-driven migration has never been and probably never will be a preplanned human activity.

So, again, looking backward, we learn that the entire human race outside Africa owes its existence to the survival of a single tribe of around 200 people who crossed the Red Sea 70,000 years ago. It took 70,000 years for our hunter-gatherer forebears to go from Africa to all the other continents of the Earth, and from primitive tools to the complex cultures, technologies, and capabilities that comprise our current civilization.

If survival-driven interplanetary migration turns out to be more a question of when rather than if, it becomes imperative to find out how long it will take to go from a life on Earth to a life on Mars, or another planet or moon in our solar system. For this migration, we have no pattern to follow, but if global efforts to curb greenhouse gases and reduce global warming serve as a model, it could take a very, very, very long time to get prepared—longer perhaps than our species might have.

Did evolution leave a few bread crumbs along the evolutionary trail to point the way, or do Homo sapiens have to figure it all out on their own? Perhaps there is nothing as specific as the laws of chemistry and physics, but are there any clues? Is there such a thing as a Species Survival Maturity Model (SSMM)?

As one begins to explore this question it again helps to look back at what enabled early humans to migrate across the face of the Earth, to go from hunter-gatherers using primitive tools to modern man using computers and emerging autonomous robots, from ape-man to spaceman.

Due to no effort on our part and equipped with all things necessary to thrive and survive, we find ourselves as a lone surviving emergent hominin species that came into existence some 300,000 or so years ago on a pretty remarkable planet that came into existence about 4.8 billion years ago. Our remarkable planet happens to be rotating in the habitable zone around a not-so-average star, one of billions, in a distant neighborhood of the Milky Way Galaxy, far, far away from its central black hole, numerous supernovae and gamma ray bursts; a galaxy that is itself just one of billions and that has existed almost as long as the universe itself.

We also find that the remarkable laws that govern the birth, life, and death of planets, stars, and galaxies enable us to peer far back into the past and look forward into the future to understand their formation, evolution, composition, and eventual demise. This orderliness creates predictability, which has allowed us to send space probes and satellites to survey or rendezvous with numerous planets, and now comets, within our solar system.

Can the clockwork laws that pervade the universe and order the rotation of the Earth around the sun and the solar systems around the galaxy offer a clue as to whether naturally occurring chemical reactions that are a subset of these universal laws will also continue indefinitely if the initial conditions that jump-started life can be maintained? One answer might be: Why not?

Because life's basic processes are naturally occurring chemical reactions that take place given the right set of conditions, the eventual appearance of life on Earth was probably as inevitable as the birth of new stars in the galaxy. Like the formation of stars and planets, everything a life-form emerging from these chemical reactions needs to survive is an integral part of the environment from which they emerge. These life-giving processes suggest that species once formed can continue, if not indefinitely, for very long intervals if those life-engendering and sustaining conditions and accompanying habitat can be maintained. The dinosaurs emerged and continued for 165 million years, until their habitat changed abruptly and was no longer able to sustain them.

Predicting outcomes and trajectories in the universe is all about initial conditions and the ability to maintain them. So an essential, if not indispensable, part of maintaining initial conditions when it comes to species survival is the preservation of the habitat and environment that gave rise to the species. Survival, then, is inextricably bound to the integrity of the sustaining environment that embodies those initial conditions. Species can thus continue as long as the environment whence they arose remains available. Otherwise, they will have to pick up and migrate elsewhere to survive, or, go extinct.

It is unlikely that our species can survive for any meaningful length of time on the other planets in our solar system—not on Mars or Venus, though these share the sun's habitable zone with the Earth, nor on gas giants Jupiter, Saturn, Uranus, and Neptune, though some of the many moons that orbit them might turn out to be marginally hospitable to life as we know it. It is equally unlikely that our species can continue to survive on Earth if the sustaining environment that gave rise to our emergence is destroyed. This is among the reasons Homo sapiens must look for Earth-like planets as we scour the galaxy and the universe for intelligent life, even as we set about restoring Earth's ability to sustain life.

The history of Earth is replete with examples of what happens to species when their sustaining habitat is destroyed. However, these extinctions occurred through no fault of the species themselves. Our species is the only species we know of capable of destroying its own habitat and is well along the way towards doing so.

Species emerge and seem to come equipped with the potential to survive almost indefinitely through the reproductive ability of individual members of the species. This ability is probably evolution's way of ensuring species continuity and avoiding species extinction. It is probably also a hint about which side of the species survival versus extinction question evolution is on. Why is it that in a universe where everything is born, lives, and dies, evolution equips species with the innate, extinction-defying ability to ensure survival through reproduction? Why would evolution go on to jump-start life again and again in the face of at least fifteen major and minor extinction events? Not only do these actions suggest species survival might well be the default evolutionary outcome but also hint at the existence of a built-in evolutionary bias for survival, one we are acquainted with—the so-called fight-or-flight response.

Species survive but individual members do not. When human parents consider having children, ensuring the continuity of the species is hardly among their considerations. Yet, like so many animal, fish, bird, reptile, amphibian, and insect species, for which bringing forth the next generation is their greatest, and often only, goal, each child that is born to a human parent contributes to the continuation of the species. It seems almost as if evolution has ensured species continuity in a way that neither depends upon nor requires conscious, individual consent.

But species survival is not the same as national, regional, or world-population survival. While it may have been just a small tribe of 200 or so that set off out of Africa 70,000 years ago, there might have been several other tribes and clans, consisting of perhaps a few hundred or so earlier humans, who never left, either because they could not make it out or because they chose to remain in Africa.

There are, no doubt, some who would want to be among that early group of Mars settlers, but those who survive to continue the human species on other planets would necessarily be a tiny fraction of the Earth's population. So, if our species could survive by some migrating to another planet, many would remain on Earth to perish in any potential extinction event that drove an interplanetary survival migration.

If the experience of people affected by recent major disasters serves in any way as a guide (consider events such as refugee flight from wars in Syria and parts of Africa; economic migrants fleeing unsustainable habitats in many regions; the earthquake in Nepal; Hurricanes Katrina, Harvey, Irma and Maria; and various floods and tsunamis such as in India, Indonesia, and Japan), one can only imagine the chaos that might ensue if our species were aware of an impending extinction-level event, one in which the only hope for survival is to get to another planet.

So, with evolution appearing to be on the side of species survival, the question then is what else does a species need to do to ensure it goes on? As one attempts to identify what might comprise a Species Survival Maturity Model, "survival must-haves" appear to fall into two broad categories: extinction-causing events that *can* be controlled by a species and those that *cannot* be controlled by a species.

Depending on their stage of development, there are things and events with potentially devastating outcomes, including extinction, that a species will be largely powerless to do anything about. Items in this broad category might include:

Location, Location, Location. As noted above, emergent life and consequent evolutionary species survival really does depend upon where in a galaxy that life comes into being. Earth is, luckily, located on a spiral arm of the galaxy, some 30,000 light-years away from the cluster of stars near the central regions, far away from the interstellar violence that can potentially block the emergence of or destroy life. However, being in a quieter neighborhood of the galaxy does not grant safety from phenomena like quasars, supernovae or gamma ray bursts. Indeed, scientists believe the sun and the solar system may have formed from the remnants of a type 1a supernova, including the iron in Earth's core. Scientists know this because the Earth is radioactive and because the heavy elements comprising its core, like iron and silicon, comprise the ejecta of these heavy elements synthesized in an exploding supernova.

Supernovae radiate very large amounts of energy over the course of a few weeks, and if one were to occur within thirty light-years from Earth, it could result in the destruction of most of the ozone layer, exposing all life to the sun's ultraviolet rays. Combined with the gamma ray radiation generated by the supernova, it could end all life on the planet's surface. Supernovae that close to the Earth have been calculated to occur once every 200 million years, but new evidence suggests they may occur more frequently, and one may have occurred in our neighborhood of the galaxy as recently as 2.2 million to 2.6 million years ago in the Scorpius and Centaurus constellations some 400 light-years away.

Gamma rays, on the other hand, are thought to be capable of putting out more energy in seconds than our sun will emit over its entire life, and gamma ray bursts are believed to be the most powerful event in the known universe. They are estimated to occur in our galaxy roughly once every 100 million years, and even if far, far away, might be capable of causing the same destruction as a nearby supernova.

Even though most gamma ray bursts we've detected have occurred nearly halfway across the galaxy, sooner or later Earth could be within range of an exploding supernova or the horrendous hellfire explosions of a gamma ray burst. For any emerging intelligent species to survive in the universe, then, it must develop the means to detect and escape from such galactic and universal species-extinction events—it has no choice but to become a multi-planet species as fast as it possibly can.

Timing and, maybe, lots of luck. Being in the wrong place at the wrong time along the planetary formation cycle or species evolution cycle, as the dinosaurs might have been, can result in extinction. Even though Jupiter has been known to repel or block errant asteroids and meteorites drawn into our solar system, there are no guarantees that one with or without our name on it might not sneak past Jupiter's watchful eye and our own vigilance, as Oumuamua, the first observed interstellar visitor, did on October 19, 2017.

Non-man-made planetary events. Supermassive earthquakes, tsunamis, superstorms, hurricanes, typhoons, floods, landslides, and droughts can all cause widespread, survival-threatening disasters. While there is a timing aspect to these as well, it bears noting that these are well beyond the planetary-formation time period. There are other potential solar/planetary events, such as rogue planets (as many as 200 billion are estimated to be in the Milky Way Galaxy), untethered from any particular star, cruising through the blackness of space, and perhaps there are other dangers yet unknown.

Glaciation and other non-man-made environment and/or climate-driven changes. These might include very rapid changes in subsistence patterns like the ones Homo sapiens experienced during the last ice age and apparently just barely survived: changes in water, moisture, vegetation, the kinds of animals and plants available (food chain sources), and the survival challenges such changes might unleash.

Inability to adapt and change in the face of survival-threatening events. Some believe that the Neanderthals, who had migrated some 250,000 years earlier to parts of Europe and Asia, in addition to being in much, much colder regions of the Earth, might have had their extinction hastened due to a loss of habitat caused by the impossible challenge of adapting to and surviving protracted periods of glaciation.

Habitat destruction. Our species is entirely dependent on the continued existence of the planet and the supporting ecosystems that gave rise to and continue to sustain life. But Earth comes with its own set of life-extinguishing mechanisms that can help or hinder survival, depending on the timing of intelligent species evolution.

Plate tectonics. Over geologic time, plate tectonics has conducted and continues to conduct an unceasing ballet of the Earth's crust, sculpting and rearranging the continents, merging them into supercontinents (Pangaea, 320 million to 200 million years ago, being the third so far, with one more in formation) and breaking them apart again, forming the now familiar pattern of continents we recognize. Some scientists believe life would have been unsustainable on many of these earlier supercontinents, so it is no small coincidence that hominin evolution, which only began some 6 million to 5 million years ago, occurred well after the breakup of Pangaea and the extinction of the dinosaurs.

And yet it was plate tectonics, combined with the end of the last ice age, that may have facilitated the migration of our forebears out of Africa at the point where the Red Sea meets the Gulf of Aden. The breaking apart of the African and Arabian plates resulted in the formation of a land bridge, at what is now known as the strait Bab-el-Mandeb, between Africa and Arabia, at a time when, due to the last ice age, the Red Sea was significantly lowered. And as Walter Alvarez tells it in his book *A Most Improbable Journey: A Big History of Our Planet and Ourselves*, plate tectonics, like ice ages (aka evolution), has continued to set the scene and choreograph numerous significant events that have shaped and will continue to shape Homo sapiens' history and evolution.

On the other hand, ice ages and the consequent loss of habitat, as noted above, may well have driven Neanderthals to extinction, and nearly ended Homo sapiens as well.

All of the above are capable of causing extinctions, and on Earth some already have; elsewhere in the galaxy, some may have prevented life from emerging in the first place. Yet, thanks to evolution, life persists, and here we are talking about it.

It would almost seem like, despite the awesome arsenals for death that the universe, like Thor, has hurled at life over the eons, in spite of all the many ways intelligent life across the universe can and probably has been prevented from emerging or outright been destroyed, like Harriet Tubman, or like Anne Frank, evolution found a way to squirrel life away. And did so on a remarkable planet with just the right environment and habitat, at just the right time, and in just the right place. Here, in the habitable zone of a long-lived, relatively benign star, tucked far away out in the galactic backwaters, in a distant part of a spiral arm in the Milky Way Galaxy, life found a way. And yet, the journey of intelligent life across the galaxy, as far as we know, is only just beginning.

As threatening and frightening as the items in the above category are, evolution found a way to navigate these galactic rapids and started life, and thereby our species, on this remarkable planet.

What comes next, however, might be even scarier. In this category, evolution appears to have turned species' survival and habitat sustainability over to an alpha species, Homo sapiens, and a successful outcome is not at all guaranteed. Homo sapiens is the first and only species—as far as we know—with which evolution seems to have chosen to share control of an outcome that can potentially bring an end to more than 4 billion years of evolution on Earth. If successful, though, it may well be the beginning of an even more exciting phase wherein, for the first time, evolution would partner with a species to chart the next course of this evolutionary journey. But why would evolution take such a risk? That is a question we will explore in a later chapter.

What generally comes to mind when considering survival-impacting factors that, unlike those listed above, can be controlled by a species are things such as climate change (global warming, air-pollution); food-source poisoning; habitat destruction; species extinction; and the seemingly endless wars and conflicts that result in the death, displacement, and impoverishment of millions of people as well as the continued destruction of Earth's habitats.

However, one might want to think of the above as symptoms rather than fundamental root causes, some of which are listed below.

The ability to remain alive and stay clear of extinction-causing events. At the risk of stating the obvious, members of a species must first exist to have any chance at surviving. There has been no lack of planet-wide extinctions of entire species during the history of the Earth. In fact, there have been no less than five major mass extinctions, and there are those who argue that we are in the Anthropocene Epoch, and it is already leading to a sixth extinction, driven by man.

Habitat. Habitat viability and species survival are inseparable. The evolutionary pattern appears to be habitat first; emergent life second. It cannot be overstated that whenever evolution has given rise to life of any form or complexity, it has invariably first established a natural, sustainable habitat with everything necessary; then, out of that habitat, life emerges, not the other way around.

From this perspective, the current excitement about establishing a significant human presence on Mars appears to be following an anti-pattern. While there is lots of excitement and chatter about establishing an alternative human habitat on Mars, and a significant amount of money being spent, it is highly unlikely that such settlements can become real, permanent alternatives to life on Earth, short of terraforming the planet. There are those who seriously question whether our species would have evolved, let alone survived, if Earth had at the time remotely resembled the current conditions on the surface of Mars. For instance, it took more than 3 billion years before enough oxygen was in Earth's atmosphere to support the rise of complex life in the form of the Cambrian explosion. There is no naturally occurring oxygen on Mars.

However, while we might one day be able to migrate to and settle on other potentially habitable planets, it does not seem as though that will happen anytime soon. Until then, our species survival is entirely dependent on the continued preservation and maintenance of the planet on which we evolved and its supporting ecosystems that gave rise to and continue to sustain life.

Migration. Migration is a built-in evolutionary survival response and is common to all species. Our species ensured its survival by migrating out of Africa at a time of habitat loss due to climate change caused by the last ice age. The urge to flee and migrate to safer climes—from country to country, across oceans, from continent to continent across the planet, or eventually even to other habitable planets—in the face of existential threats is as much built in to our species as is the evolutionary compulsion to reproduce.

Reproduction. Most if not all species appear to come prepared to perpetuate their survival by the production of offspring through the fertility of individual members of the species. It would appear that unless interrupted by uncontrollable events such as habitat destruction, most species would otherwise continue. Why is this so?

Critical mass. Any species must maintain a critical mass of fertile individuals to survive. A critical mass of male and female, childbearing and non-childbearing members of a species are an essential component of continuous survival. There is some minimum number, varying by species and habitat, below which a species begins to risk extinction simply due to the attrition dynamics of deaths and births; the interval before new offspring can themselves reproduce; predation and natural disasters; depletion and variations in food sources; and destruction of habitat. To survive, a species needs to maintain at least that minimum number of individual members of the species.

Desire to survive. The desire of individual members of a species to survive appears to be built in. When that desire is aroused and becomes pervasive, individuals collectively act as one to survive. It was the desire to survive that drove early Homo sapiens to migrate out of Africa. Survival today, however, does not appear to be a major concern of Homo sapiens as a species; neither does it preoccupy us as individuals. How important is it to the survival of Homo sapiens to cultivate, develop, and maintain a species-level survival awareness and empathy?

Survival-driven ability to adapt, invent, and innovate. Survival-driven adaptations and innovations are closely associated and comprise another of those species survival patterns we have noted. Most of the tools and technologies our early ancestors came up with were driven by their need to adapt to climate and, hence, habitat and food-source change in order to survive. In contrast, most of our adaptations and innovations today are profit and market-growth driven, the results of which often turn out to be antithetical to survival or habitat preservation. While the ability to invent, innovate, and adapt is necessary for species survival, it is not by any means sufficient, as we shall soon see. The ability of Homo sapiens to migrate from Africa across the Red Sea may have been due, not to innovations and technologies, but to the massive sea-level fall caused by an ongoing ice age.

Maturity. Species maturity or species extinction is a stark choice. Avoiding extinction due to intraspecies conflicts may well turn out to be the most difficult rung of a Species Survival Maturity Model.

In addition to being intelligent and capable of amazing inventions and innovations to survive, a species must be able to overcome and eliminate all survival-threatening intraspecies conflicts, as well as other activities that can eventually lead to self-extinction. In other words, to survive, a species has to mature and make choices that reflect that maturity. The dinosaurs' ability to survive for 165 million years was clearly not dependent upon either the size of their brains or their ability to invent and innovate, and we may find that, notwithstanding our many scientific and technological achievements, our ability to survive isn't either.

On the contrary, if a species can muster the courage and will to nurture and grow a species-wide empathy awareness that transcends ethnic, racial, religious, social, economic, political, national, and regional sympathies and allegiances, and if that awareness can mature faster than the species' ability to destroy itself and grow to a sufficient level that it begins to prioritize its survival over all else, then, and perhaps only then, can it prevent self-extinction and go on to fulfill its evolutionary purpose, a subject we will explore in ensuing chapters.

Unlike all other adaptations and innovations, which tend to be double-edged swords that endow a species with the potential to do both good and evil, to create as well as destroy, species maturity stands alone in its intrinsic and singular nature as a force for good. This is the aspiration around which all aspects of species activities—economic, political, scientific, and religious, to name a few—must coalesce in order to reach the final rung on the Species Survival Maturity Model.

But maturity does not come easy. Unlike species survival through reproduction, where continuity of the species appears to be almost autonomic, if not instinctive, species survival due to maturity is not. On the contrary, and as we shall also explore, species self-extinction due to immaturity is almost guaranteed.

It is probably a safe bet that the 200 or so early humans that fled Africa were collectively concerned about their own survival, and I guess it won't be much different when—not if—our turn comes.

Freedom to choose. The freedom and ability to choose at the individual as well as the species level may well become indistinguishable when choice is necessary to avoid self-extinction, mature, and survive. Evolution is obviously betting that, given the built-in desire to survive, individuals will choose survival over all else, even if that means choosing to become mature.

Should the survival of an entire emergent intelligent species come down to the choices of just a few individuals—Khrushchev and Kennedy, say? Should the decision to engage in global conflicts be turned over to political leaders?

Should the pollution, poisoning, and destruction of the Earth's atmosphere, habitats, and environments, which puts at risk the survival of all members of an entire species, be left to governments, corporations, and their shareholders, all of which prioritize economic growth and profits over life and a sustainable environment? Is survival of greater value to an individual and to a species than profits? A sustainable habitat more valuable than nuclear power, oil, gas, and coal? Is maturity a better choice than life-destroying technologies and innovations? With hindsight, was it wise to develop chemical agents such as anthrax or VX, weapons of mass destruction?

Just as species continuity is built in and assured by an individual member's freedom to choose to reproduce, so its maturity might depend upon the freedom of each individual to choose survival over extinction.

Control of its own destiny. The freedom and ability to choose at the individual as well as the species level is closely aligned with the control, ability, and willingness of a species to act and rapidly change course in the face of impending potential species-extinction events. It is becoming increasingly and distressingly common for the wishes and concerns of a people to be arbitrarily set aside by modern-day rulers and governments. It is this increasing loss of control by individual members of a species that places their survival as individuals at risk and threatens the species with extinction.

So even as galaxies, stars, and planets persist, even in a universe ensnared in birth, life, and death, similarly, species persist, even though they, too, evolve, thrive, and go extinct. Individual species extinction, then, is not the end of all species any more than the birth, life, and death of an individual celestial body is the extinction of all celestial bodies. Species extinction might not be inevitable, but rather, survival might turn out to be the default evolutionary outcome.

But even as evolution appears to have weighed in on behalf of species survival, that support is not unconditional. Rather, it depends upon numerous factors over which a species may or may not have control, including its ability to reproduce, or the location and timing of its evolution, or whether it evolved in a sustainable habitat. Also, in the case of an intelligent species, evolution's support seems to depend on the species' ability to adapt and innovate in response to climate change or self-driven causes of subsistence-pattern failures; its ability, freedom, and willingness to choose survival over extinction; and its ability to mature and migrate to survive when all else fails.

In subsequent chapters, we will turn our attention to exploring in greater detail these potential survival-impacting conditions and hopefully tease out those evolutionary survival patterns that might comprise a Species Survival Maturity Model.

4.

An Introduction to Intelligent Species Survival Patterns

ARE THERE EVOLUTIONARY SPECIES SURVIVAL patterns, and, if so, what are they and what might they imply about why we are here and the spread of intelligent life across the galaxy?

In view of the preceding summaries of survival-impacting factors, whose outcomes in some cases can be controlled by a species, while in other cases they cannot, the next step is to choose from among these factors and build a Species Survival Maturity Model. Next, we will try to ascertain whether or not Homo sapiens is aware such a survival model might exist and, if not, has nevertheless been pursuing a course largely in alignment with this model. In subsequent chapters, we will examine the potential consequences of a species making it all up as it evolves.

However, before going down that path, it is useful to briefly highlight parts of the intelligent species evolutionary process as it occurred on Earth and consider how we got to now. One soon realizes that evolution, having jump-started the entire process, didn't just abandon Homo sapiens on the plains of the Great Rift Valley in Ethiopia. Indeed, evolution may not have been an absent, disengaged, and unmindful planetary process after all. On the contrary, it seems to be more of a guardian and companion, a counter process to the universe's endless processes of birth, life, and, death, in a way that turns out to be quite surprising, yet pleasing.

In this chapter we begin the process of identifying some of these repeating evolutionary species survival patterns as they become evident during a brief review of the emergence and migration of our species across the face of the Earth.

It is quite remarkable that intelligent species evolution was more than 4 billion years in the making. It took evolution 4 billion years to prepare the planet to enable the transition from single-celled prokaryotes all the way to Homo sapiens, and that was only after going through multiple extinct hominin species along the way. Evolution fought against the chaotic forces of an apparently disinterested universe that seems to mindlessly destroy life, but despite it all, still found a way for life to thrive and persist. Remarkably, evolution won out, even over the universe's endless cycles of birth, life, and death; its death-dealing black holes, supernovae, and gamma ray bursts; its potentially world-destroying rogue planets, asteroids, and meteorites; and its planetary orbits and solar system itineraries that pass through potentially dangerous galactic neighborhoods. No, it was not stopped even by the nearly 700 million years of the mother of all bombardments raining down on Planet Earth during the so-called Late Heavy Bombardment period, which may have annihilated earlier evolutionary attempts. In the end, nothing in the arsenal of the universe was able to prevent the emergence and persistence of life.

Evolution, and life, had found a way, in effect saying that order can exist in the midst of this universal chaos; that the universal processes of birth, life, and death are not inconsistent with the evolutionary processes of species evolution and survival; that extinction is not inevitable; and that life and survival will have the last word.

Because when it comes down to the question of species extinction versus species survival, evolution appears to have always sided with species survival, by endowing species with the ability to evolve and persist through reproduction and preserve their survival through migration. And, when even that turned out not to be enough—as the extinction of the dinosaurs, even after surviving for 165 million years, demonstrated—evolution went on to evolve a new kind of complex animal species. This time around, though, evolution evolved an intelligent species, hominins, with the ability to do what the dinosaurs couldn't—adapt, innovate and migrate to survive, to choose its own evolutionary outcome.

Some scientists think the potential for intelligent life evolving separately and independently on some of the millions of Earth-like planets being identified across the galaxy is high; if life can spontaneously emerge and evolve on Earth, there is a pretty good chance it can also emerge on some of these planets, and probably already has.

There are others who argue that the evolutionary process it took to jump-start life on Earth, and which led to complex intelligent life (i.e., from single-celled prokaryotes to multi-celled eukaryotes, and to complex animal species and hominins), appears to have occurred just once. Not only does it appear that this process happened only once but it has been so fraught with luck and chance that one wouldn't have wanted to place a wager on the probability that complex life would have ever occurred. For these scientists, then, it seems extremely unlikely that this process might have occurred again on some Earth-like planet somewhere in the galaxy. They conclude that while simple life might exist in abundance, hominins—and, hence, Homo sapiens—might be the first and only intelligent species to have evolved, not only in the Milky Way Galaxy but potentially across the entire universe.

In their book, *Rare Earth: Why Complex Life Is Uncommon in the Universe,* authors Peter Ward and Donald Brownlee base their arguments upon the Earth's optimal distance from the sun, the positive effects of the moon's gravity on Earth's climate, plate tectonics and continental drift, the right types of metals and elements, ample liquid water, maintenance of the correct amount of internal heat to keep surface temperatures within a habitable range, and a gaseous planet the size of Jupiter to shield Earth from catastrophic meteoric bombardment.

One potential outcome implied by what this group of scientists is asserting might well be that if intelligent life is to eventually overspread the galaxy, then it must be accomplished by Homo sapiens, or a successor species that might evolve on Earth, and hopefully, a more mature one. And that is an earth-shattering thought.

If the history of intelligent life on Earth can serve as a guide, evolution might be taking this latter approach. The evolutionary pattern would dictate something such as first establishing evolutionarily mature intelligent life on Earth, then using that successful species to spread life to other planets. In other words, habitable worlds around the galaxy would be populated with mature intelligent life utilizing a single mature intelligent species reared on Earth as seedlings.

On Earth, all intelligent life seems to have emerged from Africa. This appears to be the pattern employed by evolution in spreading intelligent life across the Earth, beginning first in Africa with a single hominin species, then on to all the continents across the rest of the Earth. Relying upon a species' innate survival instincts (a species survival evolutionary pattern), evolution appears to have used climate change in the form of the last ice age, and the resulting habitat loss, to drive early humans to migrate (another intelligent species survival pattern) out of their comfort zone and away from the Great Rift Valley in Ethiopia, and eventually across the entire Earth.

It is worth noting that intelligent life does not appear to have evolved independently or simultaneously on any of Earth's other continents, even though suitable habitats probably existed there. Neanderthals left Africa and migrated to parts of Asia and Europe thousands of years before Homo sapiens did, and were contemporaries with Homo sapiens, but the Neanderthal species was not the species that eventually overspread the entire Earth. The intelligent species survival pattern here appears to be: first establish life on one continent using a single species; then use that species and continent as the base from which to launch migrations that eventually populate the entire planet.

So, might that same pattern be at work once again, perhaps this time around to begin populating the solar system and the galaxy? Could climate change and resulting habitat loss once again be the catalyst that drives Homo sapiens to migrate, this time to other planets and eventually across the stars, or will it become the cause of its extinction?

Does this potential for an apparent pattern repetition imply an evolutionary master plan? Certainly not in the sense of an architect's blueprint to build a complex structure, but the perennial interplay of the universal processes of birth, life, and death on one hand and the apparent counterbalancing evolutionary processes of species evolution and survival on the other could lead to an outcome that would be hard to distinguish from one that had been deliberately planned and executed.

The combination of climate change and habitat loss that led to the spread of life across the Earth was not just happenstance but more than likely multiple repeating intelligent species survival patterns at work across geologic time scales. So, as scientists begin to observe untoward climate patterns emerging again, one should be able to recognize these patterns at work in our times and view these growing changes in the climate for what they might be—a probable repetition of species survival evolutionary patterns. When executed this time, these patterns have the potential to drive our species to migrate to other planets as surely as they drove early humans out of Africa to find safety on other continents.

If it turns out that such an evolutionary survival pattern is at work, then might Homo sapiens be evolution's "Proto-Intelligent-Species Seedlings" that it will use to populate the rest of the galaxy? And if so, then for the first time Homo sapiens can begin to understand how they fit in and can possibly even begin to answer that perennial question: Why are we here?

Perhaps our forebears in Africa asked themselves a similar question. Today, because of this intelligent species survival pattern their migration established, Homo sapiens may be better positioned to see the answer and identify what our potential part might be in what seems to be a grand evolutionary pattern, which when fully executed, may well result in the spread of mature intelligent life across the galaxy.

If Homo sapiens are indeed the only extant intelligent life in the galaxy, it would explain the deafening silence being heard by SETI, as well as answer Enrico Fermi's question: If the universe is teeming with life, where is everybody? The answer will have to be right here on Earth—at least for now.

Based on the above, there are at least two observations one might make. First, if the Homo sapiens species is positioned to potentially spread life across the galaxy, what happens if it fails to attain the level of maturity needed to fulfill this species survival evolutionary pattern? More than likely, as the dinosaurs gave way to mammals, Homo sapiens in turn would give way to another—one would hope—more intelligent and mature species. But until then, the spread of intelligent life across the galaxy, viewed from any perspective, will have suffered a huge setback and will have to await a successor species, one capable of going boldly where no Homo sapiens has gone before.

Second, evolution's bid to overspread the galaxy with intelligent life would seem to depend on an evolving intelligent species attaining the necessary maturity to avoid self-extinction. This realization brings one face-to-face with what appears to be yet another species survival evolutionary pattern, one we will call "species-driven evolution."

It would seem that evolution has dealt intelligent species a different kind of hand, the ability to participate and even control the success or failure of the evolutionary outcome—intelligent-species-driven evolution. For the first time, evolution has endowed a species with the ability to control, within limits, not only its own survival but the survival of all other species and of the sustaining habitat itself. One potential outcome is that this species might fail. The 4-billion-year-old question is: Why? Why would evolution risk this outcome?

Not only was a very, very long time, 4 billion years or so, required for life on Earth to evolve an emergent intelligent species and support an appropriate set of sustaining environments and habitats, but an additional very, very, very long time might be required for that emerging intelligent species to attain evolutionary maturity and then develop the technologies to become not only an intelligent communicating species, but a spacefaring one as well.

What if it turns out that Homo sapiens is the only extant intelligent species in the galaxy? What if the evolutionary pattern to populate the galaxy with intelligent life resembles the spread of Homo sapiens across the Earth? And what if intelligent species really do need to attain that specific level of evolutionary maturity to avoid self-extinction? Might this not explain the vast distances between planets, between stars, the galactic distances that seem intended to make sure intelligent species are truly ready for that next step before they can begin migrating to other planets, and on across the galaxy?

To qualify for the opportunity to populate the galaxy, evolution appears to be saying that an intelligent species must be able to survive and, in order to survive, must attain evolutionary maturity. Otherwise, regardless of how smart, how inventive and innovative it becomes, without attaining evolutionary maturity, it is nothing. It will do itself in and disappear into self-extinction. And, like all other evolutionary adaptations that fail to bestow a distinct species survival advantage, its one-of-a-kind ability to control its own evolutionary outcome might, in its current form, vanish with it.

No optimistic member of our species would be rooting for that outcome, but one can't fault the evolutionary process for placing that maturity roadblock in our way in order to ensure that an intelligent but immature species—as some think humans currently are—doesn't get to export and propagate that immaturity across the galaxy. But Homo sapiens still have time and, with time, who knows. Time, however, will not change the evolutionary equation: mature to survive or go extinct. It is Homo sapiens who must change in order to survive.

As a fan of science fiction, I often wonder why so many authors, directors, and producers envision a dystopian future often populated with a galaxy of warring species. I am certainly not wishing for a galaxy full of warring spacefaring nations (are you?) with the likes of Klingons, Cardassians, Ferengi, Vulcans, and other imagined species perpetuating, on an unimaginably grander scale, the failures and immaturity we now experience on Earth.

Is it because we, but for the teachings and sometimes-nebulous visions painted by some religious texts, have never had a concrete model of evolutionary maturity on Earth? Maybe. But that's our challenge. Maybe those who wish for an evolutionary mature human species should perhaps start imagining what that might be like, and sharing that new vision, creating the myths, writing the books, and making the movies depicting an evolutionary mature human world. I challenge moviemakers and science-fiction writers to create visions of what an evolutionary mature Homo sapiens world might look and feel like.

Notwithstanding the evolutionary safeguards against an immature intelligent species spreading its immaturity across the galaxy, it still doesn't quite answer the question fully. Why would evolution risk what seems to be a high probability of failure on the part of an intelligent species? Is this the only evolutionary path forward? Is sharing control of the evolutionary outcome with a proven mature intelligent species what extraterrestrial evolution requires? Perhaps.

To begin to think of potential answers one must again search the species survival evolutionary patterns for possible clues. One that comes readily to mind, and it may not be the best one, is that of the parent-offspring pattern that pervades nature and the universe. This is a well-understood pattern: the parent gives birth to and nurtures its offspring, enabling that offspring to go forth into the world on its own. The parents in many species remain available, but the offspring must now continue to mature so it can in turn assume the responsibilities of raising the next generation.

Nobody ever thinks of evolution this way, yet evolution bears all the hallmarks of a parent: a maker of species and habitats for them to thrive and survive in, forgiving of their growing pains to a fault. But the time comes when that offspring must mature sufficiently not only to make a life of its own and raise its own offspring, but also to take its place in the world, supporting, adopting, and expanding on what is good and wholesome, and discouraging what is not. If the offspring survives to adulthood, there inevitably comes the time when the offspring must put all that it has learned into surviving to maturity, or else it will perish.

And so it is with our species. Evolution, like a parent, appears to have done all that was necessary to get us to this stage, first sharing control of our destiny with us, eventually handing control entirely over to us. We must now continue to mature or perish. The Earth, the solar system, and the universe will still be here if Homo sapiens species go extinct, and other species will continue to evolve until one matures and goes on to spread intelligent life across the stars. But wouldn't it be great if that successful species turns out to be Homo sapiens? If Homo sapiens is the only extant intelligent species, can we also attain evolutionary maturity to become the species to spread intelligent life across the stars?

That opportunity might well be ours to lose.

We conclude this chapter by drawing attention to how some of these evolutionary species survival patterns noted earlier appear to collaborate to achieve an evolutionary outcome. We have seen how patterns such as change in the form of climate change initiated a loss of habitat that in turn forced species to adapt and innovate, and when all else failed, to migrate to survive, which in turn drove early humans out of Africa and eventually across the Earth.

We also noted that that migration began in Africa and then spread across the Earth, and wondered if this could also be a pattern, one that evolution might repeat, beginning on Earth with a single surviving hominin species, Homo sapiens (or a subsequent intelligent, but more mature species), and going on to spread life across the rest of the galaxy.

We have also seen how the control and outcome of the evolutionary process for the first time appears to have been shared with an evolving intelligent species, exposing yet another pattern, species-driven evolution, and wondered whether species-driven evolution might be not only a prerequisite for Homo sapiens' survival, but perhaps, just as momentous, a new kind of evolution and a prerequisite for the spread of life across the stars—extraterrestrial evolution.

In the next section we go on to explore how evolution may have set about the making of an intelligent species. We will examine the role species-driven evolution might have played in intelligent species' survival and might continue to play in the potential spread of mature intelligent species life across the galaxy. We also look back at how evolution may have used climate change to guide and shape our hominin forebears and how it might yet shape the evolutionary outcome of us, Homo sapiens, their descendants.

PART TWO

The Making of an Intelligent Species

From Slime to Ape-man to Spaceman

UNLIKE ALL OTHER SPECIES, INTELLIGENT species like Homo sapiens evolved with latent capabilities that allow them to choose their own evolutionary outcome. The question is, will Homo sapiens use these amazing capabilities to choose survival, or use them instead to drive its own extinction?

This section examines the amazing but necessary transformation of a basic, instinct-driven, tree-hugging, ape-like creature on all fours into a bipedal, erect, self-aware, nascent intelligent species. It explores its struggle to choose between evolutionary maturity and survival, or extinction, and how evolution "guides" such species through this amazing transformation.

This section also looks at how an intelligent species' use of these capabilities engenders a new kind of evolution—species-driven evolution—and how species-driven evolution may eventually give rise to a new, re-engineered, and enhanced Homo sapiens (Homo machina?), a transition not just from ape-man to spaceman, but eventually from spaceman to Bicentennial Man.

It examines, too, how, much like a chaperone, evolution utilized climate change to guide intelligent species in the discovery and use of these newfound capabilities to adapt, innovate, mature, and migrate to survive.

In the end, it is the success or failure of this transformation that will determine whether an intelligent species such as Homo sapiens used its abilities to adapt, innovate, mature, and migrate to survive, or will have used them instead to make, create, and innovate its way to its extinction.

The Making of an Intelligent Species
From Slime to Ape-man to Spaceman

CHAPTER 5

Species-Driven Evolution
A Grand Evolutionary Experiment

Did Evolution Deal Intelligent Species a Hand in Its Own Evolution?

CHAPTER 6

The Role of Climate Change in the Making of an Intelligent Species

Evolution's Tool to Prod Intelligent Species to Greater Maturity and Survival or Extinction

5.

Species-Driven Evolution

A GRAND EVOLUTIONARY EXPERIMENT

Did Evolution Deal Intelligent Species a Hand in Its Own Evolution?

THE 4-BILLION-YEARS-LONG QUESTION WE HAVE TO ask is what might have been the "evolutionary motive," if there can be such a thing, for evolving a species with the ability to control its own evolutionary outcome? Why evolve a species with the ability to freely adapt and innovate in ways that can enable its survival, as well as its extinction, along with all other species on the planet? Why would evolution choose to bet 4 billion years of evolution on such a species making the right call?

It is perhaps not too difficult to decipher why evolution would give a species control of its own evolutionary outcome in a way that allows for choosing survival. If the "evolutionary goal" is to evolve a life-form capable of surviving in the universe, despite ceaseless existential threats and the universal pattern of birth, life, and death, clearly that life-form has to be capable of doing what the dinosaurs could not—adapt, innovate, mature, and migrate to survive. The question of why evolution would make a bet of it, we will get to in a later chapter.

So, one can identify at least one reason these exceptional adaptive and innovative capabilities were necessary: to have the ability to find a way to survive and enable life to continue whereas all prior species could not.

But while survival might have been the choice evolution "may have hoped" an intelligent species would ultimately choose, Homo sapiens seems to have chosen instead to have a blast destroying species and habitats like there is no tomorrow, and apparently couldn't care less about survival.

We'll discuss how Homo sapiens has made use of these capabilities so far, and whether, as a species, we will figure out what this freedom to choose our own evolutionary outcome might have been all about.

As a species we dreamt and wondered what, if anything, might follow Homo sapiens. Are we the end of the evolutionary chain of intelligent species on Earth? More than likely not, and here's why.

Each phase of the evolutionary process has been followed by a more complex phase, often punctuated by an extinction event, with organisms in the prior phase going extinct and/or becoming literally absorbed by or replaced, in whole or in part, by organisms of the follow-on phase, emerging in new and often different habitats and requiring different adaptations and innovations to survive.

Thus, noncellular organisms evolved into single-cellular bacteria, prokaryotes, the most famous of which is LUCA (an acronym for last universal common ancestor and is the ancestor of all life on Earth, and existed some 4 billion years ago). Single-cellular bacteria evolved into nucleated single-cellular bacteria that evolved into nucleated multicellular microbes—eukaryotes. These in turn became the foundation upon which all complex life is based, becoming the organs and adding to the nanomachines developed by prokaryotes, Earth's early and sole inhabitants during the first 2 billion years of the planet's existence. Eukaryotes evolved into plants and animals that 2 billion years later evolved into the complex animal species of the Cambrian explosion (such as the mighty dinosaurs and tiny mammals that filled the niches vacated after their extinction). Then about 6 million years ago, intelligent species (hominins) evolved and eventually (300,000 years ago) Homo sapiens, the last of the hominins—so far.

One can't help but notice the step-wise increase in species complexity and sophistication, the change in habitats and levels of adaptations and innovations, which seems to suggest that evolution never intended to stop at evolving just bacteria; rather, complex animals and intelligent species might have always been on the menu. The question we are considering is: Now that an intelligent species like Homo sapiens has arrived, what else might be on the menu? What new species, habitats, adaptations, and innovations might these need to survive? What, then, is the next level of evolutionary species complexity?

How might our species evolve, and does that evolution include the kind of cyborgs seen in many science-fiction novels, such as Isaac Asimov and Robert Silverberg's *The Positronic Man,* or movies like *The Bicentennial Man,* based on their story? Perhaps it is the android Data, of *Star Trek: The Next Generation,* that best captures what some imagine that transformation might bring about. If such transformations are possible, how do Homo sapiens get there from here? Is species-driven evolution a prerequisite for survival and hence a nascent capability built into the evolution of intelligent species?

If an intelligent species has no choice but to become a multi-planet species, as noted in Chapter 3, and must develop the capabilities to migrate from planet to planet and from solar system to solar system, as early humans had to migrate from continent to continent to survive, then intelligent-species-driven evolution might be a prerequisite for Homo sapiens survival.

Our forebears needed nothing more than their human anatomy and any continuing biological and mental evolution to migrate out of the Great Rift Valley, in what is now Ethiopia, to the ends of the Earth. However, from what little experience humans have had with spending protracted periods in space—and NASA's Twins Study has done little to allay potential fears—migration from Earth to other planets in the solar system, or to other solar systems, suggests that our current biological bodies might be in need of an urgent upgrade. Everything from morphing hearts, to reduced blood oxygen content, to shriveling bones and muscles, to increased eye pressure as fluids rush towards the head are just a few of the findings, and many of these seem to worsen with protracted stay.

And since it is unlikely that our biology would naturally evolve the kinds of upgrades protracted interplanetary or inter-solar space travel might require (at least not in the time frame needed), then it is not unreasonable to conclude that intelligent-species-driven evolution might be a prerequisite for Homo sapiens' survival. If so, evolution may have dealt Homo sapiens a hand in its own evolution, a capability no other species appears to have evolved.

Just as our forebears 70,000 years ago could not have imagined themselves evolving from ape-man to spaceman, so we today may struggle to envision yet another intelligent species transformation, this time from spaceman perhaps to Bicentennial Man.

There is of course the near- to intermediate-future alternative often depicted in science-fiction movies in which starships' life-support systems include, among other sustainability features, artificial gravity, an Earth-like atmosphere, and advanced food generation and waste recycling systems.

It is highly likely that this will be the default approach as long as it can be sustained. After all, no one is yet considering fitting humans with gills for breathing in the global oceans believed to be beneath the icy crust of Saturn's moon Enceladus, or with the ability to survive on Mars without water, oxygen, and built-in resistance to UV (ultraviolet) and gamma ray bombardment.

But what is species-driven evolution anyway, and how might one recognize it? Where can one begin to look to find clues or traces of Homo-sapiens-driven evolution at work today?

> *Intelligent-species-driven evolution is the ability of an emergent intelligent species to shape and direct its own biological and nonbiological evolution in order to survive.*

It appears to be an emergent adaptation in the evolution of the mind to lift human intelligence beyond the current limitations of the physical human brain and body, a transformation that is perhaps similar to the incremental growth in the size of the brain during our evolution from ape-man to spaceman. With species-driven evolution, Homo sapiens would have been given not just the keys to the planet, but also the ability to read and program the very code of life. This may seem absurd until one begins to understand the challenge getting to evolutionary maturity and survival past the point of self-extinction will entail.

It is not surprising then to find scientists working to harness the power of evolution. When announcing 2018 Nobel Prize winners in Chemistry, the Royal Swedish Academy of Sciences declared that 2018's laureates have "re-created the process, speeding up evolution by shuffling genes artificially in their test tubes to make evolution many times faster." According to Frances Arnold, one of the three winners, "I wanted to rewrite the code of life, to make new molecular machines that would solve human problems." Homo sapiens, the Academy claimed, is in its early days of a directed evolution revolution. Enzymes produced in a lab through "directed evolution" can be used to make everything, including renewable fuels and drugs.

Similarly, earlier, in what the *New York Times* characterized as "one of the most monumental discoveries in biology" and what Jennifer Doudna and Samuel Sternberg describe with the title of their book *A Crack in Creation: The New Power to Control Evolution*, Homo sapiens has discovered the means to peer even deeper into the workings of evolution and has begun to figure out how to potentially reprogram the evolutionary code of life.

Whether it was creating a Schwarzenegger-like muscular version of the humble beagle, or creating micro-pigs no larger than cats, using an innovative gene editing tool called CRISPR, or CRISPR-Cas9 (clustered regularly interspaced short palindromic repeats-Cas9), geneticists were able to edit these animals' genome, re-engineering the very code of life. Gene-editing has also been used to produce disease-resistant rice and many other genetic changes to nature's flora. In humans, the focus has been mainly on treating many diseases. With the capability CRISPR provides to edit and repair mutated genes in humans, it is now possible to correct mutations responsible for diseases like Duchenne muscular dystrophy, cystic fibrosis, sickle cell anemia, and more.

These capabilities might be all new for Homo sapiens, but it's important to note that bacteria, threatened by UV radiation long before the ozone layer was formed, developed DNA repair systems and have been editing and splicing DNA and swapping genes through techniques such as horizontal gene transfer for over 2 billion years. The very techniques celebrated by the Nobel Committee as twenty-first-century breakthroughs were for the most part imitated from the way certain viruses invade and replace parts of the DNA of host cells with some of their own DNA—all part of a few milliseconds of work and not for a Nobel Prize but just to survive. While Homo sapiens view these newfound capabilities as means to speed up evolution, it seems that we are the ones who have some catching up to do—about 2 billion to 4 billion years' worth.

But as we will see in a later chapter on innovation, CRISPR, like all prior Homo sapiens' innovations, has a potential dark side. It could also be used to change the genome of Homo sapiens, to modify the human germ line and change the genetic information passed on to subsequent generations. When viewed from an evolutionary perspective, that may not necessarily be a bad thing in the hands of a mature intelligent species (after all, bacteria have been using these techniques to survive for billions of years) and may well have been the evolutionary intent, a logical outcome meant to enable an intelligent species to drive its own evolution and may well, too, be the next stage in species-driven evolution.

While there are many bad things an immature intelligent species can no doubt do with such a capability, the fact that it is now possible suggests that evolution may have concluded that intelligent species survival may well depend upon having this capability. Such innovations and adaptation endow humans with, among other things, the ability to conceive and make adaptations to human anatomy and physiology to enable survival in non-Earth-like environments, such as Mars. Perhaps more important, it could give humans the ability to evolve and replicate themselves in the form of nonbiological, emergent, super intelligent life-forms—Homo sapiens 2.0.

To find clues of Homo-sapiens-driven evolution, one will need to look beyond the process of natural selection. While the exact outcome of biological evolution often remains a work in progress, and ultimately is unknowable, the purpose and evolutionary intent of species-driven evolution could not be clearer—adaptations that enable intelligent species survival. But even as an intelligent species undergoes species-driven evolution, it still needs to attain evolutionary maturity to survive.

All previous species have been at the tender mercies, or harsh tyrannies, of the inexorable forces of nature, and remained powerless to influence in any meaningful way the outcome of their evolution. The dinosaurs had no ability to know of the asteroid headed their way; had they known, they lacked the ability to take any action to mitigate the effects. And while

NASA is busy working on solutions, Homo sapiens, despite all its intelligence, is currently not much better off than the dinosaurs.

According to the late theoretical physicist Stephen Hawking, "we are at a point in history where we are trapped by our own advances, with humanity increasingly at risk from man-made threats but without technology sophisticated enough to escape from Earth in the event of a cataclysm. Although the chance of a disaster to Planet Earth in a given year may be quite low, over time it becomes a near certainty in the next thousand or ten thousand years."

There is in fact nothing that any species has been able to do to thwart or escape such planetary, galactic, or universal forces of nature and their evolutionary consequences. None, that is, except one—hominins, and now Homo sapiens, the only surviving emergent intelligent species on the planet, which might one day be capable of escaping, if not defeating, such threats. Again, according to Hawking, "By that time [the next thousand to ten thousand years] we should have spread out into space, and to other stars, so a disaster on Earth would not mean the end of the human race," a potential evolutionary outcome possible for no other species.

Unlike any other species, humans have evolved capabilities that allow them to change the potential outcome of nature in many respects, both for better and for worse. Some uses of this ability have been beneficial, and augur great promise, while others inspire great fear and show we are equally capable of creating dreadful nightmares.

Today, within limits, we can predict aspects of the weather, warn of impending hurricanes, tornadoes, typhoons, and monsoons, and to a lesser degree, earthquakes and tsunamis. We have also mostly acknowledged that man-made global warming is in progress and we can, if we choose to as a species, take corrective action to mitigate some of its potential devastating effects, as nearly 200 countries agreed to do in December 2015 by pledging to constrain global temperature increases to less than two degrees centigrade. But today we can also launch an inter-continental ballistic missile from a submerged submarine to destroy an entire nation, underscoring the fact that while our amazing ability to invent and innovate is essential to survival, it is by no means sufficient, and can potentially become the means of our extinction.

Hence, unlike Professor Hawking, while I remain optimistic about human survival, I suspect that if it were possible for Homo sapiens to begin to populate the stars before attaining evolutionary maturity, rather than ensuring our survival it would probably merely delay our extinction. After all, evolutionary survival is not achieved by creating such threats and then fleeing from them; to ensure survival, we need to stop creating them in the first place.

Clearly, then, one place one might look to find human-driven evolution at work is in the evolution of the human mind. Our species' efforts to harness the powers of the mind for the good of all remain a work in progress. The biological component of our evolution that has resulted in the anatomy and physiology we now recognize as human might be only part of the story, a necessary step to what now appears to be a follow-on phase—the evolution of the human mind and the attainment of evolutionary maturity in order to survive.

The difference between these two components of human evolution, the body and the mind, appears to be that in the former, the forces of nature are the main arbiters of the evolutionary outcome, while in the latter, the human mind is clearly in command. The forces of nature are always present, but in the case of humans, they're complemented—in some instances subordinated, or even harnessed and replaced—by the powers of the human mind to choose a different outcome.

Given its power to thwart nature, it becomes easier to understand why, as an evolutionary safeguard, an intelligent species must first attain evolutionary maturity to avoid extinction and ultimately survive. Without attaining this level of maturity, an intelligent species will eventually destroy itself and go extinct.

To find evolution at work, then, one need look no further than the history of Homo sapiens. When viewed as the evolution of the maturing human mind, the making of an intelligent species, rather than just history, one begins to see what might be evolution at work today, before our very eyes.

For the first time, a species has evolved with the nascent ability to control not only its biological urges but also the capacity to develop the means to replace that biology, eliminating some of its limitations and frailties altogether. However, while bestowing this awesome capability on our species, evolution did not simply give away the keys to the planet but appears to have left a built-in fail-safe, an evolutionary survival roadblock—to survive, an intelligent species must attain maturity.

If one can think of the evolution of Homo sapiens as occurring in phases, with biological evolution being the first phase, then the evolution of the mind can be thought of as the follow-on phase. And it is to this second phase of Homo sapiens evolution we now turn.

A teenager behind the wheel of an enormously powerful car is severely tested by the urge to see what that car can do. Add irresponsible companions egging him or her on and the potential for disaster increases. Add self-induced impairment that can result from alcohol or drugs, and other conditions such as the physical state of the terrain, the quality of the road surface, and the weather, and one has a disaster waiting to happen. That same teenager, now a mom or dad, a few years older and a few years wiser, assuming he or she survived or avoided the disaster, would not be so anxious to hand over the keys to their teen.

So, in many respects human history can be viewed as the growing pains of the continuous evolution of the human mind.

While the Earth has been spinning daily on its axis at about 1,000 miles per hour; as it hurtles once per year around the sun at 67,000 miles per hour; even as the solar system itself travels through the Milky Way at about

43,000 miles per hour, our species, like an immature teenager, has seen fit, in an infinitesimal fraction of geologic time, to invent, develop, and perfect multiple means to destroy itself.

Yes, it's true our species has managed to survive near-total extinction, both self-inflicted and natural, and to some extent may have learned from these experiences—but have we? Looking no further back than the last ten decades, it seems as though Homo sapiens is bent on self-extinction, rushing from one potential-extinction event to the next. From the First World War, to the Second World War, to the Cold War that came within inches of engulfing the planet in a nuclear holocaust; from the Namibian genocide to the Armenian genocide to the Jewish genocide to the Serbian genocide to the Rwandan genocide, one can see no end as new genocides and crimes against humanity rage on today in such conflicts as the Syrian civil war and the Rohingya expulsion from Myanmar.

We appear to still be quite evolutionarily immature, early in the process of mastering the levers of the mind to manage ourselves as individuals and as a species. Some of the issues humans have encountered might make it possible to better estimate the probability that we'll make it to mental adulthood. Some among us have done just that and have come away not very hopeful about the potential outcome.

Nonetheless, when reviewing this history, remember the goal of the evolutionary process—species survival, not species extinction—and that to survive, an intelligent species must attain evolutionary maturity. So, if the goal of the evolutionary drive in any species is to survive, one must look at the results of the works of the human mind and eliminate patterns that have proven to be destructive and a threat to our survival, while encouraging those patterns that engender maturity and enhance the chances for survival.

The history of our species—actual, imagined, and in some cases reconstructed—has mostly been documented and continues to be studied and reinterpreted. It is not our purpose here to retrace these journeys down every blind alley into which our species has wandered. Rather, the purpose here is to identify the trends and possible destinations of these trends our forbears have followed. Today, perhaps, we ask the same questions and dream the same dreams they dared to ask and dared to dream.

Like creatures trapped in some endlessly intricate cosmic maze, early humans began this grand journey of self-discovery preoccupied with a search for means to survive and initially relied, like every other species, almost entirely on instincts to forage for food, find shelter from the elements, and defend self and family from predators. In this regard, humans were outmatched by many other species that evolved with superior natural physical capabilities. While this routine, like an endlessly looping subroutine, ran unchanged in the brains of other species, the human mind, like a supercomputer running nothing more than a few low-level interrupt-driven survival applications hard-coded by nature, became engaged in aiding humans in this quest, and the evolution of the mind soon became the thing that differentiated humans from other species.

Surrounded by a plethora of flora and fauna, yet alone, our species began what must have been a frightening journey of learning, through trial and error, how to program this computer we know as the brain. Millions of years of programming later, it has barely scratched the surface and has yet to learn just how this computer of ours really works, or how it should best be used. Every other species evolved fully loaded with all the programming they will ever need to survive, but intelligent species must figure out what programs it needs and learn to write those programs, even as it must discover along the way what the requirements for such programs might be.

Is there a way to program this computer to ensure Homo sapiens' survival rather than extinction? Has evolution left us any programming notes, an algorithm or two that can point us in the right direction? If so, can we decipher the evolutionary language?

As our ability to use the powers of the mind has evolved, so have our fortunes and so have our fears. We invented ways to grow and preserve food; organized ourselves into tribes, clans, and villages; invented gods to call on to bless and protect us and give us hope, real or imagined; invented elaborate religions to lend meaning to our existence and answer questions about our origins, calm our deepest fears and yearnings, and comfort us in death and bereavement.

Thus the mind, with its powers of imagination, has enabled Homo sapiens to create mental constructs that have served our ancestors well, providing answers that have become enduring traditions. In some cases, though, they are empty superstitious rituals, enshrined in institutions, national, regional, cultural, and religious, many of which have outlasted their usefulness.

It seems, too, that we are becoming incapable of keeping up with the accelerating rate of transformation in our world that our minds continue to generate, unable to free ourselves from the cocoon of the numerous bygone, exhausted cultural practices, failed political ideologies, and spent religious practices, so we continue to worship gods we no longer believe in yet to which we still cling.

Through the millennia, our species has conjured up numerous and diverse contrivances out of our urge to survive: tools to reduce the drudgery of finding food, clothing, and shelter; medical science and medicines to treat and, in some cases, cure illnesses; technologies to better document our existence, to explore and learn about the past, to better understand the present, and to peer deeply into the future. Along the way also came laws and institutions for organizing and governing our affairs, strategies and principles for managing economies, and, yes, bigger and deadlier mechanisms for protecting ourselves from one another, to destroy enemies, real or imagined.

As we continued to learn to use our mental computers, we made many advances in the quest for survival and the search for meaning, some of which have not been without cost. It was not that long ago (2018 marked the 50th anniversary of the famous *Earthrise* photo taken by astronauts aboard Apollo 8 near the moon) we came to realize we are hurtling through space on Spaceship Earth, a rotating planet whose spin, tilt, and direction as it orbits hold huge implications for our survival, but over which we have no control. Like a child in the pilot's seat of a B-52 bomber on autopilot, Homo sapiens has to learn how to take over and fly without being taught, and is still figuring out what levers to pull and what buttons not to push in order to avoid a disaster and touch down safely.

Yes, we did unknowingly push a few of those wrong buttons along the way. We didn't intend to punch a hole in the ozone layer, yet we did and years later are still struggling to ensure we close it and keep it closed—oops! We didn't intend to precipitate global warming, but we are finding out that we've done just that—oops! Neither did we intend in our search for food to hunt or farm species to near extinction—but we have, and are still hunting species known to be endangered, like whales. In our pursuit of survival, we have plundered and pillaged nations across continents and enslaved members of our own species in search of natural resources, wood, coal, and other fossil fuels; in many cases, we did it just to acquire wealth. In pursuit of wealth, we have strip-mined vast sections of our habitat, burned and logged pristine forests and wildernesses, polluted vast expanses of waters.

In our search to understand the fundamentals of what we see around us, we managed to split the atom, and once more—oops! Like opening Pandora's box, we find ourselves rushing madly about trying to eliminate weapons of mass destruction. We were also naïve, not realizing the Earth and its habitats form a closed system. We thought we were being clever in avoiding the cost of safe disposal of industrial waste, but we are only now finding out the consequences of all the chemicals we have dumped into aquifers, rivers, lakes, and oceans. We have yet to learn the full extent of the damage we have already done to ourselves by poisoning the food chain. Have you had some tuna lately? Oops!

Though we carry around a very powerful computer between our ears, we are by no means a stand-alone computing system, disconnected from everything else. As a member of the set of species on this planet, we are, like all the others, part of an intricate, global computing network the scale of which makes the internet seem like a home computer network. Everything we do affects something else, and these effects ripple through this system, around the globe. It may take a while, as global warming has, but in one form or another, sooner or later those effects come back full circle to bite us.

To our credit, we are trying to correct some of these missteps in numerous international treaties, as in the Montreal Protocol on substances that deplete the ozone layer agreed upon September 1987 and the United Nations Climate Conference COP24 rulebook agreed upon by nearly 23,000 delegates in Katowice, Poland, but only time will tell whether we will succeed and mature. Otherwise, though we may not have thought about it this way, we may well be writing the "How Not To" chapter of the programming manual for another species yet unborn, as the dinosaurs, in the fossil record, unwittingly did for us.

So where are we headed? Is there some grand scheme for our species? Religion, soothsayers and futurists would like to think there is. I'm sorry to disappoint, but from an evolutionary perspective there appears to be just one—survival. The one thing we know for certain is that everything our species did had initially been driven by the urge to survive, but it seems that survival urge has since been replaced by a growing desire for wealth and profit. Hopefully the time will come when Homo sapiens will return to being driven by this urge to survive.

While each incremental increase in technology brought an improved ability to understand our past, present, and future, many have also come with intoxicating and addictive qualities and continue to survive parasitically, exploiting certain human proclivities that enabled them to be developed in the first place. Similarly, while many evolutionary mental constructs served a purpose at a particular stage of human development, many appear to have survived long past their usefulness (we will explore some of these in a later chapter) and may well turn out to be serious impediments to taking the next evolutionary steps, or even surviving. Indeed, as Einstein is believed to have said and the rest of us are beginning to realize, our technology has far outgrown our humanity.

All species evolve due to natural selection, but not all species evolve at the same pace, or respond in the same way to identical natural stimuli. This selection is driven by a species' need to survive, in response to complex interactions with natural phenomena. So if these interactions call for a species to change the color of its skin (like Mediterranean octopuses) to camouflage its appearance in the presence of a predator, or to develop the biological components to transition from land dwelling to water dwelling, these species are forced to adapt to survive.

In humans, the process of natural selection that shaped our species is still at work at a biological level, even if it is occurring at too slow a pace to be recognized. However, as suggested, human evolution is not confined to biology but also continues in the mind; forms of species-guided selection may have replaced natural selection in the realm of the mind.

Using the mind, humans have been able to do many things that are natural for other species but biologically unnatural for our species. Today humans are able to take flight as birds do; explore the deep like fish for considerable lengths of time with the aid of a scuba-diving outfit; remain submerged for weeks at a time in a submarine; see in the dark with night-vision equipment; navigate like a bat by bouncing sound waves off objects to identify their presence; and hear sounds from parts of the spectrum known to be audible to dogs. Humans have done all of these things out of necessity.

In short, humans have had to adapt, not by changing human biology (at least not in these examples), but by augmenting it. And while some species are known to make simple tools, the scale and the extent required to produce the examples cited above is a capability entirely unique to humans.

If the kind of natural selection and adaptation experienced by nonhuman species result in changes mainly to their physical biology, the adaptations experienced by humans today—in addition to subtle biological changes we may not readily perceive—appear to occur not by growing gills to breathe under water or sprouting wings to fly, but by using the mind to create nonbiological extensions that enable near-similar, equivalent, or superior capabilities.

This ability of our species to continually adapt to the world by using the mind to fashion the physical extensions needed to survive is nothing less than evolution at work today. But it is not the kind of evolution for which one may have been looking. To differentiate it from the kind of evolution one has been accustomed to, the kind that results in biological changes due to natural selection and adaptation, this kind of evolution is emergent, exists only as constructs in the mind, and, unless shared with others or recorded on some medium outside the body, vanishes when the brain ceases to function.

One might want to regard this emergent property of human evolution as a form of species-guided evolution.

Now that we have donned our evolutionary lenses and can recognize that evolution is alive and well, it becomes easier to understand why as an evolving species we are engaged in some of the anatomical adaptations we are making today. Let me mention just a few from medical science: organ transplant and replacement; blood transfusions; numerous forms of prosthetics including hip and joint replacements; bypass surgery; heart transplants, Lasik eye surgery; and more. These efforts indicate that Homo sapiens, like any other species, is driven to find ways to prolong survival. These adaptations also point in a direction that, if continued, may well result in the eventual replacement of a substantial part of the human anatomy, voluntarily at first, but eventually out of necessity to survive.

Homo sapiens cannot grow replacement limbs the way some species do, but because our minds continue to evolve, they enable us to adapt non-biological components to serve the same purpose, albeit less than perfectly, at least for now. Many today can readily understand the need for these adaptations, as we are also driven by the same urge to survive.

But again, where might this all lead? The evolutionary process appears to be guided by a very simple rule—the urge of species to survive. Entire species have changed their habitats from land to sea and vice versa in response to this urge to survive in the face of natural phenomena that threatened their existence. We may have started such a phenomenon in the form of global warming. The Dutch have made an art out of living below sea level, an ability and adaptation that may soon be needed by many coast-dwelling members of our species.

We now also know that there are asteroids intersecting Earth's orbit that might collide with Earth and potentially end all life as we know it. We were able to witness the collision of the comet Shoemaker-Levy 9 with Jupiter in July 1994 and saw the horrific explosions and geothermal repercussions. Hopefully we will have learned either how to destroy these in space or, as the late Professor Hawking hoped, will have had the time to develop the ability to relocate ourselves to another planet before such an event occurs.

If indeed the cosmos functions with the regularity of a precision timepiece, then it might be a question not of if but of when. It is probably no accident that our species is attempting to innovate the means for interplanetary travel, as that could turn out to be the only form of adaptation that might work.

Whatever the adaptation, whether it's living in floating homes as the Dutch propose and the Vietnamese already do along the Mekong Delta, developing underwater cities as in the legend of Atlantis, establishing human settlements on space stations, or founding new habitats on other planets as Elon Musk proposes to do on Mars by 2022, it is likely to be as radically life-changing as it must have been for those species that made radical habitat adaptations on Earth (e.g., from land to marine habitats) to survive.

All of this argues for not less but more experimentation with and integration of nonbiological adaptations for our species, which can eventually lead to a nonbiologically based life-form, aka artificially intelligent robots. If our species is driven to migrate to other planets in order to survive, it is likely that preparation for the rigors of space travel will lead to adaptations that give rise to the first bionic humans.

It is entirely possible that our need to survive may result in such extensive nonbiological adaptations that there could come a time when what few biological components might remain are vestigial, more for nostalgia than survival.

Whether or not we go there might well be driven entirely by our need to survive. We have already demonstrated our ability and willingness to enhance or replace vital biological organs with nonbiological components, like pacemakers, to prolong life. It is not so hard to imagine a future in which events precipitate such a scenario. A question one might ask is: Will our species be ready? The dinosaurs weren't.

Homo sapiens history reveals a species that at first appears to be racing off in almost every conceivable direction—literally and figuratively. At times there appears to be no rhyme or reason to some of the paths we've traveled as a result of our immature wanderings—until we begin to view that history of Homo sapiens exploring different species-driven survival strategies as part of the evolution and maturing of the human mind.

Over the course of our existence, our species has generated many strategies to facilitate its survival. Many of these strategies evolved to serve the needs of a specific evolutionary era and apparently could never have become permanent fixtures. Just as other species maintain elements of their surface-dwelling biology long after they have become underwater creatures and no longer need them, so some misguided human survival strategies (or rather, extinction attempts: wars, greed, crime, should I go on?) will hopefully come to be seen as lessons learned and serve more as vestigial appendages, paths to extinction rather than steps toward ultimate survival. Like scaffoldings used during the erection of an edifice, many of these must be dismantled to give way to new scaffoldings born of evolutionary maturity, capable of leading to our ultimate survival.

Unlike the paths of the planets and galaxies in an expanding cosmos that are driven unchangingly by the forces unleashed by the initial conditions established in the Big Bang, our species, from its first sentient moment, has been driven by an inexorable urge to survive and has used the mind to chart a very different course—a species-driven evolutionary course, its own course—than might have resulted if we had remained stuck in biological automata.

But where has this course led Homo sapiens thus far? By all appearances, our collective mental supercomputer has led us down paths that appear to increase the chances for self-extinction rather than survival. It would seem that each new generation of technologies, each new generation of innovations, serves to make it easier and easier for fewer and fewer disaffected members of our species to set off a global, species-wide extinction event.

Like many, you are probably thinking there must be a better way to program our mental computers. Indeed, as the runaway cybersecurity issue bedeviling physical computing systems, making a mockery of the current computing approach and of the way we write computer programs, cries out for a rethink of the entire programming and computing paradigm, so too, perhaps, does the current approach to life and survival Homo sapiens appears to have chosen. And as we will see in later chapters, the current microcode driving Homo sapiens is more than due for a rewrite.

Whether or not a different computing paradigm can be found to lift the scourge of cyber-insecurity is not yet known, but luckily for Homo sapiens, evolution appears to have left not only the source code to enable a reprogramming and reboot of our collective mental computers but also all the programming notes one would ever need, written in the evolutionary language of survival patterns. Whether Homo sapiens is going to adopt the evolutionary, life-sustaining code and programming patterns for survival that evolution has no doubt left, only time will tell.

Whatever one may think of survival, until about 10,000 years ago, it was the prime motivating force guiding early hominins' history and direction. It persists as the overarching rationale integrating all the disparate strands of our actions. It should ultimately be the root cause of all we do, because attaining evolutionary maturity, and thereby survival, and perhaps even the opportunity to populate the stars, will depend upon this survival urge overcoming other urges that, left unchecked, would lead to self-extinction.

In the final analysis, our survival will depend, not on the species-driven evolutionary enhancements we will have conjured up; not on our entrepreneurs, startups, technologies, and innovations; and not on the abilities of our governments, economic systems, or corporations. Nor will survival come from our racial, ethnic, or national distinctiveness or our religions, philosophies, cultures, art, fashion, or lifestyles. Our survival will come only from Homo sapiens' desire to survive and our ability to mature as a species past the point of self-extinction.

Of all the things we have used our minds to make, create, innovate, and invent, conspicuously lacking has been our ability to engineer a clear path to intelligent species maturity. That remains the as-yet-unfulfilled task species-driven evolution might well have been intended to fulfill.

As one contemplates the rungs of a potential Species Survival Maturity Model that a species must ascend, like climbing the rungs of a ladder, to avoid extinction, it soon becomes clear that attaining evolutionary maturity is among the most difficult survival challenges an aspiring intelligent species must meet.

If it turns out that an intelligent species on Earth were to mature past the point of self-extinction, might any of the potential evolutionary outcomes include exporting intelligent life across the stars, or does evolution start over again from scratch on other planets?

When we began this chapter we mused out loud whether species-driven evolution was a prerequisite for intelligent species' survival and concluded that it must be, if only because survival requires interplanetary and inter-solar migration, which poses a severe challenge to human biology and technology, possibly for a couple of hundred years to come.

It is indeed amazing that written into the code of our biology are the keys to the next phase of Homo sapiens evolution—species-driven evolution. Science-fiction writers and movie producers could not have imagined that bionic, super-intelligent life-forms like Data of *Star Trek* may be incarnations of what might well be the next phase of hominins', and hence Homo sapiens', evolution; they could not have known that evolution had already embedded within Homo sapiens the potential to evolve not only from ape-man to spaceman, but, potentially from space-man to Bicentennial Man.

Because survival in the universe implies intelligent species must become capable of inter-solar and interplanetary migration, there is an implied need for intelligent species-driven evolution, the ability of an emergent intelligent species to shape and direct its own nonbiological, survival-driven evolution. As an emergent adaptation in the evolution of the mind, survival-driven evolution may result in turbocharging human intelligence beyond the current limitations of the biological mind and brain.

Human biology is unlikely to evolve anytime soon the kinds of upgrades human anatomy and physiology might require for protracted interplanetary and inter-solar space travel or to survive on planets lacking some factors critical to complex animal and human life as explored in the chapter on habitats, Chapter 7.

Species-driven evolution appears to endow humans with the ability to conceive and make adaptations to human anatomy and physiology that could enable survival in otherwise hostile environments and, perhaps more important, to evolve and replicate itself in partly biological or completely nonbiological, super-intelligent life-forms.

It is entirely possible that species-driven evolution, when triggered by an interplanetary or inter-solar survival-driven migration, could result in the evolution of super-intelligent Homo sapiens that have more in common with *Star Trek*'s Data than with Captain Kirk. The current acceleration in fields like robotics and artificial intelligence is a sure sign that Homo sapiens is already headed that way.

But the ability to survive in non-Earth-like environments or survive protracted space travel cannot ensure species survival of either biological or super-intelligent versions of Homo sapiens unless and until evolutionary maturity is attained. And scientists are right to fear that a superintelligence created by Homo sapiens might well inherit the same flaws and weaknesses that have plagued Homo sapiens civilization throughout history, and still do.

Without attaining evolutionary maturity, all that super-intelligent Homo sapiens will do is supersize the problems that plague humanity today and thereby supersize the effort it will take to get to evolutionary maturity. Our history has shown that as Homo sapiens grew smarter, our ability to destroy grew greater while requiring inversely fewer of us to trigger a species-wide mass-extinction event.

The purpose of species-directed evolution is not to obviate the need for evolutionary maturity but to enable Homo sapiens to embark upon an even grander scale of evolution in conjunction with it—one that will enable a mature Homo sapiens, biologic or positronic, to spread intelligent life across the stars.

6.

The Role of Climate Change in the Making of an Intelligent Species

EVOLUTION'S TOOL TO PROD INTELLIGENT SPECIES TO GREATER MATURITY AND SURVIVAL OR EXTINCTION

WHAT EXACTLY IS EVOLUTIONARY MATURITY? WHILE there are other definitions, it can be useful to think of evolutionary maturity as a survival plateau, a state that results when an intelligent species has matured beyond the likelihood of self-extinction.

If evolutionary maturity can be thought of as a state that enables an intelligent species to resist and overcome all urges that could lead to species extinction, then attaining evolutionary maturity must be the highest achievement level on the Species Survival Maturity Model. As we will see, among the many "tools" deployed by evolution to prod an intelligent species toward evolutionary maturity, climate change will no doubt come to be seen as the master tool.

During the evolution of intelligent species, climate change appears to either drive these species to extinction or drive them to adapt and move up to a higher level of maturity. Evolution and climate records have shown that our species has adapted to and survived climate change so far, but it has now gone from being shaped by climate change to inadvertently precipitating climate change. It remains to be seen whether the evolutionary prod from twenty-first-century climate change will push Homo sapiens to reach for a survival plateau or over the edge into extinction.

Climate change appears to act like the ultimate species survival chaperone. It took over 4 billion years for intelligent life on Earth to reach this point, during which numerous climate events drove millions of species, including many hominin species, to extinction. By the look of things, that might be the fate of Homo sapiens, too, unless our species can get to the survival plateau of evolutionary maturity. This brings up the question of how long a species might have to attain evolutionary maturity before going extinct.

Perhaps there is no fixed expiration date, but one can glean a sense of how long a species has by looking at the approximate dates below during

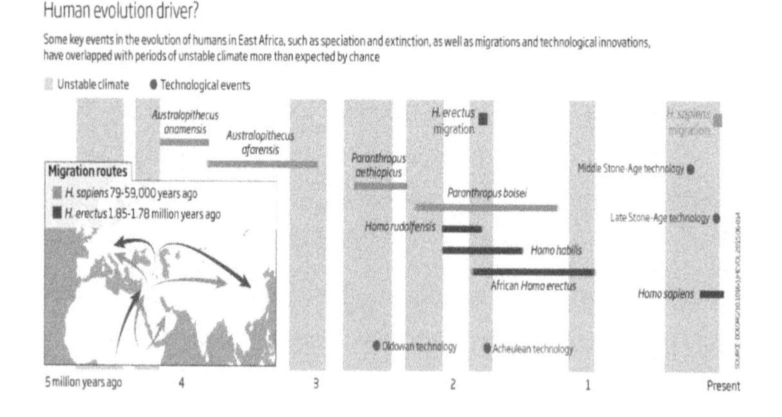

Fig. 6-1 Climate change and Homo sapiens evolution

Swings between wet and dry landscapes pushed some of our ancestors toward modern traits—and killed off others.

(Source: https://skepticalscience.com/humans-survived-past-climate-changes.htm)

which various African hominin species evolved, thrived, migrated, and went extinct, and how these may have coincided with successive climate events. Swings between freezing cold, wet, and arid landscapes pushed some of our ancestors toward modern traits—and killed off others.

Table 6-1 Longevity of some prior hominins

Species	Years
Australopithecus afarensis. Existed between 3.9 and 3.0 million years ago	900,000
Australopithecus africanus. Existed between 3 and 2 million years ago	1,000,000
Homo habilis. Existed between 2.4 and 1.5 million years ago	900,000
Homo erectus existed between 1.8 million and 300,000 years ago	1,500,000
Homo neanderthalensis. Existed between about 400,000 and 40,000 years ago	360,000
Homo sapiens, by comparison, have been around so far for only	**300,000**
Dinosaurs: Dinosaurs lived on Earth for about	**165,000,000 - 175,000,000**

(Source: Author)

It would seem from this incomplete hominin sample that the duration of prior hominin species ranged between the 360,000 years of the Neanderthals and the 1,500,000 years of Homo erectus, with 900,000 years, an approximate average, coming up twice in the table above. One cannot infer from any of the above data how long a particular hominin species like Homo sapiens might survive, but one can note the low and high ends of the survival range in the small sample. The takeaway here is that regardless of their duration, all prior hominin species went extinct. In contrast, it is at once remarkable and humbling to think that with all of hominins' increasing capabilities, the dinosaurs outlasted all hominin species combined. It is also worth noting that neither dinosaurs nor earlier hominin species appear to have brought extinction upon themselves. A more complete chart of hominins' evolution, survival, and extinction follows.

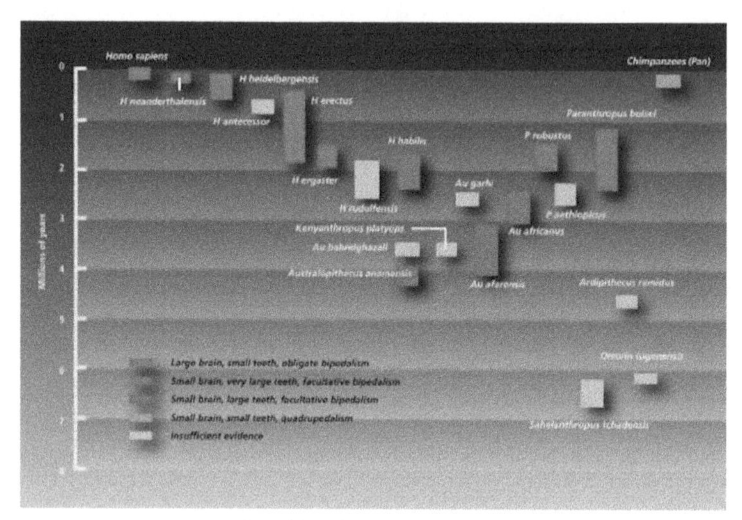

Fig. 6-2 Hominin species life-span
(Source: Pratclif.com)

Viewed on the scale of geologic time, there is no steady state a species can achieve that means it has it made and can now kick back and relax, if only because the universal treadmill of birth, life, and death has placed an expiration date on all stars, planets, and galaxies. It rolls inexorably on, altering and destroying old worlds, habitats, and species and building new ones. Just as an expert surfer rides oncoming waves, jumping from one to another in order to keep moving and not get pulled under, an emerging intelligent species will need to mature enough to jump from planet to planet, from solar system to solar system, possibly even from galaxy to galaxy. Such, I am afraid, might be what our future holds, and could be the fate of our species should we survive the looming, but this time self-inflicted, climate change.

Assuming an intelligent species survives and gets to that level of maturity, it will probably have only a few million to a billion years left before the star around which its home planet revolves begins to commit hari-kari by expanding into its habitable zone, incinerating everything in its path, and burning the planet to a crisp, extinguishing all forms of life. Moreover, that time window could turn out to be much, much shorter as with aging of the star, the solar heat increasingly intensifies, encroaching upon and destroying former habitable zones around the star and thereby destroying sustainable habitats on the planet.

So even though the sun won't die for another 5 billion years, it will grow into a red giant so large that it will engulf the planet and make Earth uninhabitable much sooner than its expiration date. After about a billion years the sun will become hot enough to boil the oceans. There are those who think that life as we have come to know it now at the beginning of the twenty-first century may well become unsustainable before the end of the century as existing habitats and ecosystems will have changed radically enough to make it difficult, if not impossible, for our species to continue to survive on parts of the planet.

Our sun is currently classified as a "main sequence" star. This means that it is in the most stable part of its life, converting the hydrogen present in its core into helium. For a star the size of ours, this phase lasts a little over 8 billion years. Our solar system is just over 4.5 billion years old, so the sun is slightly more than halfway through its stable lifetime.

Fig. 6-3 The sun becomes a red giant

As the sun matures into a red giant, the oceans will boil and Earth will become uninhabitable. (Credit: Fsgregs, CC BY-SA. Read more at: http://phys.org/news/2015-02-sun-wont-die-billion-years.html#jCp)

The good news is that the Earth may have at least another billion years before the sun begins to render the planet even more uninhabitable than Venus. Given how far our species has come in less than 300,000 years, a billion years should afford Homo sapiens a comfortable interval to get its evolutionary maturity act together, or failing that, enough time for a replacement intelligent species to emerge on Earth to reach that survival plateau.

So, barring an extinction event, natural or self-inflicted, evolution appears to have built in enough time before the Earth's expiration date for our species to evolve to maturity and become capable of migrating to planets with stars still much earlier in their life cycle. If that were the end of the story, there'd be much reason for hope, because with time, our species could make it. But I am afraid it's not going to be that simple.

The bad news is that the current rate of climate change, whether driven by humans or due to natural causes (much of the evidence points to humans), may well reduce that billion years that evolution so generously gave us to less than a few hundred, if that, unless our species can find the will and then the way to stop and reverse the activities that are at the root of global warming. By increasing the accumulation of greenhouse gases in the atmosphere, trapping heat from the Earth in the atmosphere and in the oceans, our species may have inadvertently sped up the Earth's expiration date, cutting short its ability to sustain life, including the life of our species.

While twenty-first-century Homo sapiens might be the only intelligent species to have precipitated global climate change, it is by no means the only one to have lived during a period of climate change; rather, it is the last in a long line of hominins, our forebears. Indeed, one only has to take a look at Fig. 6-1 above to see how climate change shaped intelligent species evolution, showing the overlap between climate change events and hominin species' evolution, survival, migration, and extinction.

Overall, the hominins fossil record and the environmental record show that hominins evolved during an environmentally variable time. Higher variability occurred as changes in seasonality produced large-scale environmental fluctuations over periods that often lasted tens of thousands of years.

In this chapter we will examine how climate has shaped hominins' and thereby Homo sapiens' evolution and continues to do so. With signs of global warming manifesting now, climate change can further drive our species to reach even greater levels of adaptations and evolutionary maturity—or to extinction.

"The geologic period of human evolution has coincided with environmental change, including cooling, drying, and wider climate fluctuations over time. How did environmental change shape the evolution of new adaptations, the origin and extinction of early hominin species, and the emergence of our species, Homo sapiens?" is the question explored by Rick Potts and answered below in excerpts from his writings.

"Earth's climate has always been in a state of flux. Ever since our ancestors branched off the primate evolutionary tree millions of years ago, the planet has faced drastic swings between moist and dry periods, as well as long-lived glacial freezes and thaws. It's clear that early humans were able to survive such changes—our existence confirms their success," so writes Brian Handwerk in a *Smithsonian* article published September 30, 2014, which draws on the works of Potts.

Potts, head of the Human Origins Program at the Smithsonian National Museum of Natural History, writes that "hominins experienced large-scale shifts in temperature and precipitation that, in turn, caused vast changes in vegetation—shifts from grasslands and shrub lands to woodlands and forests, and also from cold to warm climates. Hominins' environments were also altered by tectonics—earthquakes and uplift, such as the rise in elevation of the Tibetan Plateau, which changed rainfall patterns in northern China and altered the topography of a wide region. Tectonic activity can change the location and size of lakes and rivers. Volcanic eruptions and forest fires also altered the availability of food, water, shelter, and other resources. Unlike seasonal or daily shifts, the effects of many of these changes lasted for many years, and were unexpected by hominins and other organisms, raising the level of instability and uncertainty in their survival conditions, driving many to extinction. Survivors were able to adapt to these climate changes."

To scientists like Potts, it has become clear from the evolutionary record that climate change has accompanied, and may well have choreographed, many major advances in our species' evolution.

"Fossils of hominids—all two-legged species related to human beings—document a history of human evolution from the ape-like Lucy (the first known Australopithecus afarensis) to the hand axe-carrying Homo erectus to the climate-changing masters of the planet that we are today. Now, at the onset of another climatic change event believed to have been triggered by human behavior, scientists say, this past about our species' ability to survive climate change and adapt over millions of years can offer clues," writes Gayathri Vaidyanathan, in *ClimateWire* on April 13, 2010, in an article titled "Scientists attempt to understand how human ancestors adapted—or not—to previous periods of climate change," and published in *Scientific American*.

Earlier climate events were indeed drastic, as recorded in layers of the Earth. Scientists correlate them with a sometimes-sparse fossil record to draw correlations between climate and evolution.

Currently, the human fossil record shows a correlation with climate patterns in Africa. About 5,000 to 10,000 years ago, the Earth's axis of tilt shifted (a process called precession), which changed the amount of rain that Africa received.

"It is not true that the Sahara Desert has been a permanent feature for millions of years," said Peter deMenocal, a professor at the Lamont-Doherty Earth Observatory at Columbia University. Africa oscillated between wet and dry every few thousand years, and each shift induced adaptation in the creatures that lived in the region. "Civilizations and populations can be very plastic that way," said deMenocal. "Climate change alters ecological landscapes, creates unnatural selection pressures, and promotes genetic selection to fit the pressures." Swings between wet and dry landscapes pushed some of our ancestors toward modern traits—and killed off others.

About 3.35 million years ago, Ethiopia was forested and Lucy's species thrived with its ape-like features. The climate changed and the habitat switched to woodland, and then to the African savannah. Then, about 2.95 million years ago, it switched back to woodland. Unable to adapt, Lucy's species went extinct around this time after 900,000 years.

About 2 million to 2.5 million years ago, an intense dry period led to the first migration of Homo erectus out of Africa into Southeast Asia, according to scientists.

About 5,000 years ago, with the creation of the Sahara Desert, humans migrated into the Nile Delta, creating an urban settlement, according to deMenocal.

Rick Potts observed that "many organisms have habitat preferences, such as particular types of vegetation (grassland versus forests), or preferred temperature and precipitation ranges. When there's a change in an animal's preferred habitat, they can either move and track their favored habitat or adapt by genetic change to the new habitat. Otherwise, they become extinct. Another possibility, though, is for the adaptability of a population to increase—that is, the potential to adjust to new and changing environments. The ability to adjust to a variety of different habitats and environments is a characteristic of humans."

"People think we're such a successful species, nothing can happen to us," observed Potts. But, he pointed out, most of our ancestors sooner or later went extinct. Homo erectus, the forerunner of modern humans, lived for 1.5 million years, he said. Homo sapiens, by comparison, have been around for only 300,000 years. Yet even they decreased in population size to between 600 and 10,000 breeding pairs when hit with mega-droughts, heavy monsoonal rains and the eruption of a volcano near Sumatra about 70,000 years ago.

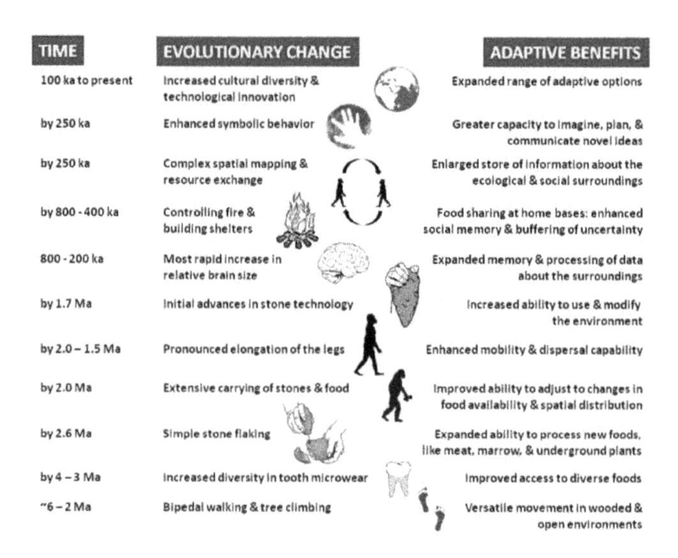

Fig. 6-4 Evolutionary change and adaptations over geological time
Ma = million years ago; ka = thousand years ago.
(Smithsonian diagram)

Overall, the evidence shows that hominins were able to adapt to changing environments to different degrees. The genus *Homo*, to which our species belongs, had the capacity to adjust to a variety of environmental conditions, and Homo sapiens is especially able to cope with a broad range of climatic conditions, hot and cold environments, arid and moist ones, and with all kinds of varying vegetation. Our species uses resources from a vast variety of plants and animals and uses many specialized tools. We have many social contacts and means of exchanging resources and information to help us survive in a constantly changing world.

Fig. 6-4 illustrates how the emergence of human characteristics from 6 million years ago to present conferred benefits that improved the ability of our ancestors to survive unpredictable and novel environments.

Summarizing the effects of climate change on intelligent species evolution over geologic time, we find that climate change altered ecological landscapes, created unnatural selection pressures, and promoted genetic selection to fit the pressures. Hominins experienced large-scale shifts in temperature and precipitation that, in turn, caused vast changes in vegetation—shifts from grasslands and shrub lands to woodlands and forests, and also from cold to warm climates.

A growing number of scientists are beginning to think that major climate shifts may have also forged some of the defining traits of humanity; in particular, a few large evolutionary leaps, such as bigger brains and complex tool use, seem to coincide with significant climate change. Swings between wet and dry landscapes pushed some of our ancestors toward modern traits—and killed off others. All prior Hominin species were driven to extinction due primarily to climate change.

Fast-forwarding to more recent periods of Homo sapiens history, 10,000 to 5,000 years ago, our species changed as it transformed from small communities of hunters and gatherers to settlements supported by farming and domestication of animals, to increasingly larger populations dwelling in villages, towns, and eventually cities. Throughout this interval, numerous Homo sapiens civilizations developed, thrived, and then collapsed. Many scientists now believe these repeated collapses of civilizations were driven primarily by repeated, extended, and severe bouts of climate change and reveal that our species struggled to make sense of Earth's climate machinery and its devastating cycles of monsoons, El Niños, La Niñas, and protracted multidecade droughts.

While the fall of a great empire is usually due to a complex set of causes, drought has often been identified as either the primary culprit or a significant contributing factor in a surprising number of such collapses.

"Drought is the great enemy of human civilization. Drought deprives us of the two things necessary to sustain life—food and water. When the rains stop and the soil dries up, cities die and civilizations collapse, as people abandon lands no longer able to supply them with the food and water they need to live," writes Jeff Masters in a March 21, 2016, article published in *WunderBlog* titled "Ten Civilizations or Nations That Collapsed From Drought," in which he lists many one might not have known were casualties of drought due to climate change.

Two significant takeaways: Whereas climate change drove our hominin forebears to extinction, it has driven more recent Homo sapiens civilizations to collapse, and the primary climate change factor turned out to be drought.

Collapse #1. The Akkadian Empire in Syria, 2334–2193 BC. In Mesopotamia 4,200 years ago.

Collapse #2. The Old Kingdom of ancient Egypt, 4,200 years ago. The same drought that brought down the Akkadian Empire in Syria severely shrank the normal floods of the Nile River in ancient Egypt.

Collapse #3. The Late Bronze Age (LBA) civilization in the Eastern Mediterranean. About 3,200 years ago, the Eastern Mediterranean hosted some of the world's most advanced civilizations.

Collapse #4. The Maya civilization of AD 250–900 in Mexico. Severe drought killed millions of Mayan people and initiated a cascade of internal collapses that destroyed their civilization at the peak of their cultural development, between AD 750–900.

Collapse #5. The Tang Dynasty in China, AD 700–907. At the same time as the Mayan collapse, China was also experiencing the collapse of its ruling empire, the Tang Dynasty. Dynastic changes in China often occurred because of popular uprisings during crop failure and famine associated with drought.

Collapse #6. The Tiwanaku Empire of Bolivia's Lake Titicaca region, AD 300–1000. The Tiwanaku Empire was one of the most important South American civilizations prior to the Inca Empire.

Collapse #7. The Ancestral Puebloan culture in the Southwest U.S. in the eleventh and twelfth centuries AD. Beginning in AD 1150, North

America experienced a 300-year drought called the Great Drought. This drought has often been cited as a primary cause of the collapse of the ancestral Puebloan (formally called Anasazi).

Collapse #8. The Khmer Empire based in Angkor, Cambodia, AD 802–1431. The Khmer Empire ruled Southeast Asia for over 600 years but was done in by a series of intense, decades-long droughts interspersed with intense monsoons in the fourteenth and fifteenth centuries, which, in combination with other factors, contributed to the empire's demise.

Collapse #9. The Ming Dynasty in China, AD 1368–1644. China's Ming Dynasty—one of the greatest eras of orderly government and social stability in human history—collapsed at a time when the most severe drought in the region in over 4,000 years was occurring, according to sediments from Lake Huguang Maar analyzed in a 2007 article in *Nature* by Yancheva et al.

Collapse #10. Modern Syria. Syria's devastating civil war that began in March 2011 has killed over 300,000 people, displaced at least 7.6 million, and created an additional 4.2 million refugees. While the causes of the war are complex, a key contributing factor was the devastating drought that began in 1998. The drought brought Syria's most severe set of crop failures in recorded history, which forced millions of people to migrate from rural areas into cities, where conflict erupted.

"There is a strong correlation between unusual climate shifts and exceptional events in human history," writes Brian Fagan in his book *Floods, Famines and Emperors*. He, too, lists a number of civilizations whose collapse coincided with huge climate events that may have hastened their collapse. Some of these include the ancient city of Ur, the Ancient Egyptian Kingdom, the Moche and Maya civilizations, and the Ancient Pueblo people of the American Southwest. They are discussed at length in Fagan's book and, because of their relevance to our discussion, some of these and a few others are referred to below.

The Harappan civilization. The mysterious fall of the largest of the world's earliest urban civilizations nearly 4,000 years ago in what is now India, Pakistan, Nepal, and Bangladesh and extended across the plains of the Indus River from the Arabian Sea to the Ganges now appears to have been due to ancient climate change, researchers claim, notes Charles Choi in a *Live Science* article.

Choi writes: "Nearly a century ago, researchers began discovering numerous remains of Harappan settlements along the Indus River and its tributaries, as well as in a vast desert region at the border of India and Pakistan. Evidence was uncovered for sophisticated cities, sea links with Mesopotamia, internal trade routes, arts and crafts, and as-yet un-deciphered writing."

"They had cities ordered into grids, with exquisite plumbing, which was not encountered again until the Romans," Liviu Giosan, a geologist at Woods Hole Oceanographic Institution in Massachusetts, told *Live Science*. "They seem to have been a more democratic society than Mesopotamia and Egypt—no large structures were built for important personalities like kings or pharaohs."

Like their contemporaries in Egypt and Mesopotamia, the Harappans, who were named after one of their largest cities, lived next to rivers. The Harappans had complex trade routes and a system of writing. This civilization built up in a "goldilocks" period when the rivers flooded often enough to support agriculture. As the climate changed, so did the monsoon season, lowering the floods and support for their cities.

Initially, the monsoon-drenched rivers the researchers identified were prone to devastating floods. Over time, monsoons weakened, enabling agriculture and civilization to flourish along flood-fed riverbanks for nearly 2,000 years. Eventually, these monsoon-based rivers held too little water and dried up, making them unfavorable for civilization. This change would have spelled disaster for the cities of the Indus, which were built on the large surpluses seen during the earlier, wetter era. The dispersal of the population to the east would have meant there was no longer a concentrated workforce to support urbanism. Eventually, over the course of centuries, Harappans apparently fled along an escape route to the east toward the Ganges basin, where monsoon rains remained reliable.

"Cities collapsed, but smaller agricultural communities were sustainable and flourished," said co-author Dorian Fuller, an archaeologist with University College London. "Many of the urban arts, such as writing, faded away, but agriculture continued and actually diversified."

"Our research provides one of the clearest examples of climate change leading to the collapse of an entire civilization," Giosan said. It remains uncertain how monsoons will react to modern climate change.

To better understand what this might have been like for the Harappans when the monsoons failed and the droughts followed one need only look at the impact of the 2015–2016 El Niño across Sub-Saharan Africa.

The United States Geological Survey (USGS) and other agencies report that warmer ocean waters in the Pacific Ocean have triggered drought

Fig. 6-5 Drought-driven peanut farm failure in Mozambique
A woman in a field amid a failed peanut harvest in Mozambique (Source DW)

in Africa. There have been abrupt changes in the weather. Rainfall in southern Africa from October to December 2015 was at the lowest level for those three months in thirty years. The World Food Program (WFP), which released those figures, blamed the weather pattern El Niño, which reoccurs every two to seven years. El Niño's impact varies with location. In some regions, the changes in warm water currents associated with El Niño cause an increase in rainfall and flooding, as in parts of Kenya and southern Ethiopia. In other regions, the opposite happens and rainfall ceases.

The drought on the Horn of Africa and in southern Africa—in Zimbabwe, Malawi, South Africa, Mozambique, and Ethiopia—had reached devastating proportions. The ground was parched, the harvest ruined. Severe drought had brought hardship to tens of millions of Africans. Aid agencies feared for the worst. "Many people no longer had any stocks of food left and were trying to sell their chickens, goats, and cows in order to provide for their families," said Marc Nosbach, country director for the aid organization CARE in Mozambique. But once they have sold their livestock, they have nothing to fall back on. Mozambique was experiencing its worst drought since the 1980s, and around 35% of families faced chronic food insecurity, according to CARE.

Fig. 6-6 Drought-driven cattle die-off in Tanzania

These cattle died of lack of water in western Tanzania early this year.
(Source DW)

Somalia had been hit the worst by the drought. In the north of the country, the first deaths from hunger had been reported. The United Nations

(UN) says more than 40% of the population depended on food aid. At least 360,000 children were malnourished, and about 70,000 of them were in danger of starvation. Tens of thousands of families had left their homes in search of water and grazing land for their livestock.

Ethiopia suffered drought conditions in 2015 that were comparable to the severe drought and ensuing famine of 1984, during which hundreds of thousands of people perished. Like the case in the 1980s, the 2015 Ethiopian drought was related to a strong El Niño. Unlike that terrible episode, widespread acute food insecurity was avoided in 2015–2016 due to effective climate services, early warning of potential food insecurity, and social safety nets, particularly through the Famine Early Warning Systems Network (FEWS NET).

The UN Framework Convention on Climate Change (UNFCCC) believes that such emergencies will recur frequently in the future: "On average global temperatures have risen by about 1°C since the start of the Industrial Revolution. But in parts of Africa, temperatures are rising much faster than the global average. . . . By the end of this century, temperatures in Africa could have jumped by between 3 and 6°C. The effect on harvests and the population would be horrendous." It is important to bear in mind that it was just a 3–5°C change in global temperature that exacerbated the Permian-Triassic mass extinction, when more than 95% of all species went extinct. At such high temperatures all animal species would die.

The Classic Maya Civilization of Southern Mexico, Guatemala, Belize, and Honduras flourished between the time of Christ and AD 900 and then suddenly collapsed, leaving survivors to disperse into small self-sustaining villages where their descendants live to this day, abandoning the ruins that stretch across the peninsula.

Computer models of deforestation in the Yucatan and its impact on rainfall published in *Geophysical Research Letters* suggest that a severe drought, exacerbated by widespread deforestation (logging), might have triggered the mysterious Mayan demise.

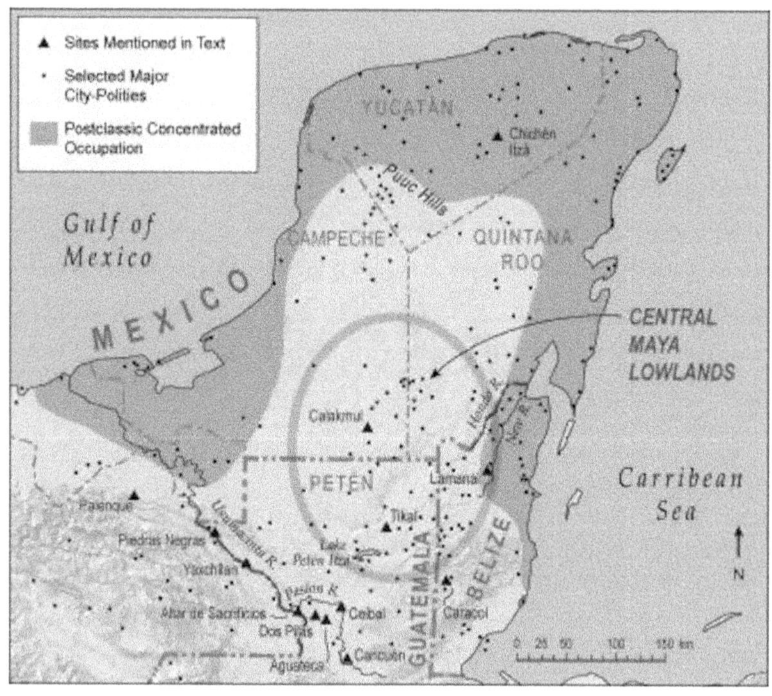

Fig. 6-7 Locations of abandoned Mayan cities
(Source: Geophysical Research Letters)

In an August 2012 article titled "Why Did the Mayan Civilization Collapse?" written by Joseph Stromberg and published on *Smithsonian.com*, Stromberg explains:

"Because cleared land absorbs less solar radiation, less water evaporates from its surface, making clouds and rainfall more scarce. As a result, the rapid deforestation exacerbated an already severe drought—in the simulation, deforestation reduced precipitation by 5 to 15% and was responsible for 60% of the total drying that occurred over the course of a century as the Mayan civilization collapsed. The lack of forest cover also contributed to erosion and soil depletion."

In a time of unprecedented population density, this combination of factors was likely catastrophic. Crops failed, especially because the droughts occurred disproportionately during the summer growing season. This forced peasants and craftsmen into making a critical choice, perhaps necessary to escape starvation: abandoning the lowlands, leaving the ruins we see today.

Fig. 6-8 The Mayan city of Tikal
The Mayan city of Tikal was abandoned in the ninth century.
(Source: Getty Images)

Researchers observed that one of the lessons here is that reshaping of the environment can often have unintended consequences and one may not have any idea of what they are until it's too late.

The Moche civilization comprised farmers and anchovy fishermen who flourished along the north coast of Peru between AD 100 and 800 and were known for their farming expertise, their cotton, and their expert weaving. They depended on runoff from the Andes to irrigate their fields until a thirty-year drought between AD 563 and 594 drastically reduced the amount of runoff reaching coastal communities. Then El Niño struck. Torrential rains swamped the Andes Mountains and the coastal plains, turning the arid rivers into raging torrents.

The arduous work of years vanished in a few weeks. Around AD 700, the Moche abandoned their cities. El Niño flooding and drought had broken the back of a wealthy and powerful state. According to Fagan, their brilliant technology and irrigation could not guarantee the survival of a highly centralized society. There were limits to the climate shifts the Moche civilization could absorb, and they eventually ran out of options. Much of their ideology comes down to us in their remarkable clay vessels, in their textiles, wall art, and metallic objects.

Angkor Wat, the capital of the Khmer Empire, the major player in southeast Asia for nearly five centuries, was a thriving metropolis of 750,000 before a series of mega-monsoons made it unlivable and, according to scientists, led to its collapse in spite of valiant but ultimately failed efforts to battle drought.

Fig. 6-9 Monks in Angkor's Ta Prohm temple
(Source: Alamy)

At its peak in the twelfth and thirteenth centuries, the Khmer capital of Angkor sprawled over 1,000 square kilometers and possessed a complex network of channels, moats, and embankments and reservoirs known as *barays* (large reservoirs) to collect and store water from the summer monsoons for use in rice paddy fields in case of drought. Recent evidence suggests that prolonged droughts might have been linked to the decline of Angkor.

Scientists studying tree rings from Vietnam suggest the region experienced long spans of drought interspersed with unusually heavy rainfall.

The researchers deduced a 1,000-year-long climate history of Angkor from the barays. They found at around the time Angkor collapsed the rate at which sediment was deposited in the baray dropped to one-tenth of what it was before, suggesting that water levels fell dramatically also.

Some scientists view Angkor as an example of how technology isn't always sufficient to prevent major collapse during periods of protracted climate change. Angkor had a highly sophisticated water-management infrastructure, but this technological advantage was not enough to prevent its collapse in the face of extreme environmental conditions. In the end, scientists note that while failure of the water-management network was not the sole reason for the downfall of the Khmer Empire, it might have been insufficient to cope with sudden and intense variations in climate and protracted drought.

The Ancestral Pueblo (aka Anasazi) civilization rose from complete obscurity to build some of the largest towns in ancient North America about 1,000 years ago. Then, suddenly, they moved out of their great pueblos and vanished from history. They simply dispersed from their homelands in the face of untenable environment conditions during two major drought cycles identified from ancient tree rings: AD 1130–1180, and the great drought of AD 1275–1299. Their response to climate change lay in their ideology, in which the notion of movement played an important part. They inherited the cumulative knowledge of 10,000 years of foraging in an unrelenting landscape where survival depended on a close knowledge of alternative foods in times of hunger. They became farmers and developed water-conservation technology.

Theirs was a land of unpredictable rainfall and frequent droughts. They survived and flourished in a diverse environment of desert, plateaus, and mountains between 1,200 and 2,400 meters above sea level with rapid changes in elevation and accompanying environmental changes for centuries because they knew their homeland intimately. Anyone farming this environment knew the value of water conservation and the importance of mobility.

Stories of the past told by Pueblo Indians share one common element—movement. By movement they meant short migrations from one place to another, from one mountainside to another. Movement is one of the fundamental ideological concepts of Pueblo thought, because mobility perpetuates human life. Says the Tewa Tessie Naranjo, teacher of the Tewa language and culture: "Movement, clouds, wind, and rain are one. Movement must be emulated by the people." Keeping on the move was the way to survive. The Pueblo believed that without movement there was no life. Their history has been a continual process of moving, settling down, then moving again. People have moved, joined together, and separated since long before farming began in the Southwest more than 3,000 years ago.

When fifty years of intense drought caused by a prolonged La Niña settled over the Colorado Plateaus in AD 1130, the Pueblo deserted their great pueblos and dispersed into small villages and hamlets after farmers and traders stopped bringing goods and food and instead kept scarce supplies for themselves. The massive grand pueblos they had built were abandoned and stood empty. By early AD 1200, they had all gone.

This would be akin to most of the towering skyscrapers and elegant apartment buildings in Singapore or Hong Kong and Shanghai being abandoned and the populations fleeing to villages in Malaysia and smaller villages and towns along the Pearl River Delta, respectively. This is not that unimaginable a scenario for most major cities that depend on food supplies from external sources if years of protracted drought were to leave supermarkets, farmers markets, and stores devoid of food and water, with former trading partners from food-exporting countries hoarding their food supplies for their own populations.

According to Franz Broswimmer in his book *Ecocide*, "Human history is replete with accounts of the early ecocidal activities of great empires such as Babylon, Egypt, Rome, ancient China, and Maya, all of which destroyed their forests, and the fertility of their topsoil, and killed off much of the original fauna through a combination of their linear thinking and their insatiable drive for material wealth."

The most flourishing lands of antiquity were sites of civilizations that remained powerful and wealthy for great periods of time, but they are today among the poorest regions of the world. Large parts of these areas are now barren deserts; most of the ancient cities are abandoned and have become targets for religiously inspired terrorist groups like ISIS; and local people now often have little to no historical awareness of their social or ecological past. To be sure, famine, warfare, and disease contributed to the demise of ancient civilizations, but one of the primary causes of their decline was the depletion of their environmental resources.

And the destruction of forests that happened back then is continuing today on an industrial scale. Most of the world has been fighting the deforestation of both the Amazon and the Congo for nearly thirty years. The Ivory Coast has lost more than 80% of its forests in the last fifty years mainly to cocoa plantations and farms. In addition are the illegal timber trade that is destroying Myanmar's forest and the illegal timber trade in Laos—only Indonesia and the EU seem to have come up with a sustainable approach.

In fact, consumption of tropical timber by the U.S. and other industrial countries plays a significant role in tropical deforestation. The pillaging lumber industry has taken very little heed of the world's concern that permanent damage is being done to the environment. According to the Global Policy Forum, "Rampant timber exploitation has not only destroyed the environment but also funded illegal arms deals and fueled bloody civil wars and regional instability. Extensive forest destruction throughout the world has had severe social and economic consequences for indigenous peoples, leading to rights violations, communal conflict and harsh repression by governments and timber companies."

Consequently, the world's natural forests cannot sustainably meet the soaring global demand for timber products under current forest management practices, and alternatives such as fast-wood plantations—or commercially planted forests—fail to create the ecosystem natural forests provide species dependent upon them.

Tree planting might seem to make sense as a way to make up for Earth's vanishing forests except that planting trees does not a forest make.

According to James Lovelock in his book *Gaia*, "It is insufficiently appreciated that a forest's ecosystem is an involved entity comprising a huge range of species from microorganisms, nematodes, and invertebrates to small and large plants and animals. In addition, natural forest tends to evolve with climates while tree plantations die."

Lovelock continues: "What is less well known is how destroying forests disturbs the atmosphere on a global scale and to an extent at least comparable with the effects of modern urban industrial activity. In fact, agriculture in total has climate effects comparable with those caused by fossil fuel combustion. Forest clearance has direct climate consequences, water recycling and atmospheric albedo change (a measure of the amount of sunlight reflected by Earth's surface or atmosphere) and is also responsible for much of the carbon dioxide emissions. Fires of this type inject into the air, in addition to carbon dioxide, a vast range of organic chemicals and a huge burden of aerosol particles. Some of the chlorine now in the atmosphere is in the form of the gas methyl chloride, a direct product of tropical agriculture and a substance we now know was produced in abnormal quantities as a consequence of prior primitive agriculture.

"This brutal disturbance of natural ecosystems always involves the danger of upsetting the normal balance of atmospheric gases. Changes in the natural production rate of gases such as carbon dioxide or methane and of aerosol particles cause perturbations on a global scale."

While climate events were often not the entire cause of the collapse of human civilizations, they invariably exerted huge sustainability pressures, exploiting every weakness and leaving very little room for error, and often hastened the collapse of these civilizations. Climate change and an exhausted water supply in many instances dealt the final blow.

Droughts have killed more than 11 million people worldwide since 1900 and now affect twice as much land area as in 1970, according to the UN Food and Agriculture Organization. Within fifteen years, the world water supply will fall short by at least 40%, a United Nations report warned.

Today, drier areas, covering more than 40% of the world's land surface, are inhabited by 30% of the world's population (2.5 billion people); support 50% of the world's livestock; grow 44% of the world's food; account for the majority of the world's poor, with around 16% living in chronic poverty. Scarce natural resources, land degradation, and frequent droughts severely challenge food production in these areas.

With continued global warming, some regions are likely to get wetter, while other regions, already on the dry side, are likely to get drier. It remains to be seen what coping strategies, adaptations, technologies, and innovations our species will bring to bear and whether, unlike these earlier civilizations, we will come out of this warming era intact centuries from now.

Scientists now know that it was El Niño, La Niña, and ENSO (El Niño–Southern Oscillation), recurring climate systems, that caused much of the climate chaos that drove the collapse of the above civilizations.

Fagan points out that while scientists have always known that recurrent climate events drove temperature extremes, raging wildfires, droughts, floods, and extreme human suffering, especially of the poor, and that these have in the past put civilizations under great stress and hastened the end of more than a few of them, they have come to understand that these events were not confined to local areas or specific countries, but were part of a global weather machine that simultaneously impacted multiple countries across oceans and continents.

"For the first time," writes Fagan, "scientists were able to infer, albeit crudely, the existence of climate anomalies in one part of the world if scientists also knew of simultaneous (but not necessarily similar events) occurring half a world away." Until recently they lacked the scientific tools to appreciate just how profoundly short-term climate change has affected the rise and fall of civilizations.

According to USGS, El Niño is an irregularly occurring and complex series of climatic changes affecting the equatorial Pacific region and beyond every few years, characterized by the appearance of unusually warm, nutrient-poor water off northern Peru and Ecuador, typically in late December. This occurs when the normal trade winds weaken (or even reverse) and lets the warm water that is usually found in the western Pacific flow instead toward the east. The effects of El Niño include reversal of wind patterns across the Pacific, drought in Australasia, and unseasonal heavy rain in South America.

El Niño and La Niña are opposite phases of what is known as the El Niño-Southern Oscillation (ENSO) cycle. The ENSO cycle describes the fluctuations in temperature between the ocean and the atmosphere in the east-central equatorial Pacific.

La Niña is sometimes referred to as the cold phase of ENSO, and El Niño as the warm phase. These deviations from normal surface temperatures can have large-scale impacts not only on ocean processes, but also on global weather and climate. El Niño and La Niña episodes typically last nine to twelve months, but some prolonged events may last for years. While their frequency can be quite irregular, El Niño and La Niña events occur on average every two to seven years. Typically, El Niño occurs more frequently than La Niña.

El Niños often mean droughts, famines, forest fires, and floods, and the world has seen many devastating El Niños that resulted in millions of deaths and great destruction. India until recently experienced repeated droughts and famines due to El Niño climate events, when the monsoon rains failed. Famine was endemic in India for thousands of years, until railroads and improved communications made shipment of food supplies to hungry villages practical. Going as far back as 1344–1345, and again in 1629–1631, 1685–1688, and again in 1780, 1896–1897, and yet once more in 1899–1900, Indian farmers died by the tens of thousands, sometimes millions. No one knows for sure exactly just how many people died during the great famine of 1899, but it could have been as many as 4.5 million. The population declined for ten years between 1895 and 1905.

There were major El Niños in 1957–1958, in 1972–1973, and as recently as 2015–2016. The devastations of El Niños and La Niñas continued for centuries, leaving death and destruction in South America, Africa, Australia, America, and Europe, until scientists began to study the world's climate machine, understood how it works, and developed the innovations that have significantly reduced the impact of these climate anomalies.

Climate-Change-Driven Adaptations and Innovations. With the ability to migrate becoming increasingly less of an option for most of our species today, the only remaining options available is to adapt, innovate, and mature. In response to these short- to intermediate-term climate-change pressures, our species has come up with numerous innovations to enable evacuation, disaster-mitigation planning, and advanced notification of populations potentially in harm's way.

Over the past quarter century, scientists have blanketed the world with monitoring devices in an effort to make forecasts about emerging El

Niños and other large-scale atmospheric interactions. Scientists deployed a network of tidal gauges, drifting and moored buoys, many far offshore, and devices for measuring the amount of water vapor above the oceans. Satellites were deployed to transmit buoy data and measure sea-level heights. In addition, computer models have been developed, implementing climate-prediction models used to generate long-term weather forecasts. Together, these allow climate scientists and weather forecasters to track weather anomalies from inception, forecast their progress, and follow their development, enabling populations and governments to take necessary action to prevent and/or reduce disasters.

There is no question that these innovations and adaptations save lives. Nonetheless, people (especially the poor) continue to suffer and die due to floods, mudslides, and forest fires, and billions of dollars are lost in property, livelihood, and infrastructure damage, suggesting these innovations and adaptations are imperfect replacements for migration—the ability to permanently move oneself and one's belongings out of harm's way.

Often the weak link in the effectiveness of a solution is the seeming inability of governments to take adequate and appropriate action in the face of impending climate anomalies. Even developed regions of the Western world like the EU have been able to mount only a fragmentary response to the current climate-refugee crisis, with Germany making most of the effort, while countries like the USA and Australia erect barriers, banish migrants to other countries, and impose extreme vetting. In addition, while the patchwork of humanitarian and religious charities makes a difference, more and more people living in failed states are likely to experience less or no assistance, and will be forced to deal with the effects of these climate anomalies on their own.

The limitations of our adaptations to climate change, compounded by the inadequate or total lack of effective action by governments and the limited resources of humanitarian organizations have become the drivers of modern-day migrations. Unless a permanent, comprehensive, and coordinated global response mechanism is established, with the resources and authority to act to save lives, our species will be in for a very rough ride as climate change further shrinks the Earth's habitable areas due to the gradual rise in sea levels alone.

Unlike short-term climate change caused by El Niños and La Niñas, long-term climate change driven by global warming is expected to have a punishing impact on most species' and Homo sapiens' survival. Among potential reasons, scientists now believe that Earth's climate machinery appears to have hot and cold stable states that geologists refer to as greenhouse and icehouse, respectively, and in between are the metastable periods like the current interglacial in which twenty-first-century Home sapiens civilization exists.

The last known hothouse occurred 55 million years ago around the geologic period known as the Eocene. The Eocene was already warm by today's standards, but a geological accident caused the release of 2 tera tons, or 2 million, million tons, of carbon dioxide that caused the temperature in the Arctic and tropical zones to rise between nine and fifteen degrees. This release may have taken 10,000 years and as long as 200,000 years for temperatures to return to previous states. In comparison, Homo sapiens have injected as much in just the last 200 years, and, in addition, the Earth itself through normal process would have released as much again. So, if one is still wondering how real global warming is, just reflect upon what the Eocene hothouse, without our help, was like.

It is neither unreasonable nor alarmist, then, to expect global temperatures to rise comparatively quickly and potentially (some scientists believe) much faster than projected by the IPCC (Intergovernmental Panel on Climate Change) climate models, and, unlike those models, potentially in non-continuous discrete jumps of several degrees at a time, over brief intervals, before stabilizing at one of its hothouse stable states. And that could potentially drive most species and Homo sapiens to extinction.

Meanwhile, "According to an ongoing temperature analysis conducted by scientists at NASA's Goddard Institute for Space Studies (GISS), the average global temperature on Earth has increased by about 0.8° Celsius (1.4° Fahrenheit) since 1880; two-thirds of the warming has occurred since 1975, at a rate of roughly 0.15–0.20°C per decade." The year 2016 has gone down in history as the hottest year since our species has begun keeping global records. And if a picture is worth a thousand words, a NASA rendering of what the Earth's temperature profile might have looked like between 1885 and 1994 and how it has changed between 2005 and 2014 tells the whole story and is reproduced in the diagrams below.

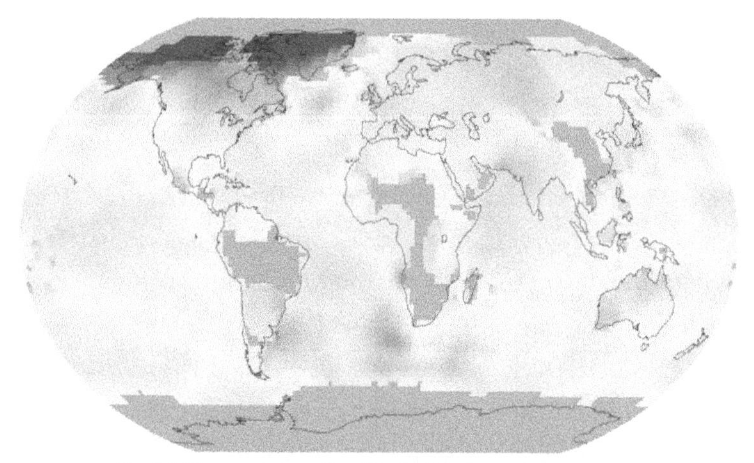

Fig. 6-10 Temperature profile of the Earth between 1885 and 1994

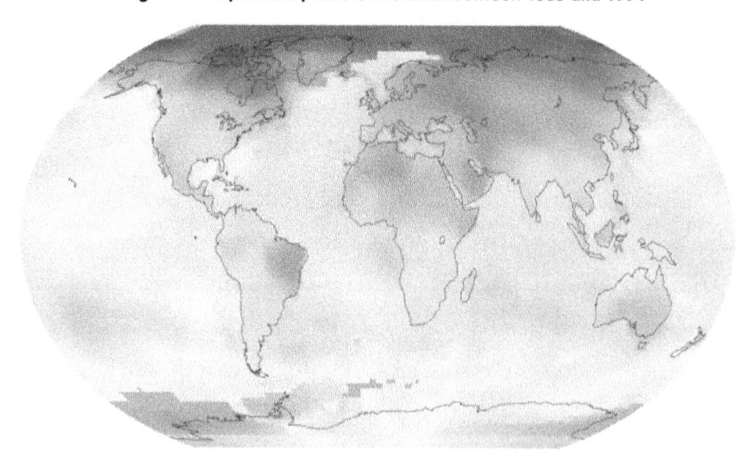

Fig. 6-11 This 2005-2014 profile shows how the temperature profile of the Earth has warmed since 1885

A one-degree global change might sound small, but it takes a vast amount of heat to warm all the oceans, the entire atmosphere, and the landmass by that much. In the past, a one- to two-degree drop was all it took to plunge Earth into the Little Ice Age. A five-degree drop was enough to bury a large part of North America under a towering mass of ice 20,000 years ago. According to NASA, "The global temperature mainly depends on how much energy the planet receives from the sun and how much it radiates back into space—quantities that change very little. The amount of energy radiated by the Earth depends significantly on the chemical composition of the atmosphere, particularly the amount of heat-trapping greenhouse gases."

In addition, the rapidity of carbon dioxide and other pollution would have given the Earth climate machinery little time to self-adjust and may well have already engendered several positive feedback systems that could result in a fast buildup of global heat in the atmosphere and particularly in the oceans.

Some scientists differentiate between *global temperature* and *global heat*, considering the latter a better indicator of how rapidly the Earth is heating up by how rapidly ice caps melt and ocean levels rise. The rise in global temperatures might seem to stall at times, but the rise in global heat, being absorbed by melting of glaciers and expansion of the oceans as it warms, is considered a better indicator, and indeed is shown by the continuous rise in sea level, compared to global temperature fluctuations since the 1970s. Some scientists believe that once the glaciers are melted and ocean temperature reaches some form of dynamic equilibrium, global temperature increase will accelerate and might accelerate precipitously during the second half of the twenty-first century.

In the coming decades, climate change will increasingly threaten humanity's shared interests and collective security in many parts of the world, disproportionately affecting the globe's least developed countries. Climate change will pose challenging social, political, and strategic questions for the many different multinational, regional, national, and nonprofit organizations dedicated to improving conditions worldwide. Organizations as different as Amnesty International, the U.S. Agency for International Development, the World Bank, the International Rescue Committee, and the World Health Organization will all have to tackle the myriad effects of climate change.

Unlike local climate change, where individual countries have been able to respond and make some difference, global climate change is going to require a global response.

An impartial observer, considering how our species, living in a crowded world with no unclaimed areas left for migration, responds to short-term climate change brought on by weather anomalies like El Niños and La Niñas, might conclude two things. First, that while innovations and adaptations have mitigated loss of life and limb, these tend to come into play only after millions have already suffered; second, built in to these responses appears to be a willingness to tolerate loss of life as inevitable and necessary—a clear but unacknowledged trade-off between dollars and human life, with a greater willingness to forfeit those lives when they belong to the poor and voiceless.

Given the current response of many nations, including an unwillingness to open their borders to migrants while increasing global warming exacerbates these climate events, one despairs for the future, as such conditions can only get worse.

It's one thing to respond, however imperfectly, to the short-term climate changes discussed above, but it is going to be quite another to respond effectively when the fallout becomes several orders of magnitude worse than we have experienced so far.

Most of the civilizational collapses discussed above were due to natural causes. These civilizations did not trigger global warming. The El Niños, monsoons and other weather patterns that devastated their cities, towns, and villages were not due to their warming of the oceans. They were simply overtaken by what scientists now know to be cyclical changes in the global climate machinery. They had limited generational history and little if any awareness of these climate cycles. While their survival efforts were indeed valiant, their adaptations and innovations were overpowered by the intensity and duration of climate shifts on the one hand and exacerbated by the often corrupt political, cultural, military, and religious worldview and practices of their leaders on the other.

Given the above track record, the open question is how would modern Homo sapiens manage and survive climate change beginning in the twenty-first century? Will climate change drive the collapse of yet another Homo sapiens civilization or the extinction of our species? How will our widespread alterations of Earth's landscapes, forests, atmosphere, and oceans, rivers, and coastal waters interact with the tendency of Earth's environment to shift all on its own as we have seen above?

While twenty-first-century Homo sapiens are no doubt better prepared than those earlier civilizations, and our innovations and ability to adapt may have improved by several orders of magnitude, we still suffer from some of the same deficiencies; in some cases, we are confronted with versions of the same problems that are potentially worse. Here are a few of these issues:

Regional versus global: The problem is now bigger because the world is flat. Homo sapiens have grown from small, separate, regional centers to a global civilization. Most disasters now, whether due to warfare, climate, or pandemics, tend to have ramifications across the entire globe, affecting survival, trade, travel, economics, and finance worldwide.

Population growth: By 1300 BC, the world's hunter-gatherer population was approaching 8.5 million. Today the world's population exceeds 7.8 billion and is expected to reach 9 billion by 2050. Continued population growth exacerbates the already-diminished carrying capacity of the Earth.

Population density in metropolitan areas: In 1800, only 3% of the world's population lived in urban areas. By 1900, almost 14% were urbanites, although only twelve cities had 1 million or more inhabitants. In 1950, 30% of the world's population resided in urban centers, and the number of cities with over 1 million people had grown to eighty-three. For the first time in history, humans are now predominantly urban. It is expected that 70% of the world's population will be urban by 2050. Cities occupy less than 2% of the Earth's land surface but house almost half of the human population and use 75% of the resources we take from the Earth.

Because most of the food, water, electricity, and fossil fuels that supply these cities is shipped in from distant (often international) locations, this high-population-density way of living exposes modern Homo sapiens to a greatly intensified version of the failed supply chain that drove the ancient Pueblo people to desert their great pueblos. Modern Homo sapiens might end up abandoning their cities as the Pueblo people did when confronted with climate change in the form of major droughts, such as a fifty-year drought or the Great Drought.

Loss of freedom to migrate. For millennia, writes Fagan, "countless Stone Age peoples of remote prehistory relied on mobility (migration) and well-developed social networks for survival. Mobility, low population densities, and flexible ways of living allowed foragers to adjust easily to droughts, floods, and other vagaries of climate. They were opportunistic, accustomed to sudden climate change, and able to adapt to it in ways that became impossible when population rose rapidly after the last ice age. Experience, low population densities, and the sheer flexibility of human existence allowed Stone Age foragers all over the world to ride with the punches of the global climate machine."

Today, migration for some of the same reasons that drove early Homo sapiens is no longer an option, as can be seen in the plight of migrants from Africa trying to enter Europe and Central Americans trying to enter the USA.

Droughts, water shortages, and forest fires: Due to increased population densities and the lack of freedom to migrate, short-term climate change (hurricanes, monsoons, typhoons, floods, forest fires, droughts) has become an important delimiting factor in human existence and survival.

We still don't know for sure, and probably won't until it's too late, how anthropogenic global warming and widespread habitat destruction will affect the global climate machine and its already deadly cycles, but if our response to short-term climate-change disasters so far is any indication, Homo sapiens is definitely not ready for protracted climate disasters like those that drove earlier civilizations to collapse.

One need only look at the devastation caused by Hurricanes Andrew, Katrina, Harvey, and Irma; the massive earthquakes in Nepal and Mexico; and the typhoon-driven storm surges and floods across Southeast Asia and India to see how ill-prepared some countries are, as well as Homo sapiens in general.

It was habitat loss resulting from climate change that drove early Homo sapiens out of Africa; climate change and habitat loss contributed to the collapse of many ancient civilizations as mentioned above; and it is habitat loss yet again, due in part to climate change, but exacerbated by civil wars, regional conflicts, and failed states, that are the triggers of migration today.

It wasn't that these ancient civilizations were without innovations and technologies; in fact, in almost every case the evidence shows that they were quite innovative and resourceful. Nonetheless, they collapsed, forcing surviving peoples to desert their cities and migrate to more sustainable regions. Climate change and habitat loss overcame their resourcefulness and best technologies.

Today, despite our advanced technologies and innovations, we still lack adequate coping mechanisms for repeated and sustained El Niños' and La Niñas' floods and dry spells, and the resulting droughts that have been known to last forty, fifty, and even 300 years across continents, conditions that brought down many of these early civilizations. Just as they did, we still depend on nature for water for ourselves, our crops, and livestock, and our technologies won't prevent the widespread disaster that such droughts will bring to vast regions of the planet as they did before and are still doing across Sub Saharan Africa.

Today, we live in a world with several orders of magnitude more people who will potentially be affected by similar climate events, and those events are increasing in intensity, possibly due to global warming, according to scientists. The world is warming more rapidly than at any previous time, and melting glaciers will put at risk the water supply that feeds rivers (like the Nile) and streams bringing water to many population centers. With drying streams and reduced rains come droughts, death of livestock, failing crops, a shrinking food supply, famine, stressed and inadequate humanitarian aid, disease, conflicts, and widespread death.

In his book *The Great Warming: Climate Change and the Rise and Fall of Civilizations*, Fagan says, "We are entering an era when extreme aridity will affect a large portion of the world's now much larger population, where the challenges of adapting to water shortages and crop failures are infinitely more complex." And it is a world without room for the migrations early humans engaged in to escape what might have been similar climate events.

The journey along the road to intelligent species maturity and survival is littered with the remains of extinct hominins, forbears of modern Homo sapiens. While the collapse of recent Homo sapiens civilizations did not result in the end of our species, these events nonetheless show Homo sapiens have not learned to survive in habitats continuously rearranged by short-term climate change, and remain insensitive to how rearranging the Earth's ecosystems and habitats in pursuit of profit and growth can exacerbate climate-driven disasters and civilization collapse.

The challenges confronting modern Homo sapiens are not only the short-term climate-change events these earlier collapsed civilizations experienced and not only the short-term climate change now exacerbated by global warming, but also the many potential disasters, both forecasted and unknown, of global warming. Because of this, what modern Homo sapiens face may not be just a potential global civilization collapse, but a potential extinction event.

No intelligent species has experienced a mass extinction event. Yet even the worst of Earth's mass extinctions saw survivors, invariably a small percentage of the prior extant population. Prior mass extinctions have resulted in more than 50% of all species dying; in the case of the "Great Dying" (Permian-Triassic mass extinction), as much as 95% of all species on the planet died. Populations in most early hominin species may have never exceeded 1 million, compared to the 7.8 billion that exist today. As noted above, Homo sapiens decreased in population size to between 600 and 10,000 breeding pairs when hit with climate-driven megadroughts, heavy monsoon rains, and the eruption of a volcano near Sumatra, about 70,000 years ago.

Applying those percentages to the current human population would suggest that between 3,700,000,000 to 7,030,000,000 could potentially perish, leaving approximately 3,700,000,000 to 370,000,000 surviving members of our species—if any. "If any" because survival of a species is not a numbers game, but a question of sustainable habitats (a topic we will get to in the next chapter), or alternate habitats to which survivors can migrate. Most of our species today, leaders in particular—climate-change believers as well as climate-change deniers—have not begun to wrap their minds around the chaos, carnage, and resultant civilization wreck a looming extinction could mean long before it actually happens.

Many are looking to emerging technologies and a retreat to Mars in case of an extinction event on Earth. But some extinction events in the past weren't preceded by multidecade, let alone multi-century, warning. The dinosaurs had no warning whatsoever. In the case of our species, it is not an inability to recognize the signs of a potential extinction event but a deliberate choice to stay the course and remain on a path that might lead to what has come to be known as the "Sixth Mass Extinction." Our species may have lost the ability and freedom to act to save itself—an important rung on the Species Survival Maturity Model that must be scaled if a species is to avoid self-extinction.

Subsequent to most prior extinctions, the emergent species tended to be almost entirely different from those that had died off. The extinction of the dinosaurs is believed to have made it possible for mammals, and thereby hominins, and thereby Homo sapiens to emerge.

In this section we took note of the emergence of species-driven evolution enabled by the amazing powers of the human mind; capabilities that will enable our species to respond to the pressures of habitat loss due to climate change; and our option to choose survival over self-extinction and overcome the existential threats of the universe.

In the following section we learn why a sustainable-capable habitat is indispensable to species survival and whether Homo sapiens will use those emerging powers of the human mind to adapt, innovate, mature, and migrate to survive or, instead, make, create, and innovate the means of its extinction.

PART THREE

Evolutionary Rules for Intelligent Species Survival

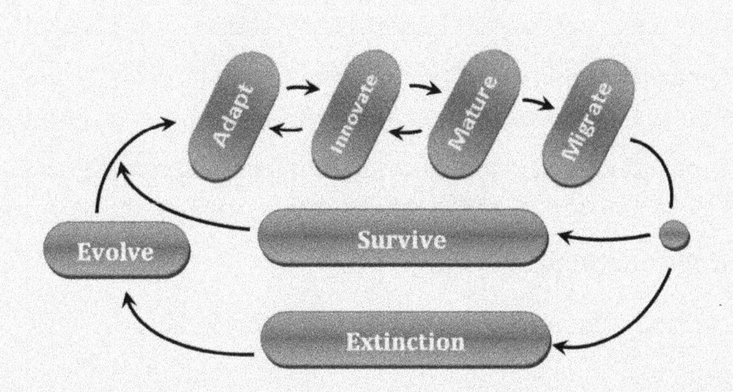

UNLIKE ALL PRIOR SPECIES THAT EVOLVED a priori in a state of evolutionary maturity, intelligent species like Homo sapiens need to mature to evolutionary maturity just to survive and to avoid self-extinction.

The very attributes that enable intelligent species to transcend the limitations of prior species—intelligence and freedom to choose its own evolutionary outcome—are simultaneously its greatest strengths and weaknesses.

Homo sapiens survival or extinction will be the outcome of the race in progress between its ability to adapt and innovate versus its ability to mature and migrate.

If our ability to adapt and/or innovate outruns our ability to mature, it becomes all too easy for a vanishingly small few humans to drive our species to extinction. If, too, we are unable to migrate to escape a species-wide extinction event, self-inflicted or by the universe, once again our species will end in extinction. Homo sapiens' challenges, then, are to mature much faster than it adapts and innovates—a not-so-easy challenge as we will see—and/or, like our forebears, must remain free, willing and able to migrate to escape that extinction.

But if a bacteria that's barely visible in a microscope, without a nervous and endocrine system—just a bag of chemicals inside a membrane—has been endowed by evolution to know how to make the right enzyme for whatever sugar is available, should not an emergent intelligent species also evolve with the capabilities to know how to make the right choices to survive?

Just as there are rules, patterns and models that govern the universe, and rules that regulate the different kinds of cells in the body, and rules that regulate the number and kinds of animals in a given place, so in this book we ask what are the rules, patterns and models that also must exist to guide an intelligent species to survival. The diagram above illustrates the evolutionary survival patterns—Adapt, Innovate, Mature and Migrate to Survive or go Extinct. Three of these—Adapt, Innovate and Migrate—we will explore in this section. But first, we must look at habitats, where these patterns play out, and why habitat sustainability is indispensable to survival.

Evolutionary Rules for Intelligent Species Survival

CHAPTER 7

Habitats

The Conditions Necessary for the Evolution and Sustainment of Life

CHAPTER 8

Sustainability

Species Survival Ensures Habitat Survival, and Habitat Survival Ensures Species Survival

CHAPTER 9

Adaptation

Must be Survival-driven to Preserve Habitats and Biodiversity

CHAPTER 10

Innovation

From Ape-man to Spaceman in Three Million Years—to . . . ?

CHAPTER 11

Migration

A Built-in Evolutionary Survival Response

To Survive, an Earth-based Emergent Intelligent Species Must Mature Enough to Avoid Self-extinction and Become Capable of Interplanetary and Intersolar Migration

7.

Habitats

THE CONDITIONS NECESSARY FOR THE EVOLUTION AND SUSTAINMENT OF LIFE

"The most intellectual creature to ever walk Earth is destroying its only home."—Jane Goodall

THE HISTORY OF LIFE ON EARTH HAS SHOWN THAT A sustainable habitat is a prerequisite for the evolution of life and is indispensable for the long-term survival of a species. Habitats, in this context, are the set of galactic, solar, and planetary preconditions necessary for the emergence and continued survival of complex animal life, and specifically emergent intelligent species. And, while from a human perspective the Earth might seem to abide forever, the habitats that sustain life on Earth do not.

The history of life on Earth also shows that whenever evolution gives rise to life of any form or complexity, it invariably first establishes a natural survivable habitat, and, only then, out of that habitat life emerges, not the other way around. This pattern of habitat first, emergent life second, appears to be universal and is consistent with the universal pattern of birth, life, death noted earlier.

There could be no galaxies before there first existed a universe; no stars before there first existed a galaxy; and no planets before there first existed a star around which most revolve. Similarly, there could be no life before there first existed a sustainable habitat from which life could evolve. Hence a pre-existing, sustainable habitat and species survival are inseparable.

Widespread and pervasive loss of sustainable habitats in the past has invariably resulted in mass or near-mass extinctions. Earth has seen fifteen and counting, five big ones that wiped out most complex life and one in particular referred to as the "Great Dying" 250 million years ago, when nearly all life, an estimated 95%, went extinct. Our species is entirely dependent on the continued existence of the planet and the ecosystems that gave rise to and continue to sustain life. Snowball Earth ice ages, global interglacial warming events, and the consequent loss of habitat, as noted earlier, may well have driven Neanderthals to extinction and nearly ended Homo sapiens, our species, as well.

What constitutes a sustainable habitat and what needs to exist to maintain the sustainability necessary to support emergent intelligent species, on Earth or elsewhere in the galaxy?

In this chapter we note the patterns evolution appears to have used to establish a sustainable habitat for the emergence and continued sustenance of complex animal life on Earth; we examine what other planets might need to sustain complex animal life—what planetary preconditions and evolutionary preparations are necessary for the evolution and survival of an emergent intelligent species; and we explore what damage to or loss of a sustainable habitat on Earth might imply for the continued existence of Homo sapiens as a species, given that prior pervasive habitat loss has invariably led to species extinction.

Animal life has survived on Earth for nearly a billion years, complex animal life for over 500 million years, beginning just before the Cambrian explosion. The earliest hominins can be traced back to at least 5 million to 6 million years, and Homo sapiens, the single remaining emerging intelligent species, a little over 300,000 years. So, what has made Earth, unlike Mars and Venus, the other terrestrial planets in the continuously habitable zone of our solar system, a habitat for the evolution and survival of complex animal life and intelligent species? And how have such species fared in the past, and how might they in the future, if those preconditions are interrupted or, as may have been the case on Mars and Venus, completely eliminated?

The fossil trail is littered with the remains of extinct hominins, forbears of modern Homo sapiens. It is a record of an emergent intelligent species' struggle to attain evolutionary maturity and thereby survive. Yet despite an average hominin species existence of about a million years each, the main reason they all went extinct was loss of sustainable habitats, often driven by climate change. To survive, an emergent intelligent species must come to grips with the need to maintain a sustainable habitat, even on a planet like Earth that comes with all the habitability preconditions checked, even if that planet is predisposed to sustaining complex animal life, even if it has done so for almost 1 billion years, and even if it has repeatedly demonstrated its ability to jump-start life all over again after numerous near-total extinctions.

In Chapter 3 of their book *Rare Earth: Why Complex Life Is Uncommon in the Universe*, Peter Ward and Donald Brownlee write: "Most of the Universe is too cold, too hot, too dense, too vacuous, too dark, too bright, or not composed of the right elements to support life. Only planets and moons with solid surface materials provide plausible oases for life as we know it."

So, what exactly is a suitable habitat?

If a suitable habitat is one that is capable of initiating and supporting the evolution and survival of complex animal and intelligent life for billions of years, then we have been living on a planet that provides just such a habitat. We now know that Earth is the only planet in the solar system where such life exists and probably ever evolved. As far as we know, Earth is probably the only planet with complex animals and an emerging intelligent species in the entire Milky Way Galaxy. If so, that would be a big, big, big (leaving out here the expletive that a certain U.S. vice president is reputed to have said) deal. And unless you are Luke Skywalker, things like galaxies and how big they are are probably not what you tend to worry about. Even so, it's not a bad idea to get to know the galactic neighborhood where our species evolved.

According to NASA, the galaxy has been around for about 13.6 billion years, almost as long as the universe itself. It measures some 120,000–180,000 light-years in diameter and is home to Planet Earth, the birthplace of humanity. Our solar system resides roughly 27,000 light-years away from the Galactic Center, on the inner edge of one of the spiral-shaped sections called the Orion Arm. Earth's star, the sun, is just one of approximately 200 billion stars in this galaxy.

If Earth turns out to be the only planet with complex animal life and an emerging intelligent species, as some scientists are beginning to suspect, I think one will have to agree with the vice president. It is a *very, very, very big deal.*

Fig. 7-1 The Milky Way Galaxy

(Source: *shutterstock.com*)

Indeed, when looking back from the moon as the Earth rose into view (See Fig. 1-1), astronaut Bill Anders recalls: "We came all this way to explore the moon, and the most important thing is that we discovered the Earth." It is past time that we as a species begin to discover the Earth before it's too late.

Because we have always had Earth's extensive sustainable habitats and ecosystems, we haven't really had to give much thought to what a sustainable habitat really consists of and requires to continue supporting life. Our species grew up on Earth and simply took all it offered for granted. But if we were forced to find another planet in this or another solar system (the nearest is Alpha Centauri—a binary system about 4.2 light-years away), would you or I know what "survival must-haves" our species should include on its shopping list?

It's similar to packing for an extended field trip. It's only when we begin to think about those must-have items we have to take along that we appreciate the comforts and conveniences of home that we've been taking for granted.

In a recent poll to find out where people are happiest, expatriates were asked the best and worst places to live. Taiwan came out on top, followed by an aspiring Malta, while Ecuador only just retained its place on the list. Kuwait, Greece, and Nigeria remain at the bottom of the pack. *InterNations,* a network for expatriates, reported that in addition to claiming first place out of sixty-seven countries in the overall ranking, Taiwan is in the top ten for every individual index. Taiwan also holds first place in the quality of life and personal finance indices; especially impressive are the quality and affordability of its health care and the enviable financial situation for expats living there.

So what should members of our species be looking for if the question is instead which planet or moon in the solar system is best suited to becoming a replacement habitat for Earth, capable of supporting complex animal and human life, in order to escape potential extinction or as an alternative home world?

When looking for a new home today, one would consider the location—the particular neighborhood, convenience to shopping, desirable schools, health care facilities, transportation, ethnic composition (yes, one is sorry to say), crime, noise, and other types of pollution and environmental considerations. And, of course, the type of house or apartment as well as real estate prices and taxes are important. Similarly, a search for a new habitat for our species, Earth 2.0, must encompass not only the planet but also the home star around which the planet orbits, and the region of the galaxy where it's located.

That's just for starters. There are a few additional items not on an Earth-based house-hunting list that you literally won't be able to live without on most, if not all, planets and moons in our solar system and probably many, if not most, exoplanets in the galaxy as well. These items include many basics we take for granted here on Earth, such as an oxygen-based breathable atmosphere; a suitable temperature range; shielding on the surface from ultraviolet and gamma-ray bombardment from space and from the home star; running water on the surface; food of any kind; fertile soil; animals, birds, reptiles, insects, and marine life; oceans, rivers, and lakes; and of course plants of any kind. Without plants there'd be no photosynthesis, and without photosynthesis there wouldn't be enough oxygen in the atmosphere to support complex animal life.

When choosing an exoplanet, as when choosing a home on Earth, "location, location, location" can be even more important. The following is an incomplete list of some of the "must-have" requirements for a replacement for or alternative to Earth. Much of the following is drawn from recent scientific research reports and books, such as *Rare Earth: Why Complex Life Is Uncommon in the Universe*, by Ward and Brownlee.

The home star of a potentially life-evolving and sustaining planet will preferably be a single star (not a binary or multiple star system) in the galactic habitable zone: one that does not pulse or rapidly change its energy output, or give off too much ultraviolet radiation, but emits a near constant energy output for billions of years; one that is located in a safe region of the galaxy, far from the galactic center and from sources of gamma-ray bursts or supernovae; and one that is moving relatively slowly on its journey through its galaxy.

To be a habitat for the evolution and sustainability of complex animal life, a planet must be terrestrial (rocky and with a hard surface), stably in a circular orbit (with a fixed spin rate—Earth has a twenty-four-hour spin rate), with a fixed tilt of its axis of rotation (aided by the gravitational effects of the sun and moon, Earth has had, for most of its recent geologic existence, a 23-degree tilt angle that drives long-term stability of its surface temperature). It must also be in a continuous habitable zone and far enough away from its home star to avoid being tidally locked by its gravitational force (where one side of the planet always faces the star and the other always faces the coldness and darkness of outer space, like our moon does), and it must be shielded from frequent comet and meteor bombardment by larger

Fig. 7-2 Potential habitats

A comparison of terrestrial planets and moons thought to be able to sustain complex animal life (Source: European Space Agency http://www.universetoday.com/42782/where-could-humans-survive-in-our-solar-system/)

outer planets in its solar system, the way Jupiter shields Earth, as it did in July 1994 from the comet Shoemaker-Levy.

In addition, it must possess a stable climate, a range of temperatures (15°C–45°C) that allows complex animal life to evolve and survive and liquid water to exist (0°C–100°C), and oceans with an average temperature of about 50°C, with a chemical composition conducive to complex animal life and a salinity and pH factor favorable to the formation and maintenance of proteins. The atmosphere must have the right mix of gases. The planet must be capable of supporting plate tectonics to help maintain suitable surface temperatures and formation of continents, thereby engendering biodiversity. It must also have a magnetic shield to protect the surface from cosmic, gamma and UV ray bombardment. In addition, since it could take more than 4 billion years for an intelligent species to emerge (that's how long it took to get to Homo sapiens), it must be able to do all of the above and maintain these conditions for billions of years. Whew! Got all that? That's quite a list, and that's not even being picky. The diagram above shows the relative size of some potential Earth 2.0 habitats.

When a planet and a supporting solar system fail to maintain most if not all of these requirements, life fails to evolve, or if life evolves, it fails to develop into complex animal life, or worse, it evolves into complex animal life and intelligent species only to have it all end when one or more mass extinctions occur. Too much heat or cold, not enough food and other nutrients, too little or too much water, oxygen, or carbon dioxide, excess radiation, incorrect acidity in the environment, environmental toxins and other organisms, or some combination of these has been known to repeatedly drive many of Earth's species to extinction.

A quick perusal of the table below, which lists the habitat preconditions the above worlds must have to be potential Earth replacements, would give even the most intrepid of our species pause and, despite the movie, may permanently flush one's mind of any thoughts about volunteering for the first human habitat on Mars.

Some potential planets and moons compared with Earth for "must-haves" are shown in the table. Each line item contributes a single point to a total score for each world. Earth's score is 15. To qualify as an Earth 2.0, a planet or moon must equal or exceed Earth's score.

Table 7-1 A comparison of Earth with Mars, Venus, and the moons Titan, Ganymede, and Europa

Sustainable habitat requirements	Earth	Titan *(Saturn's moon)*	Ganymede *(Largest moon of Jupiter)*	Europa *(Smallest moon of Jupiter)*	Mars	Venus
SCORE	15	?[16]	?	?	4	4
In the sun's habitable zone [3]	Yes	No	No	No	Yes	Yes
Plate tectonics[2] stable	Yes	Unlikely	Unlikely	Unlikely	No	No
Abundance of elements - 26 needed for life[1]	Yes	No rocky, metal elements such as silicates and iron	?	?	Yes	Yes
Size	1 sol mass	0.163 of Earth's	8.72×10^7 km^2 (.171 of Earth's)		≈ -5 sol	≈ Same as Earth[4]
Gravity[12]	9.8 m/s^2	.14 of Earth's	1.428 m/s^2 (0.146 g)	1.315 m/s^2	3.711 m/s^2	???
Surface	Rocky granite	Rivers, lakes of liquid ethane and methane	Mostly ice with a fair amount of rock in the ice near the surface	Frozen—possible warm-water interior	Dusty and red	Basalt, granite
Atmosphere[5] oxygen[15]	Yes	No. 95% nitrogen and 5% methane	Thin, small amounts of oxygen, not enough to support any form of life	Tenuous, composed solely of oxygen	None[6]	None[7]

Avg. temp[13] range for plants, animals, humans	Yes 15°C to 45°C	No −180°C, or −292°F	No −113°C (−171°F) to −183°C (−297°F)	−260°F −160°C	No[8]	No 465°C
Earth-like magnetic fields[14]	Yes	No	Has a magnetic field	No	No[9]	No
Surface oceans, rivers or lakes—liquid or frozen	Yes	No water. Rivers of ethane and methane	No surface. Salty oceans 100 miles below	?	None so far[10]	No
Running water on the surface	Yes	No	No	No	No[11]	No
Food of any kind	Yes	No	No	No	No	No
Life of any kind	Yes	? TBD	? TBD	Ecosystems in hydrothermal vents	None	None
Animals, birds, reptiles, insects, and marine life	Yes	No	No	No	No	No
Fertile soil and plants	Yes	No	No	No	No	No

(Source NASA historical pages. See also http://www.universetoday.com/22603/mars-compared-to-earth/)

Notes

1. And heavy metal (needed for magnetic field and heat sources). Metal-poor planets may not be able to support or maintain animal life.
2. Helps keep Earth's temperature.
3. As well as in the animal habitable zone.

4. Almost the same size as Earth.
5. Nitrogen (77%), oxygen (21%), argon (1%), water vapor, carbon dioxide (0.038%)
6. There is no naturally occurring oxygen on Mars, mostly carbon dioxide (95.32%), nitrogen (2.7%), argon (1.6%), oxygen (0.13%), water vapor (0.03%), nitric oxide (0.01%).
7. Mostly carbon dioxide: 96% CO_2, 3% N_2
8. -46°C (–51°F) Extremely cold. Mean surface: Equator –50°C; Poles: – 130°C.
9. Mars may have lost its oceans and atmosphere due to lack of such a magnetic shield.
10. Mars oceans? None so far but signs of dry past river beds.
11. Mars. Very small amounts of water in the atmosphere; ice water might be beneath much of the surface.
12. Mars gravity: (approximately 1/3 Earth's).
13. Earth's average temperature range for plants, animals, humans—14°C (57.2 °F) 2°C – 45°C.
14. Earth's magnetic field shields the surface from (cell-destroying) UV, cosmic, and other rays.
15. Earth's atmospheric oxygen–rich breathable air.
16. ? Unknown.

Habitats are about survival. There are those who think, and the science shows, that one or two items on the above table of life's "must-haves" might exist below the surface of some planets in spent subterranean lava tubes; for instance, frozen or liquid water might be on Mars. But, wouldn't it be extremely ironic if our species took 4-plus billion years to go from "slime" to cavemen and from cavemen to spacemen only to have to go back to being cavemen all over again on Mars or other potential worlds in search of water and protection from UV and cosmic ray bombardment?

Even if there is one, or even a few, of life's "must-haves" on a potential habitat, like water on Mars, it's worth observing that man does not live by water alone. One does not get to pick and choose when it comes to the preconditions that must exist for the evolution and survival of complex life. That's not how the game of life is played; it's either all of them, or life as we know it does not evolve or soon goes extinct. There is no negotiating with Mother Nature.

Evolution, as we noted earlier, has had more than 4 billion years of practice getting the formula right for the emergence and survival of complex life. So, when it comes to the evolution and survival of complex life and intelligent species as we know them, at least for now and probably for a long, long time to come, there is going to be no way to fool Mother Nature. Given what we know now, it is probably a fair conclusion that on any of these Earth 2.0 wannabes we would either have to cut it or it will cut us out of existence.

There are those who think our species will be able to terraform a potential Earth 2.0, adding to a planet what it lacks to make it habitable. While that might be possible in a few thousand years, if not a million, it probably won't be soon enough to save our species if we somehow manage to render Earth incapable of supporting complex animal life in the interim. Terraforming, or modifying a world to be habitable for humans, would minimally require reconstructing the planet's atmosphere and biosphere practically from scratch, eradicating any native ecosystem, unless one plans to walk around in a spacesuit. That might be an assignment for the likes of Captain Kirk or Jean-Luc Picard and the Federation in the twenty-fourth century, but for now it is orders of magnitude more challenging than the not-so-minor tweaks needed to restore Earth's environment to a sustainable state.

New tools are beginning to emerge that make it easier to pay more attention to what's happening to the sustainability of Earth as a habitat for humanity and the search for other potential habitable planets. A team of astrobiologists has now proposed a Planetary Habitability Index that includes four groups of variables, each of which is weighted by its importance to sustaining life. Astronomers have often estimated the habitability of extrasolar planets and moons based mostly on their temperatures and distance from the nearest star, but as we have seen in the table above and will from the Planetary Habitability Index below, having a suitable temperature range is just one of many factors needed to sustain complex animal and human life.

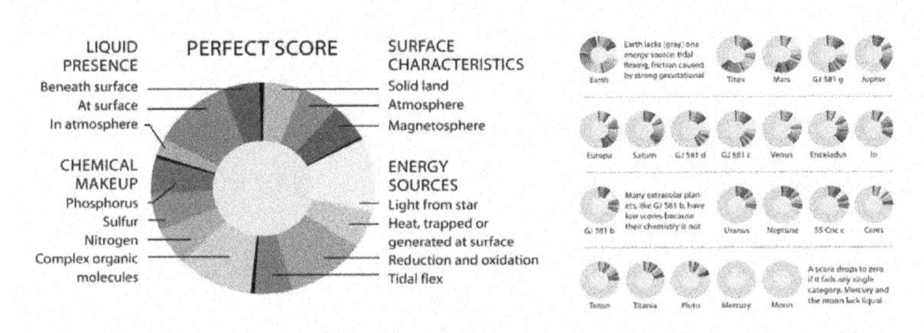

Fig. 7-3 The New Planetary Habitability Index
(Source: Popular Science
https://www.popsci.com/content/new-planetary-habitability-index)

The planetary index shown above uses atmospheric analysis to determine the presence of life. Except for the chemically unreactive rare gases like argon, helium and xenon, most gases in Earth's atmosphere (oxygen, methane, nitrogen, carbon dioxide, nitrous oxide, and others) are produced or used by organisms. Earth's atmosphere then is kept in a steady but dynamic state of disequilibrium due to the presence of life.

By comparing the major gases comprising the atmospheres of Venus, Mars and what it might have been on a potential pre-life Earth with those comprising Earth's current atmosphere (shown below), one can clearly see the difference on a planet with life versus one without. Most gases comprising a planet's atmosphere are produced by the life-forms inhabiting that planet.

Table 7-2 Gases comprising planetary atmospheres

	PLANET			
GAS	VENUS	EARLY EARTH	MARS	EARTH AS IT IS
Carbon dioxide	96.5%	98%	95%	0.04055%
Nitrogen	3.5%	1.9%	2.7%	79%
Oxygen	trace	0.0	0.13%	21%
Argon	70 ppm	0.1%	1.6%	1%
Methane	0.0	0.0	0.0	1.7 ppm
Surface temperature °C	459	240 to 340	-53	13-14
Total pressure, bars	90	60	0.0064	1.0

(Source: NASA)

Hence, unlike the gases comprising Earth's current atmosphere shown above, the atmosphere of a planet without life would be at or in a state of near chemical equilibrium. Reacting gases on the surface would show no net energy. And indeed, a spectroscopic analysis (infrared spectra) performed on the atmospheres of Venus and Mars confirmed their atmospheres were comprised mostly of carbon dioxide with trace levels of nitrogen, oxygen and other gases, and showed no evidence of chemical reactivity and were in a state of near if not complete chemical equilibrium, thus indicating complete or near complete absence of life.

It should be clear that none of these potential habitats come close to meeting the survival needs of complex animals, let alone emergent intelligent species. So in view of the slim prospects of even identifying, let alone getting to, a world with the right properties to support life, if the choice is to immediately stop and begin to repair the destruction our species is doing to the Earth's habitats, ecosystems, and environment, or attempting to go to a planet like Mars that may have once had but now no longer possesses those "must haves" for life, the choice could not be clearer. If as a species we want to survive, mature, and go on to populate the stars, then for now and a considerable while to come, there is no planet B.

To be clear, having to find a new habitat for humanity is not going to be optional. Given Earth's expiration date, referred to earlier, our species will have to find and migrate to one or more planets in the solar system or among the stars, but that time is quite a ways away. Evolution has built in enough time before that expiration date—another 500 million to 1 billion years—unless of course we manage to render the Earth unfit as a habitat before then.

Meanwhile, given all the buzz about Mars, some might begin to think our species has a near-term choice between survival on Earth or escaping to another world, but I'm afraid—and this is coming from a science buff—at least for the next few hundred plus years, the choice may come down to survival on Earth or the extinction of most of our species. Choose well.

Personally, I would definitely err on the side of restoring the Earth's ecosystems and environment over taking my chances on Mars. And, as most expats might agree, just as having a home country one can return to at any time offers one a sense of security and consolation, so would maintaining the Earth as a sustainable habitat to which our species can return might one day be equally indispensable.

Some observers, rightly or wrongly, muse that at the moment we are still too immature a species to embark on this next major evolutionary leap, and evolution appears to agree. By placing the planets and stars far enough away (as the continents must have seemed 70,000 years ago when Homo sapiens began migrating out of Africa), evolution seems to have anticipated that our ambitions might outpace our preparedness and has erected these mind-boggling inter-planetary and inter-solar distances as evolutionary roadblocks, as it were, to ensure we take the time to achieve that level of evolutionary maturity needed, to scale that Survival Plateau we encountered in the prior chapter, before embarking on this next step.

Migrating to potential new earths would require an unprecedented level of global co-operation that can only come from that level of maturity a species must attain before migration to the stars becomes possible. In this regard the International Space Station might well be one small step for Homo sapiens.

Given that prior pervasive habitat loss invariably led to extinctions, even mass extinctions, and in at least one event, near-total extinction of all species, it is reasonable to be concerned about the prospects for our species in view of our extensive and accelerating destruction of many of Earth's ecosystems, habitats, and environment. Just as looking back from the moon afforded a perspective and awareness of Earth in a way not previously possible, looking back at what early Earth might have been like can allow us to discover Earth again, this time not just as a planet but as a habitat built for humanity, possibly the only such habitat within light-years.

Have you ever wondered what Earth was like when it was first formed? Growing up on Earth today, it's difficult to imagine a time when our planet was not always as it is now; it took evolution over 3 billion years to transform the early Earth from a seething, molten cauldron into a near-perfect habitat for humanity.

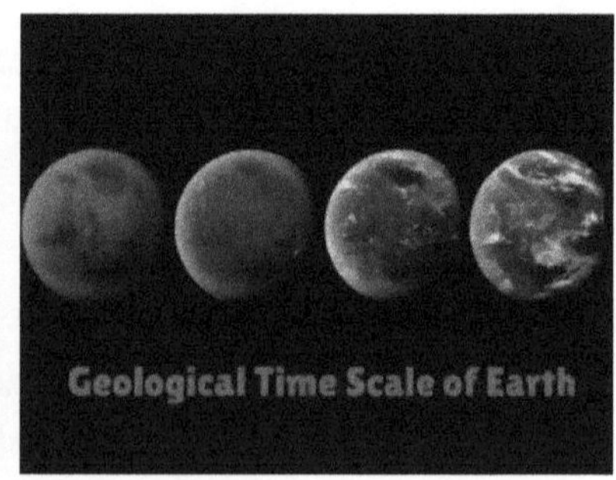

Fig. 7-4 Earth changes over geological time
(Source: *www.gainhistory.blogspot.com*)

The early Earth (shown above) was nothing like it is today. Life as we know it could not have existed on Earth 3.8 billion years ago. Life is a very complex and delicate phenomenon that is easily destroyed, as Ward said in his book *Rare Earth*. Ward noted that for the first 700 million years of Earth's existence, during the so-called Late Heavy Bombardment period, some 4 billion years ago, life in any form would have been highly unlikely, or if it had emerged it would have been destroyed repeatedly. In fact, for nearly all of Earth's history, life was limited to microbial life, small, hardy creatures, invisible to the naked eye, that seemed to have evolved soon after the Late Heavy Bombardment period concluded. And it is from these invisible microbes, extremophiles, evolution's planetary terraforming crew, all known life-forms evolved. Some scientists believe the galaxy might be teeming with planets that harbor this level of life.

While in this book we focus on intelligent species, hominins' and Homo sapiens' evolution and survival, that evolution did not just begin with the great apes in the family Hominidae, or even from other mammals. "Back before the first dinosaurs, before the first fishes, before the first worms, before the first plants, before the first fungi, before the first bacteria, there was an RNA world—probably somewhere around 4 billion years ago, soon after the beginning of Planet Earth's existence," writes Matt Ridley in his book *Genome: The Autobiography of a Species in 23 Chapters*. And, according to Lynn Margulis in her book *Symbiotic Planet: A New Look at Evolution*, "We evolved from a long line of progenitors, and ultimately from the first bacteria. No species existed before bacteria. In fact, all organisms visible to the naked eye are composed of once-independent microbes that over the course of those 4 billion years teamed up to become the large life-forms we see today."

It was to these microscopic bacteria that evolution assigned the task of terraforming Earth. Simply by living they prepared the watery, seething cauldron to become a habitat for the evolution of complex animals and intelligent species that was to happen billions of years later. Earth was populated by bacteria for most of the geologic periods known as the Proterozoic (0.57 billion–2.5 billion years ago) and the Archean (2.5 billion–4.5 billion years ago) that together comprise the Precambrian geologic period after which complex animals and intelligent species life appeared. It is mostly accurate to say that most of Earth's crusts, ecosystems, oceans, and atmosphere are the products of these tiny life-forms. In fact, we live on a planet built by our bacterial ancestors that are deeply integrated into the ecosystems that comprise its habitats.

As we have seen above, most of the gases comprising Earth's atmosphere are produced by these microbes. For example, it was cyanobacteria that began the generation of much of early Earth's oxygen. It was bacteria that joined together to form the first cell and thereby engineered larger cells, prokaryotes and eukaryotes, including all ancestors to both plants and animals. We now know, for instance, that plastids and mitochondria, organelles within the human cell but outside the nucleus, are in fact remnants of once free-living microbes. And because of this collection of remnant, independent bacteria we now know as mitochondrial DNA, we can trace all living Homo sapiens back to a single female ancestor in Africa.

Unlike our microbial extremophile ancestors, complex animal and intelligent species life, Ward notes, is quite a different matter. Too much heat or cold, or too many gamma rays, X-rays, or other types of ionizing radiation (not unlike many planets seen in our solar system and others) are not conditions conducive to the emergence of complex animal life and the evolution of an intelligent hominin species like Homo sapiens. Such lifeforms require a very delicate and narrow range of conditions to ensure evolution and survival; the early Earth had a ways to go before complex animal life could evolve.

In describing the early Earth, Ward notes that the length of the day was shorter because Earth was rotating faster than it does now. The sun might have seemed like a red orb, supplying little heat (25%–40% less luminosity) and burning with less energy than today. It would have appeared much dimmer, as light would have to penetrate a poisonous, turbulent atmosphere composed mainly of carbon dioxide (some scientists think about 90%—as much as on Venus today), hydrogen sulfide, steam, and methane. The Earth would have looked like an alien planet and required a spacesuit, as only trace amounts of oxygen might have been present. The sky might have been orange to brick red, and seas covering most of the planet, except for a few volcanic islands, would have been muddy brown and clogged with sediment. There were no trees, no shrubs, no seaweed or floating plankton in the sea. It would have seemed from all appearances to be a dead world.

It was going to take nearly another 3 billion years before complex animal life would evolve—less than a billion years ago, during the last 10% of the planet's existence. If nothing else, this might be a not-too-subtle hint that complex animal life is quite a tricky phenomenon to get right, and establishing a suitable habitat for its evolution required a rather delicate balance of galactic, solar, and planetary habitat conditions for it to emerge and survive. Several factors might have contributed to this 3-billion-year delay between the appearance of microbial life and the evolution of complex animal life.

Early Earth was essentially a watery world; a sprinkling of emerging volcanic islands were the only solid surfaces before the continents formed sometime during the first 2 billion years. Surface temperatures may have been too hot—hot enough to melt rocks. In addition to the need for the planet to cool down to a range of temperatures conducive to animal life, between 15°C and 45°C, there might have been only trace levels of oxygen in the early atmosphere. Animals require a higher atmospheric oxygen content to evolve and survive.

It took a very long time for oxygen levels to build up. The interaction of sunlight and water molecules may have produced the first atmospheric oxygen—about 1%. A further buildup followed, generated by cyanobacteria breathing in carbon dioxide and breathing out oxygen, but surface rocks would have oxidized much of that. It was going to take another billion years before oxygen buildup, accelerated by plants through photosynthesis, would take off. In fact, it took more than 2 billion years before the concentration of oxygen was high enough in Earth's oceans and atmosphere (currently 21%) to support the evolution of complex life that greatly accelerated during the Cambrian explosion.

Gaining an appreciation of what it took for Earth to transform from planetary chaos to a habitat capable of jump-starting life and sustaining for billions of years the conditions necessary to support complex animal and intelligent species is necessary if only because our species has been engaged in a rapid unraveling of it all. It is also indispensable to understanding both the geological time scales involved and the planetary, atmospheric, ecological, and biological processes evolution employed to turn Earth into the human habitat we know today.

One can hopefully be forgiven, then, for harboring a deep sense of skepticism when it comes to talk about Homo sapiens terraforming Mars and other planets for human survival anytime soon.

Table 7.2 shows that Earth's early atmosphere was not too dissimilar to that of Mars and Venus—lacking any of the reactive gases that would have indicated the presence of life. It should come as no surprise, then, to learn that our species is not the first to have changed Earth's atmosphere. That honor and (depending on your point of view) dubious distinction already belongs to cyanobacteria (blue-green algae), as noted above. Cyanobacteria,

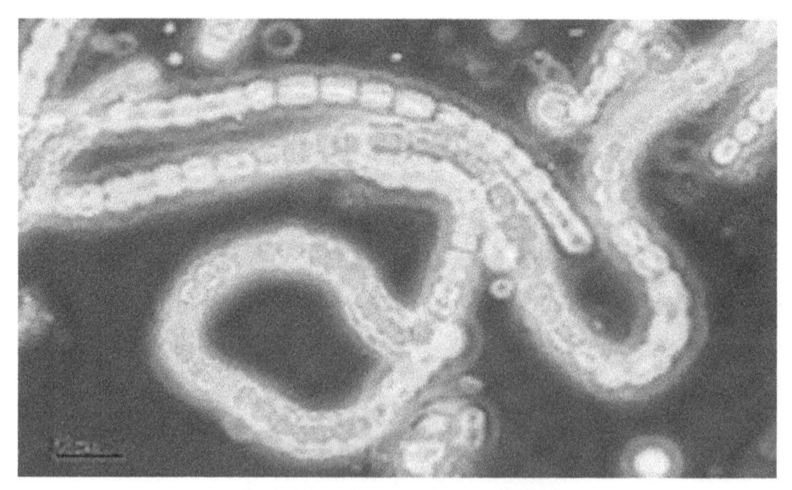

Fig. 7-5 Cyanobacteria (Phytoplankton)
Responsible for the buildup of oxygen in the Earth's atmosphere
(Source: https:/wikipedia.org/wiki/Phytoplankton)

pictured below, without detonating a nuclear bomb or setting off any other weapons of mass extinction, are thought to be responsible for one of the most significant extinction events in Earth's history (though some scientists are now questioning this interpretation of events).

In an early atmosphere dominated by carbon dioxide, ammonia, and water vapor (mostly greenhouse gases), oceanic cyanobacteria, which evolved into multicellular forms more than 2.3 billion years ago, are believed to have been the first microbes to produce oxygen by photosynthesis. The increased production of oxygen set Earth's original atmosphere off balance. Free oxygen is toxic to certain types of anaerobic organisms (obligate anaerobic organisms), and the rising concentrations may have destroyed most such organisms at the time. Indeed, cyanobacteria, as mentioned above, are thought to be responsible for one of the most significant extinction events in Earth's history, and probably the first. Like the Karate Kid, all they did was breathe in and breathe out, breathe in and breathe out, completely unaware that by so doing they were setting off an atmospheric gas exchange (CO_2 to O_2) that would end in a mass extinction as well as begin the process of getting Earth's atmosphere ready for complex animal evolution that would begin 2 billion years later.

This suggests a climate event that changes atmospheric gases from, as in this case, a predominantly carbon dioxide-based to an oxygen-based atmosphere might not be a good thing. It did not have so happy an ending for the predominantly anaerobic species living at that time, as most would have been driven to extinction or to hide in low-oxygen locations just to exist as they still do today in your and my intestines—our so-called microbiome. Just imagine what would happen to all the oxygen-breathing species if the reverse were to occur today—and that may not be so farfetched, given the reverse atmospheric gas exchange in which our species is currently engaged.

The cyanobacteria-driven extinction event offers a few takeaways for Homo sapiens today, as our species seems bent on doing the reverse. Whereas evolution flushed the early Earth's atmosphere of carbon dioxide (CO_2) and replaced it with oxygen (O_2), a gas essential to sustain complex animal and human life, our species seems intent on dredging up all the CO_2 that evolution so carefully hid away in the bowels of the Earth, to dump it all back into the atmosphere. And it seems intent to follow that with methane (CH_4)—which traps heat twenty to thirty times more efficiently than CO_2—which until recently had been buried under the Arctic and Antarctic.

To better understand how this tiny speck of a creature managed to initiate a mass extinction event, one needs to understand a little about how Earth's current atmosphere came about.

Looking at what the atmosphere was like during Earth's early history as compared with what it is today, and how that came about, helps us understand the significance of the CO_2 buildup, decline in oxygen levels, and simultaneous temperature increase, and the potential effects of these changes on complex animal and human life. In fact, the importance of the rise and fall of oxygen and carbon dioxide levels over geologic time is fundamental to understanding the history of life on Earth, and how changes in the composition of these atmospheric gases have affected and can affect our lives today.

In their book *A New History of Life: The Radical New Discoveries about the Origins and Evolution of Life on Earth*, Peter Ward and Joe Kirschvink note that "perhaps the most influential factors other than temperature that most importantly influenced life's history on Earth were the changing volumes (atmospheric gas pressures) of life-giving carbon dioxide for plants, and oxygen for animals. The relative amounts of both CO_2 and O_2 in Earth's atmosphere over time have been and continue to be determined by a wide range of physical and biological processes. Almost more than any other aspect, it has become clear that the interplay and concentrations of the various components of the Earth's atmosphere have been dominant determinants of not only what kind of life (or there being any life at all) on Earth, but the history of that life."

Getting the Earth ready, then, for complex animal and intelligent species life was no easy matter, and here is the backstory.

With the possible exception of water vapor, Earth essentially had no atmosphere until the buildup of gases emanating from violent volcanic eruptions, believed to have consisted mainly of CO_2 and a cocktail of other gases that might be ejected from a volcano, including water, sulfur dioxide, cobalt, disulfur, ammonia, methane, and more. The hydrogen and helium that might have accompanied the Earth's formation process might have been stripped away by the sun's UV rays. The high levels of CO_2 (some believe it dominated the atmosphere at this time) would imply a very warm to very hot atmosphere, but the lower amount of heat the sun was thought to be giving off at the time would have more than made up for it. And while Earth's atmospheric oxygen content is 21% today and may have been as high as 35% just before and during the period of the dinosaurs, the early Earth had no free atmospheric oxygen.

This means that the early microbes and extremophiles that emerged soon after the end of the Heavy Bombardment period some 3.8 billion–4 billion years ago would have had to be tolerant of a CO_2-dominant cocktail of atmospheric gases. And indeed that was the case. However, before there could be any chance for multicellular life to emerge, Earth needed a change from a CO_2- to an O_2-based atmosphere, as well as a UV shield.

We noted earlier that fundamental to understanding life's history is the importance of the rise and fall of oxygen and carbon dioxide over time. The trick evolution had to perform was changing a predominantly CO_2-based atmosphere to one with a level of atmospheric oxygen sufficiently high to support the evolution of complex animal and human life. Today the CO_2 content of the Earth's atmosphere is approximately .036% by volume (a huge drop), whereas oxygen content grew from essentially trace levels to approximately 21%. And because DNA is easily destroyed by UV light, blocking these harmful rays became a prerequisite for the emergence of multicellular life-forms. How did evolution pull this off?

First, the generation of oxygen in small amounts (≈1%) through a process referred to as photochemical dissociation might have started through a breakup of water molecules by ultraviolet rays. However, as noted, the sun's UV rays would have killed any early life exposed to them until there was enough oxygen in the stratosphere to permit the buildup of ozone, which formed a protective layer stopping UV radiation from reaching the surface of the Earth.

This shield permitted the emergence of cyanobacteria, which produced more oxygen, increasing levels to about 10%–12%. Cyanobacteria inhaled carbon dioxide and water, and with sunlight, produced oxygen as a by-product. The shield also facilitated the emergence of plant life that also consumed CO_2 and gave off oxygen. Both of these processes resulted in a reduction of CO_2 atmospheric content while building up O_2. This continued for almost 3 billion years, until enough oxygen accumulated in the atmosphere to permit the emergence of the first animals some 500 million years ago.

> *This atmospheric content changeover is thought to have been the key event in early Earth history that initiated eukaryotic cell development, land colonization, and species diversification.*

So, finally, the early Earth was beginning to accumulate enough oxygen for complex animals and emergent intelligent species to evolve. However, oxygen is a promiscuous, corrosive, and poisonous gas that reacts easily, and this changeover from a CO_2- to an O_2-dominant atmosphere was not without consequences. It drove one of Earth's great extinctions—the Great Oxygenation Event—and meant that most anaerobic life forms may have been driven to extinction, except for those that managed to survive in anoxic (low-oxygen) habitats.

Before leaving this section, it is worth reflecting on the massive planetary transformation evolution engineered to prepare this delicate cocktail of atmospheric gases that complex animals and intelligent species would need to evolve, survive, and thrive. It took a mass extinction no less, and 3 billion years, and like a massive propane tank dangerously close to a raging fire, Earth has skull-and-crossbones signs emblazoned all through the fossil record of extinct species that show what happens when these atmospheric gases are reversed.

Could the current CO_2 buildup trigger a reverse in that delicate balance of atmospheric gases once more? If it does, this time around, it could be to our peril.

What would happen if the current life-sustaining levels of atmospheric oxygen were to change? To understand how and why oxygen levels could change today one must consider both the sources and consumers of oxygen. The known sources of oxygen today are photochemical dissociation, or the breakup of water molecules by UV light, and photosynthesis by cyanobacteria and plants, including marine-based plants, phytoplankton alone accounting for more than 50% of the Earth's oxygen. Consumers, or sinks, include oxidation (or chemical weathering) of surface rocks, animal (including human) respiration, and, finally, burning of fossil fuels.

We now know that atmospheric oxygen content has not always been constant at around 21% but has built up over geologic time from almost none. We also now understand that atmospheric carbon dioxide appears to have gone in the opposite direction—declined over geologic time from highs possibly greater than 90% to what it is today, approximately 0.036%. We have explored briefly the effects of an atmospheric changeover from a CO_2- to an O_2-dominant atmosphere and the subsequent extinction of anaerobic species unable to adapt.

The unasked question here is: Can this oxygen-based atmosphere, thought to have been the key event in early Earth history, initiating eukaryotic cell development, land colonization, and species diversification, be reversed? If so, how?

What's not so well known is that O_2 might have been as high as 35% just prior to the time of the dinosaurs, from 135 million to 65 million years ago, and might have fallen to as low as 15% during a 120-million-year period that ended in a mass extinction at the end of the Permian period. That reduction to 15% might have contributed to the great Permian extinction of both land and sea species, the worst of all extinctions, the Great Dying, a time when almost all life died. Atmospheric oxygen content depletion during this period would have definitely led to the extinction of species that had adapted to living in an oxygen-rich environment.

If a drop to a 15% O_2 level may have contributed to species extinction during the late Permian, could a similar drop today also have consequences for our species, since we too require an oxygen-rich environment?

If you are starting to get the impression that a predominance of oxygen in the atmosphere seems to spell life whereas a predominance of carbon dioxide seems to spell death for complex animals and humans, you may be on to something and might also be wondering about the implications for species survival given the current atmosphere. According to Ward and Kirschvink in *A New History of Life*, oxygen and carbon dioxide levels (particularly oxygen) are the most important of all factors dictating animal survival, death, and diversity. Times of high oxygen give rise to an increase in species, whereas when low oxygen prevails, species die out.

They note further that geochemists have long known that CO_2 levels and atmospheric oxygen levels show trends that are inversely related to one another. "When oxygen levels rise, CO_2 is usually dropping," note Ward and Kirschvink. They also suggest that it might not be the changing CO_2 but a combination of falling oxygen levels and CO_2-driven global rising temperature.

For example, cold water holds more oxygen than warm, so in a cold world with high oxygen levels, life in the sea would rarely be compromised by too little oxygen. In a warm world, a high-CO_2 world, where there is already relatively little oxygen, most bodies of water will quickly go stagnant, affecting life.

Today the world is warmer due to increasing CO_2 levels, while oxygen levels are falling. We are also living in a time of alarming and rapid loss of species to extinction, most of which is attributed to human activity such as poaching, habitat encroachment, and overfishing. Could an atmosphere-based component have contributed to the loss of some of these species as well? Can it also be due in part to falling oxygen levels accompanied by rising CO_2-driven temperature increases (due to a different kind of human activity)? In fact, this has already begun to happen.

Because the buildup of atmospheric CO_2 also means a reduction in oxygen in the air and in the oceans, it bears asking whether we are unwittingly engaged in a transformation from an oxygen- to a carbon-dioxide-based atmosphere? Whereas evolution flushed the Earth's atmosphere of CO_2 and replaced it with O_2, our species seems intent on doing quite the opposite. We are essentially dumping back into the atmosphere more and more carbon dioxide and soon methane. The cyanobacteria-driven extinction event should serve as a warning to Homo sapiens today, as our species seems bent on doing quite the reverse but with a similar result.

Burning carbon-based fuels, including oil and gas, produces a carbon dioxide molecule—two oxygen atoms get locked up together with one of carbon. That oxygen that drives the burning process comes from the atmosphere and hence decreases the amount of oxygen available for you and me to breathe. Similarly, carbon dioxide sequestering may not be too bright an idea when viewed from the oxygen-depletion perspective because it also sequesters oxygen in the process.

So, while global warming induced by greenhouse gases (including CO_2, of course) are not insignificant, an emerging problem might be the depletion of the atmospheric free oxygen needed by all complex animal life to survive. The Scripps 02 monitoring program's Historical Global Atmospheric Oxygen Levels Graph, shown below, documents and tracks this phenomenon.

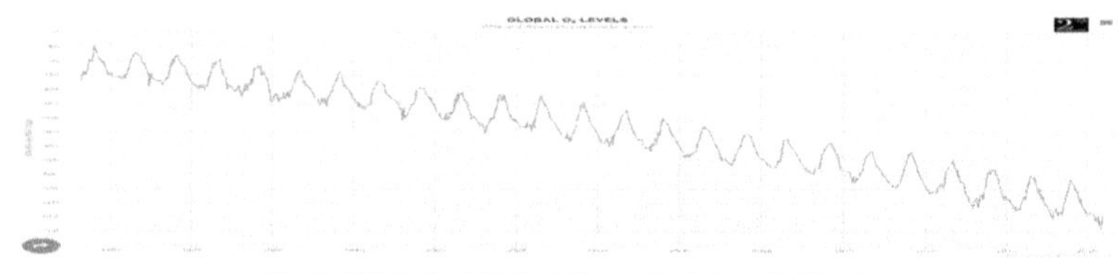

Fig. 7-6 Historical Global Atmospheric Levels Graph
(Source: Oxygenlevel.org)

In an article titled "O_2 Dropping Faster than CO_2 Rising: Implications for Climate Change Policies," geneticist Dr. Mae-Wan Ho writes: "Within the past several years scientists have found that oxygen (O_2) in the atmosphere has been dropping, and at higher rates than just the amount that goes into the increase of CO_2 from burning fossil fuels, some 2 to 4 times as much, and accelerating since 2002-2003. Simultaneously, oxygen levels in the world's oceans have also been falling."

And although today there is much, much more O_2 than CO_2 in the atmosphere (20.95%, or 209.460 ppm of O_2 compared with around .04% or .380ppm of CO_2), how low a level and for how long can oxygen levels drop before it becomes a global life-threatening issue? It's worth remembering that during the Great Permian Extinction, oxygen levels are believed to have fallen to about 15% compared with present-day levels and remained around that vicinity for nearly 120 million years and is believed to have contributed to the Great Permian Extinction.

In humans, failure of oxygen energy metabolism is the single most important risk factor for chronic diseases including cancer and death. "Oxygen deficiency is currently set at 19.5 percent in enclosed spaces for health and safety; below that, fainting and death may result. Humans, all mammals, birds, frogs, butterflies, bees, and other air-breathing life-forms depend on this high level of oxygen for their well-being," according to "Living with Oxygen," at *Science in Society*.

Many might view this as a non-issue because of the current lopsided amount of O_2 versus CO_2 levels in the atmosphere, but there was a time in Earth's history when the atmospheric content of these gases was completely reversed. It took hundreds of millions of years for the current oxygen-dominant state to be reached. Yet, in a *Scientific American* article titled "The Ocean Is Running Out of Breath" published February 25, 2019, Laura Poppick quotes Andreas Oschlies, as saying that "the past decade ocean oxygen levels have taken a dive—an alarming trend that is linked to climate change." Oschlies is an oceanographer at the Helmholtz Center for Ocean

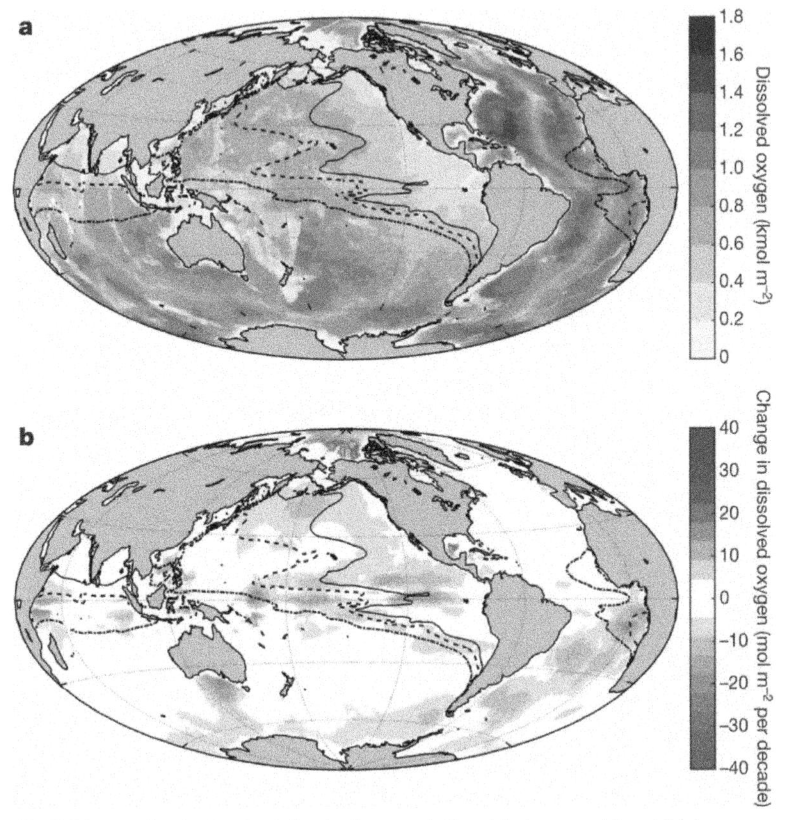

Fig. 7-7 A map showing levels of dissolved oxygen in the global oceans (a) and (b), how oxygen levels have declined or risen per decade since 1960.

Credit: Schmidtko, et al., Nature 2017

Research Kiel in Germany, whose team tracks ocean oxygen levels worldwide. "We were surprised by the intensity of the changes we saw, how rapidly oxygen is going down in the ocean and how large the effects on marine ecosystems are," he says. Oxygen levels in some tropical regions have dropped by a startling 40% in the last fifty years, some recent studies reveal.

The year 2018 was the hottest year measured for Earth's oceans compared with the 1981-2010 average and the consequences are catastrophic. The current rate of ocean warming is equivalent to five Hiroshima-size atomic bombs exploding every second.

Not only has the oxygen level dropped across the Earth but it is now at 10% and even lower in vast so-called dead zones in various regions of the oceans. There are now more than 400 coastal dead zones around the world, the most well-known being the dead zone in the Gulf of Mexico (see below), where the Mississippi River dumps fertilizer runoff from the Midwest. The Louisiana Universities Marine Consortium reported that this year's so-called dead zone covers 7,722 square miles.

Others have appeared off South America, China, Japan, southeast Australia, and New Zealand.

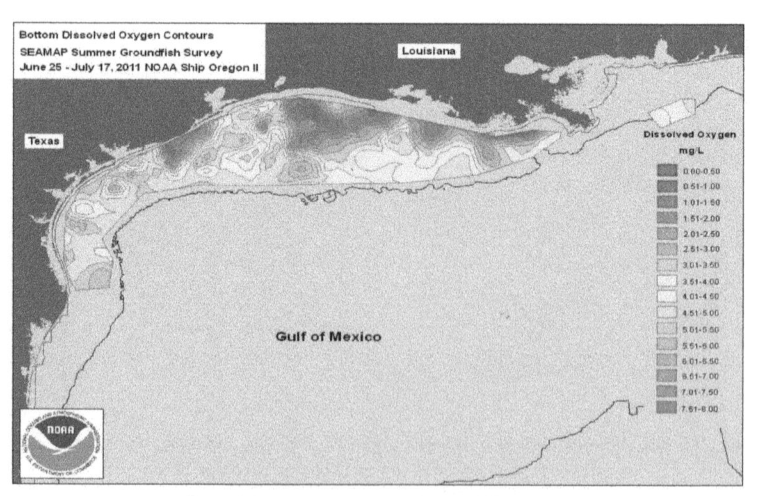

Fig. 7-8 Dead zone spanning the Gulf Of Mexico

Large fluctuations in the ocean's oxygen have occurred in the past history of the Earth. As noted above, anoxic (O_2 free) oceans at the end of the Permian period 251 million years ago were associated with elevated atmospheric CO_2 and massive extinctions on land and sea.

Repeated surveys have indicated that the upper 3 kilometers of the oceans have warmed and intermediate waters of high-latitude have freshened over the past decades. Studies with models have confirmed the suspicion that these are the consequences of global warming from burning fossil fuels and other human activities. The surveys have also detected a decrease in dissolved O_2 in intermediate waters in all the oceans, with small increases in deeper waters of the North Pacific and South Indian oceans. Strong fluctuations in oxygen have been observed in the upper 100 meters from year to year and over decades, and long-term changes are reported in the subpolar and subtropical regions.

> *In some places, the oxygen is getting so scarce that fish and other animals cannot survive. They can either leave the oxygen-free waters or die.*

The Kiel Declaration on Ocean Deoxygenation convened more than 300 scientists from thirty-three countries in Kiel, Germany, in September 2018 and is a sure sign that many are now starting to take this decline in oxygen seriously. Many, too, are starting to realize that loss of phytoplankton-generated oxygen due to increased warming of the oceans may pose a greater threat to our species' survival than flooding due to sea-level rise. While much attention has been given to the loss of islands and coastal cities as potential consequences of global warming, not much has been paid to how continued increase in ocean temperature can result in the depletion of oxygen in the oceans and atmosphere. A study conducted by scientists from Britain's University of Leicester found that an increase of about 10.8°F (6°C) in the temperature of the world's oceans could prevent phytoplankton's oxygen production by disrupting photosynthesis. "It would mean oxygen depletion not only in the water, but also in the air," said the research team. "Should it happen, it would obviously kill most of life on Earth."

Phytoplankton may well turn out to be the Canary in the Ocean.

Scientists estimate that phytoplankton contribute between 50% to 85% of the oxygen in Earth's atmosphere. According to *National Geographic*, phytoplankton are at the base of what scientists refer to as oceanic biological productivity, the ability of a water body to support life such as plants, fish, and wildlife, and are the foundation of the oceanic food chain. Fish, whales, dolphins, crabs, seabirds, and just about everything else that lives in or off of the oceans owe their existence to phytoplankton, one-celled plants that live at the ocean surface.

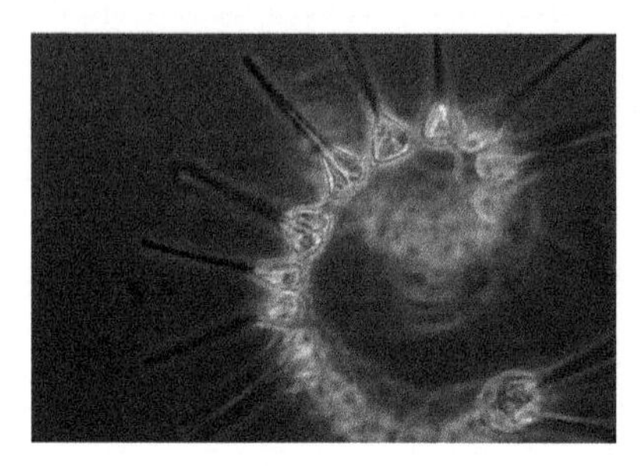

Responsible for the buildup of oxygen in the early Earth's atmosphere

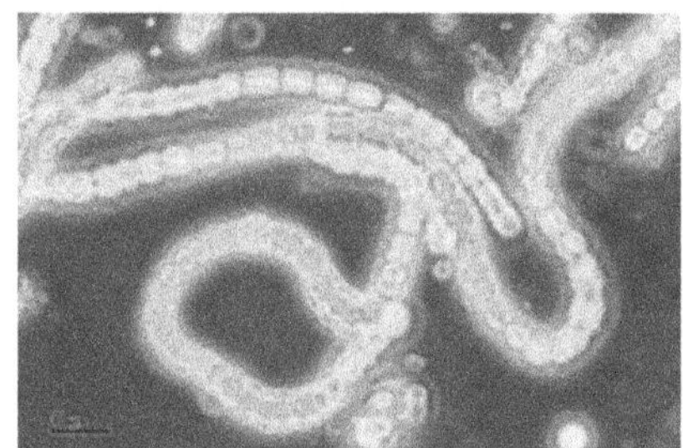

Phytoplankton – Image via NOAA

Fig. 7-9 Cyanobacteria & phytoplankton

A glance at these two urchins above affords a little insight into the ways and patterns of evolution. No, if our demise were to come from our heedless continued destruction of Earth's habitats, it won't be by Klingons or Cardassians, or even the Empire's Dark Lord of the Sith, Count Dooku, but just from an innumerable army of lowly phytoplankton—evolution's perfect replicators.

What's so uncanny about this potential cause of global mass extinction is that it was another tiny thing, a bacteria species, cyanobacteria, that was responsible for jump-starting the initial buildup of oxygen in the atmosphere that changed early Earth from a CO_2-based world to an O_2-based world, which led to our evolution. If our species persists in dumping CO_2 into the atmosphere, thereby increasing the temperature of the oceans, it may turn out that another tiny thing, this time a single-celled ocean plant, phytoplankton, responsible for between 50% and 80% of the free oxygen in our atmosphere, will no longer be able to produce that oxygen.

How strange! It was oxygen that got us into the game of life, and it might well be oxygen that takes us out. Perhaps there ought to be an oxygen-level monitor app (and there is though not an app—https://www.oxygenlevels.org/) not unlike the PM2.5 app some carry around to track air pollution.

To borrow from a phrase made famous by President Bill Clinton, when it comes to life and survival, it's the oxygen, stupid! But we didn't stop there; we went after the rest—Earth's last remaining source of oxygen, the forests.

Forests use carbon dioxide (CO_2) and produce oxygen and are thought to produce the remaining 20%-30% of the planet's oxygen. They also provide shelter and food for many different types of plants and animals. NASA reports that today 30% of the land is covered by trees and as much as 45% of the carbon stored on land is tied up in forests. But here again, our species has been doubling down on making sure that this only other remaining source of free oxygen, which simultaneously removes some of the excess CO_2 we continue to pump into the atmosphere, is being rapidly erased. Is this suicidal behavior?

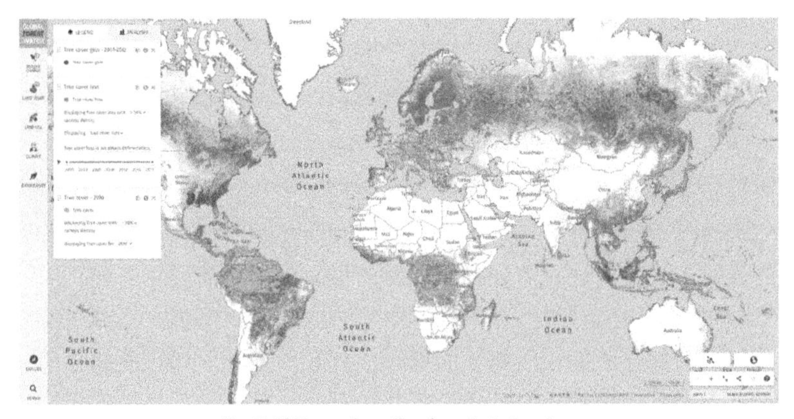

Fig. 7-10 Dynamic online forest alert system
(Source: GBL Forest Watch)

Global Forest Watch (GFW) shows a dynamic online forest alert system (map shown below) that enables one to track the gain and loss of forests around the world from 2001 to 2017 and the impact on CO_2 emissions.

According to Ward and Kirschvink, "The most important driver removing carbon dioxide from the atmosphere is the weathering of silicate rocks to granites, and sedimentary and metamorphic rocks with a granite-like chemical composition—a rock type rich in silicon. Chemically weathered silicate rock reacts with the atmosphere and removes molecules of carbon dioxide. Deep roots of trees facilitate rapid weathering and in Earth's history have caused carbon dioxide levels to plunge and plunge quickly. By lowering carbon dioxide, Earth's temperature also drops. In fact, as trees grew taller and taller, roots went deeper into the Earth, and the resultant cooling of the planet led to one of the longest ice ages that began in the Carboniferous period of Earth's history, a time when the Earth was blanketed with trees."

If there ever was one, perhaps that was the time when the Tree-hosts, the Ents of Middle-earth, carried Frodo's lost companions into battle against Mordor. Now if only we can muster the courage to take on the new lords of Mordor, such as President Jair Bolsonaro of Brazil and others, who seem keen on destroying the remaining forests across the Earth.

Evolution has left us a pattern to follow on how to remove carbon dioxide from the atmosphere while building up atmospheric oxygen, but instead of increasing the number of trees, our species is engaged in the very opposite.

A study published in the journal *Nature* says that deforestation is responsible for the removal of over 15 billion trees each year. The number of trees has dropped 46% globally since the advent of human civilization, according to an article in *The New Zealand Herald.* In fact, our species has destroyed a tenth of Earth's remaining wilderness in the last twenty-five years, and there may be none left within a century if trends continue, according to an authoritative new study in *Nature*, "Mapping tree density at a global scale" by T.W. Crowther, et al.

Impossible, you think? Not so impossible when one considers what the Rapa Nui people of Easter Island did to themselves. The way Jared Diamond tells it in his book *Collapse*, Easter Island is the "clearest example of a society that destroyed itself by overexploiting its own resources." Worldwide, researchers found a vast area the size of two Alaskas—3.3 million square kilometers—had been tarnished by human activities between 1993 and today, which experts said was a "shockingly bad" and "profoundly large number."

Michael J. Benton, in his book *When Life Nearly Died: The Greatest Mass Extinction of All Time*, draws attention to patterns observed in prior extinction events that might be playing out to some degree today. "The natural world is complex, and consequences are often unpredictable. As tropical forests are cleared and reefs are poisoned, we are losing not only species, but also whole habitats. Destroying species and habitats piecemeal can lead to a runaway crisis as seemed to have happened in the past. Low levels of extinction can turn into high levels. It could be that removing one or two species from an ecosystem does little damage. But if another few species are picked off, then another few, and then a few more, a point may be reached when that ecosystem will collapse. And, once the world becomes locked into a spiral of downward decline, it is impossible to see how any human intervention could turn it back."

For example, destruction of forests can kill ocean fish. "[What] happens in the atmosphere happens in the oceans as well," says Benton, "as changes in the gas composition or temperature of the air penetrate the upper levels of the ocean. Plants take up carbon dioxide during photosynthesis and pump out oxygen. Animals require oxygen but produce carbon dioxide as a waste product. There is a balance here and that balance can be perturbed by destroying too much of the world's forests. It is not simply enough to identify a new source of carbon dioxide; that source has to be capable of overwhelming the usual negative atmospheric feedback systems that have a tendency to cope with fluctuations and bring excesses of one input back to a standard level." Negative feedback means that a process is countered by opposite, or negative, processes, resulting in a state of equilibrium. Positive feedback, on the other hand, means that the process is enhanced by more of the same, with other processes operating in the same direction resulting in disequilibrium.

This makes one wonder whether Earth's atmospheric systems are in a state of equilibrium or disequilibrium.

The balance Benton refers to above seems to have kept atmospheric temperatures relatively stable until our species started burning fossil fuels during the mid-1800s, pumping increasingly huge volumes of CO_2 into the atmosphere. Judging from the ongoing increase in global temperatures, it is not clear that that negative feedback mechanism is keeping pace with the rapid increase in atmospheric CO_2. In fact, about half of these emissions are removed by the short-term carbon cycle each year (discussed below). The rest continue to build up in the atmosphere, thereby driving global temperatures even higher.

As CO_2 continues to build up year after year in increasing volumes, it would appear that the atmospheric system might have begun to switch to a positive feedback cycle and here is why.

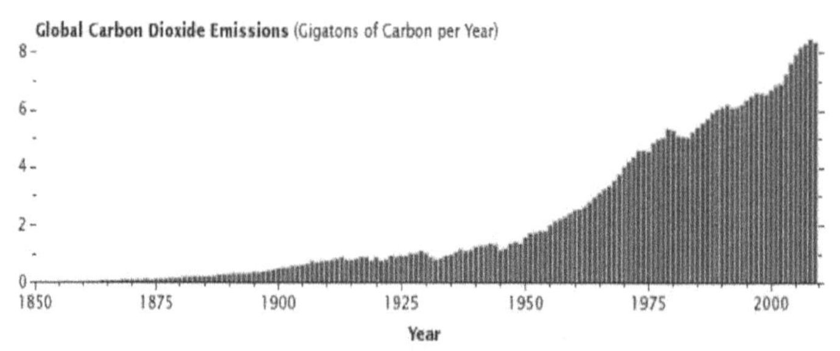

Fig. 7-11 Global carbon dioxide emissions (1850–2000)
(Graph by Robert Simmon, using data from the Carbon Dioxide Information Analysis Center and Global Carbon Project)

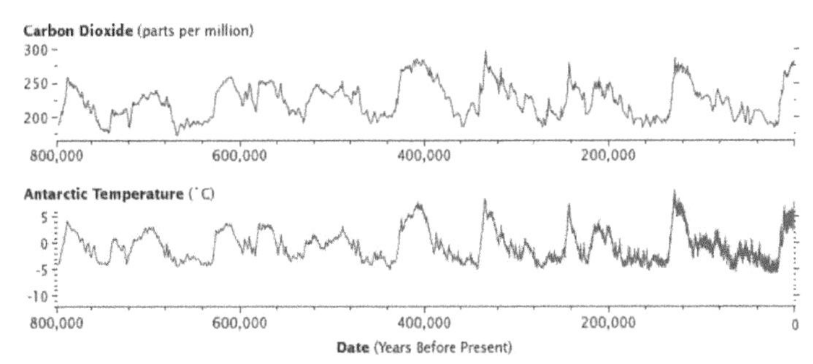

Fig. 7-12 800,000-year correlation between CO_2 and atmospheric temperature

Levels of carbon dioxide in the atmosphere have corresponded closely with temperatures over the past 800,000 years. (Courtesy of NASA based on data from Lüthi et al., 2008, and Jouzel et al., 2007)

The back-story on carbon and CO_2—the short- and long-term carbon cycles.

What we have talked about above refers mainly to the short-term carbon cycle. It is dominated by plant life and can increase atmospheric carbon dioxide by about 25%. Carbon dioxide is taken up during photosynthesis, and some of this carbon becomes locked up as living plant tissue. When leaves fall, or plants die, this carbon is transferred to the soil and can again be transformed into other carbon compounds, in the bodies of microbes, other plants, or animals. As this carbon passes through the food chain, it is then respired out of the animal as carbon dioxide, and the cycle starts all over again.

However, if the carbon remains locked up in plants or animal tissue and gets buried without being consumed, it becomes part of a large, organic carbon reservoir within the Earth's crust, and is no longer available to the short-term carbon cycle. It becomes part of the long-term carbon cycle and involves very different kinds of transformations, the most important of which is the transfer of carbon from the rock record into the ocean or atmosphere and back again. The time scale of this transfer is measured in millions of years.

The transfer of carbon to and from rocks can cause changes in the Earth's atmosphere much greater than changes caused by the short-term carbon cycle because there is a lot more carbon tied up in rocks than in all of Earth's biomass and atmosphere combined. In the past, this cycle has resulted in atmospheric carbon dioxide levels swinging both up and down more than 1,000%.

According to NASA, carbon dioxide levels reached 403.2 parts per million (ppm) in 2016., 405.12 ppm in November 2017 and 408.02 ppm in November 2018. By 2100, carbon dioxide levels in the atmosphere will have reached 935 parts per million, meaning the gas comprises nearly 0.1% of the atmosphere. If by the end of the century carbon dioxide in the atmosphere more than doubles, much of Africa, South America, and India will endure average maximum temperatures of more than 45°C. What NASA did not say was that at such temperatures most animals die.

To understand why, we must return to our earlier discussion of the negative-and-positive-feedback natural atmospheric system. We have seen how negative feedback has kept the Earth's climate system in a stable state until the mid 1800s. What began to happen after that provides some examples of how our species might have, inadvertently, already thrown that switch to a positive setting, thereby initiating an atmospheric positive feedback cycle. The following is a not-so-unlikely scenario.

Increasing global temperatures by increasing atmospheric CO_2 levels is causing glaciers to melt, exposing the permafrost and a potential release into the atmosphere of very large amounts of methane (that had been buried for millions of years). This would be a positive feedback, a reinforcement of the buildup of heat-trapping atmospheric greenhouse gases. This further increases global atmospheric and oceanic temperatures. When that increase in oceanic temperatures gets close to or exceeds 6°C, phytoplankton (responsible for 50%–80% of atmospheric oxygen) can no longer carry out photosynthesis. When phytoplankton begin to die in large enough numbers, the levels of the atmospheric oxygen we breathe begin to take a nosedive, and so does the entire food chain for all life-forms that derive a living from the sea. Similarly, as we raze the Earth's forests we are reinforcing the same positive feedback mechanism because with fewer trees, more CO_2 remains in the atmosphere while simultaneously eliminating trees' contribution to atmospheric oxygen levels. When the atmospheric oxygen drops to a low enough level, humans and animals begin to die in large numbers, and the sixth mass extinction, believed to be already in progress, begins to accelerate.

The question no one I am aware of has been able to answer definitively is: Have we tripped that temperature-raising CO_2-increase, O_2-decrease scale too far in the wrong direction, to a runaway greenhouse effect? While Earth's history shows that atmospheric balance slowly returns, it is over a period counted in millions of years, and for our species, it might as well be eternity, as we will be long gone.

With an exponentially rising world population (now about 7.8 billion, rising to an estimated 9 billion by 2050) combined with: tens of millions of fossil-fuel vehicles, power stations, farm animals pumping more and more CO_2 and CH_4 (methane) into the atmosphere; and with what can only be characterized as a mindless destruction of the world's forests and plankton due to warming oceans—that together produce all oxygen on the planet—one must ask on which side of this equation does nature come down? We already know where some members of our species stand—more CO_2, please. Where do you stand?

Nature appears set to cast her vote at the falling-oxygen ballot box, and based on what is happening in the atmosphere and oceans, it might not be too hard to tell the outcome. Whatever her vote, the result cannot be hacked by Russians and won't be up for debate by climate change deniers, either. Recent reports show atmospheric and ocean oxygen levels have begun to fall much faster than previously, even as global temperature continues to rise with the continuing increase of atmospheric CO_2 levels now exceeding 415 ppm.

And, just in case anyone still thinks this is not a level to be taken too seriously, it's worth noting that last time CO_2 levels were this high was during the Pliocene Epoch (17 million–2.5 million years ago). Back then temperatures were 3°C–4°C warmer, trees were growing at the South Pole, and sea levels were 15 to 20 meters higher, according to researchers in an April 2018 *ScienceDirect* article. At this point it is also important to note that our species is not even close to inventing a planet-size air conditioner that can reduce global temperatures, or an oxygen-generating technology that can replace all the oxygen that's being slowly sucked out of the atmosphere, nor has it begun to build floating cities.

For the diehard techies out there, let me remind you that this won't be easy—it took evolution nearly 3 billion years to produce an oxygen-dominated atmosphere at high-enough levels to jump-start and sustain complex animal and human evolution and survival.

In the absence of comparable human interventions, then, it looks like Homo sapiens is on its way to doing evolution one better. In less than 300,000 years of our existence we are pulling off a reverse cyanobacteria trick—replacing atmospheric oxygen with CO_2. Unlike the cyanobacteria, however (and they are beginning to look like the more intelligent species here), if we succeed, it will be we who go extinct. Can you imagine what the cyanobacteria historians and scientists would say if we were able to ask their advice? They would probably be screaming at us at the top of their tiny lungs: *Please, please, stop fooling with the* CO_2 *while you still can or else you will take us out, too*. It's really not nice to fool with Mother Nature.

What happens when one combines falling oxygen levels and increasing carbon dioxide levels with high temperatures? How much hotter can it get? It is probably fitting to conclude this chapter with a reminder of what's driving the destruction of Earth's habitats.

Global surface temperatures soared past previous records to make 2017 the second hottest year on record, according to two top U.S. science agencies. The year 2016 was the hottest, the third in a row to hold the dubious honor—after 2014 and 2015—completing a natural hat trick of sorts, the *Washington Post* reports. Both NASA and the National Oceanic and Atmospheric Administration (NOAA) recorded the highest average surface temperatures in 2016 since the agencies started tracking such data.

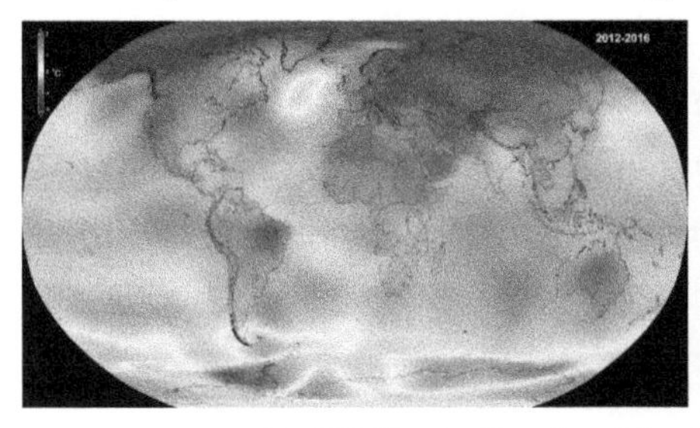

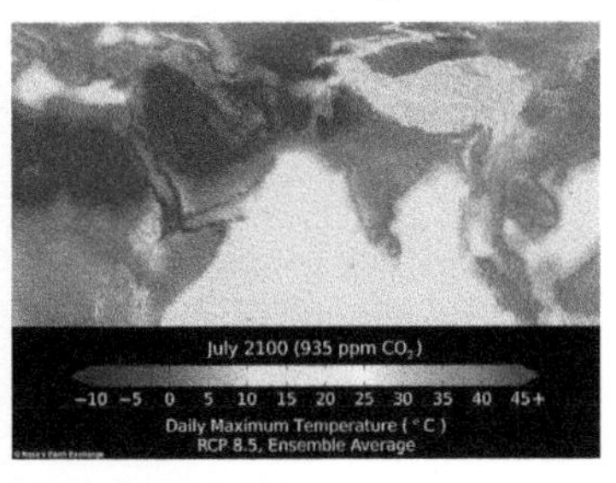

Fig. 7-13 Visualization of peak temperatures in North Africa, the Middle East and northern India

Maximum temperatures in North Africa, the Middle East, and northern India will exceed 45°C by 2100, as can be seen in the close-up of the region from the maps produced using NASA's new climate projection data set. (Source: NASA/Goddard Space Flight Center Scientific Visualization Studio - Jan. 18, 2017)

NASA notes that since the late 1800s, average global temperatures have risen by 1.1°C. And from January to August of last year, NOAA says that each month broke its own warmest record.

So, the Earth is getting hotter. Why should anyone care? Not everyone is convinced Earth's heating up is even driven by human activities. Regardless of the cause, what's not up for debate is that global climate change and warming is occurring, and rapidly. Further, increasing atmospheric CO_2 can only exacerbate increasing atmospheric temperatures.

But here are some other questions worth thinking about. How much CO_2 is too much, and how and when can this buildup be stopped? How low does the oxygen have to fall before it becomes a problem, and can anything be done to stop or reverse this trend? How high can the temperature increase before it's too high, and how can one stop or reverse this trend? Assuming there are ways to do all of this, how long might it take to begin to see a course correction?

While there are a lot of projections and models exploring all of the above, the disconcertingly obvious fact is that nobody really knows for sure exactly how this might all play out, or even whether or not we have already passed the point of no return. *This is all uncharted territory for our species.* However, what scientists do have a better handle on is what happened in Earth's early history, when similar or near-similar conditions prevailed—but I doubt this will make anyone feel any better. What these events and intervals of time do show is what happened to Earth as a habitat and to the species that depended on that habitat, under similar conditions, and how long they lasted. Scientists refer to those times as greenhouse mass extinction events, some of which lasted millions of years.

No discussion of Earth as a habitat can be complete without a consideration, however brief, of greenhouse mass extinction events, and how things worked out for the species that lived during those times. As you will see, the fact that these occurred naturally tens, and sometimes hundreds, of millions of years before our earliest ancestor looked out across the land offers no comfort when 50% to 95% of all life died. Hopefully, however, they can inform our decisions and accelerate potential course corrections.

We noted earlier that what might have been Earth's first mass extinction, the Great Oxygenation Event, resulted from a changeover from CO_2- to O_2-dominant atmospheric gases. According to Ward and Kirschvink, oxygen and carbon dioxide levels (particularly oxygen) are the most important of all factors dictating animal survival, death, and diversity. In addition, geochemists have long known that CO_2 levels and atmospheric O_2 levels show trends that are inversely related to one another. When oxygen levels rise, carbon dioxide is usually dropping. Keep this in mind as we examine these greenhouse mass extinction events.

These greenhouse extinctions shared a set of common characteristics and occurred over a vast expanse of time, from 400 million years ago to 100 million years ago, and included extinction during the Devonian geologic period, the Permo-Triassic (Permian) extinction, Triassic-Jurassic (T-J) extinction, and extinctions repeatedly during the Jurassic and Cretaceous periods, ending with the last known greenhouse extinction at the end of the Paleocene epoch, some 60 million years ago. High temperatures, high CO_2, and low oxygen levels appear to have coincided with these major mass extinctions. Research now shows that each occurred in a world of quickly rising CO_2 (and perhaps methane as well), and some of the lowest oxygen levels the world has ever seen. So as we review these events, we will pay close attention to the roles these three factors might have played in these greenhouse mass extinctions, as they seem to be starting to play a similar role today.

According to Ward and Kirschvink, these mass extinctions show strata depositions in low-oxygen conditions. It turns out that the minerals pyrite, uraninite, and the sulfur isotopes are very strong indicators of a lack of oxygen. Under such conditions the strata turn black because they contain pyrite and other sulfur compounds, reduced by chemical processes that can occur only in the absence of oxygen. However, while low-oxygen levels are characteristic of all of these greenhouse extinctions, researchers find the drop in oxygen levels during the Triassic to be especially stunning. According to Ward and Kirschvink, oxygen reached its lowest level of the past 500 million years in the late Triassic. Oxygen dropped to a minimal level of between 10% and 15% and stayed there for about 5 million years, from 245 million years ago to 240 million years ago.

This was a long, slow drop in oxygen levels, culminating in the Triassic mass extinction. Yes, from our perspective this extinction took a very long time, but what's more important here is the reason it happened—falling oxygen levels in the midst of high CO_2 in a high (above 35°C–45°C) temperature world. If it took 5 million years for oxygen to finally drive that extinction, given that atmospheric oxygen levels are falling today, and falling faster still, how long will it take to get down to a level where, accompanied by rising CO_2 and increasing temperatures, animal and human life begins to be seriously in peril? I am not sure if anyone ran this model, but wouldn't you like to know the result? I would. There is a CO_2 backstory as well.

While there is no direct way to measure the exact volume of CO_2 present in the past, plants are highly sensitive to CO_2 levels. Very high CO_2 levels during these extinctions is evidenced by the reduced number of stomata—the tiny openings on leaves used to absorb CO_2—found on fossil leaves. In the presence of high CO_2 levels, plants produce leaves with fewer stomata. Using this method, a paleobotanist could determine whether CO_2 was falling or rising at the time, and also estimate how many times higher or lower from a base-level observation.

The fossil leaves show that CO_2 was spectacularly high for all four extinctions, and the rise happened quickly, on the order of thousands, not millions, of years. Some estimates place the peak CO_2 levels between 2,000 ppm and 3,000 ppm, compared with April 2018's monthly average of 410 ppm and rising. According to NASA, CO_2 is expected to double by 2100 to 900+ ppm.

Is this scenario beginning to sound a little like our twenty-first-century CO_2 run-up? As noted earlier, CO_2 levels are far higher now than they have been for any time during the past 800,000 years. That means that in the entire history of human civilization, CO_2 levels have never been this high.

We now turn our attention to the third component in this deadly trio. Given the role rising global temperatures have played in greenhouse extinctions, it is not unwise to be concerned about how hot it can get given current conditions. If those events are any guide, very hot indeed. One Chinese-American research team found the reason it took so long for life to recover in the seas after the Permian extinction can be at least partly attributable to temperatures of 104°F (40°C) in the sea and a blistering 140°F (60°C) on land.

Today Earth is already smoldering. Heavily populated parts of the Earth have already reached these temperatures. An August 2, 2018, article in economist.com, "The World Is Losing the War Against Climate Change," said: "From Seattle to Siberia this summer, flames have consumed swathes of the northern hemisphere. One of 18 wildfires sweeping through

California, among the worst in the state's history, is generating such heat that it created its own weather. Fires that raged through a coastal area near Athens last week killed 91 [see article]. Elsewhere people are suffocating in the heat. Roughly 125 have died in Japan in 2018, as the result of a heat wave that pushed temperatures in Tokyo above 40°C for the first time."

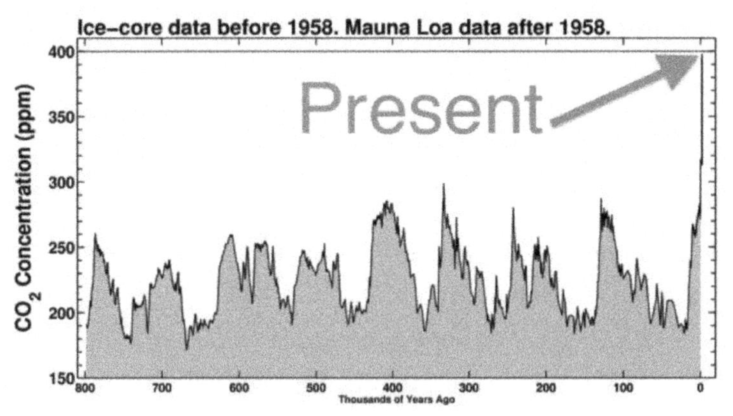

Fig. 7-14 Ice-core data before 1958; Mauna Loa data after 1958
(Source: http://scrippsco2.ucsd.edu/data/atmospheric_co2/icecore_merged_products)

Such calamities, once considered freakish, are now commonplace. Scientists have long cautioned that, as the planet warms—it is roughly 1°C hotter today than before the industrial age's first furnaces were lit—weather patterns will go berserk. An early analysis has found that that sweltering European summer would have been less than half as likely were it not for human-induced global warming.

It is tempting to think these are temporary setbacks and that mankind, with its instinct for self-preservation, will muddle through to a victory over global warming. In fact, it is losing the war.

If you were wondering how long temperatures are likely to remain this high, geochronology now shows that high temperatures lasted for at least the first 3 million years of the Triassic, and may have been climbing during that period, with a maximum temperature occurring during a time interval known as the Smithian stage (a million-year time interval around

247 million years ago), which had the highest known temperatures since animals first evolved.

None of this happened overnight. These conditions were reached after thousands of years in some cases, but most reached those levels over tens of millions of years. Nor did the impact all come toward the end for animals living at the time, as a look at the impact these "three horsemen" brought makes plain.

Above we asked what happens when one combines falling oxygen levels and increasing carbon dioxide levels with increasing high temperatures? The answer is extinction.

To understand the havoc wreaked on animals by rising CO_2, rising temperatures, and falling oxygen, one must look at metabolic rates, the pace at which oxygen is consumed. It turns out that pace is directly influenced by temperature; as temperatures rise, metabolic rates go up, and as metabolic rates rise, so does the need for oxygen, as the chemical reactions of life are oxygen-dependent. Metabolic rates can double or triple with each 10-degree rise in temperature. In a low-oxygen world, the consequences of this would have a major impact on life.

All evidence of those times points to a stark and environmentally challenging world for animals. Many reptiles returned to a life in the sea due to the hot, low-oxygen world of the Triassic. The highest temperature any animal can withstand is not even halfway to the temperature that boils water. At 40°C, most animals die off, and the rest die at about 45°C. Most marine organisms die above a 100°F (37.8°C) level in the sea. Photosynthesis stops at a temperature slightly higher. According to Ward and Kirschvink, the entire zone of the tropics would have been devoid of animals, and complex life would have hung on only at high latitudes. Land animals would have been rare even in mid-latitudes.

To sum up this brief look at greenhouse mass extinctions, we now know that if our species continues dumping CO_2 into the atmosphere and driving global temperatures up while oxygen levels are falling, Homo sapiens probably won't have a happy ending. Even if we are mature enough as a species to pull back while there may still be time to avoid triggering yet another greenhouse mass extinction, the destruction already in progress will have damaged Earth's habitats so deeply that it can only result in suffering, death, and destruction in orders of magnitude beyond any prior disaster we as a species have experienced. And we probably won't be able to escape to Mars by then either.

You might look at the above and say, *Oh well, this kind of global change has taken thousands and even millions of years in the past, so I need not worry. At least it won't happen in my or even my children's lifetime.* Such thinking fails to take into account a very significant wild card amongst the variables. The difference between what is happening today and the rate at which it is happening compared with what happened back then is us; we are here—all 7.8 billion of us—going up to 9 billion by 2050.

Back then, no one was driving CO_2 buildup at a rate faster than at any time since the dinosaurs went extinct. There weren't 7.8 billion humans drawing down the planet's atmospheric oxygen budget while pumping CO_2 into the atmosphere every time we take a breath, take to the streets in our fossil-fueled vehicles, or turn on the lights, AC units, or heat. And there was no one cutting down the forests (the lungs of the planet) that remove CO_2 from the atmosphere while contributing 20%–30% of the Earth's oxygen. And, alas, no one was pouring waste and chemicals into the oceans, driving up water temperatures and increasing acidity, thereby destroying reefs and other marine ecosystems, habitats, and food chains, and threatening the remaining 50%–80% of Earth's oxygen produced by plankton.

So, this time around it may not take nearly as long and, indeed, some of its harmful effects are already being felt across the globe. If you plan on being alive during the next eighty or so years, or if you identify as a Millennial, or any subsequent generation, depending on your location, there is a good chance you will begin to experience the consequences. So, what are you going to do about it?

At the beginning of this chapter we noted that habitat survival is indispensable to species survival and species survival is indispensable to habitat survival. We noted then, too, that if a species is no longer able to maintain a sustainable habitat it must be able to migrate to survive or it will go extinct. Given what we have done and are still doing to Habitat Earth, how would you rate the chances of Homo sapiens' survival?

Let's imagine Ambassador Spock, a member of the Vulcan High Command, orbiting in a cloaked Vulcan Galactic Cruiser in high Earth orbit. He's been tasked with making first contact when our species shows signs of having attained the right level of maturity (and that probably won't be measured by a warp engine signature—sorry, Trekkies). How do you imagine he would rate our progress?

I imagine Ambassador Spock might, with that characteristic twist of his head, say something like:

> *"Curious! This species appears to have a death wish. Why, they're depleting their planet's atmospheric oxygen by filling it with CO_2, the very opposite of what their planet's evolution did to enable life. And they are depleting the oxygen that trees generate by rapidly destroying their planet's remaining forests—sure wish we had such lush forests on Vulcan—and now they have begun threatening the oxygen produced by phytoplankton by increasing the temperature of the planet's oceans."*
>
> *"Strange! I seriously doubt anyone, not even the Klingons, could have come up with a more ingenious plan for exterminating this species than they themselves already have, except perhaps, that curious institution of theirs, Hollywood. Strange, indeed! They are definitely not ready to meet the rest of the galaxy."*
>
> *"I will check back in a few hundred years to see whether they have managed to avoid self-extinction."*

Every so often one learns of a prominent individual at the peak of their career, like the late Anthony Bourdain, who inexplicably chooses, or is driven, to end it all in an act of suicide. How unspeakably sad. What sense of utter despair and desperation could have brought this about? We struggle to grasp what drives someone to choose such an end, and yet an even greater enigma exists when we realize that our entire species seems to be bent on suicide.

Doctors can identify suicidal behavior and place an individual on suicide watch. Doctors can also recognize an immune system run amok and respond with the appropriate medication for an autoimmune disease. But whose job is it and how does one respond when an entire species begins to exhibit suicidal behavior, seems intent on attacking their own planet's life-support systems?

It took evolution nearly 4 billion years to transform the Earth from the nightmarish cataclysm it was at its beginning into a near perfect habitat capable of evolving and sustaining complex animal and intelligent species life, but alas, it has taken Homo sapiens a mere 200 to begin to turn it back into a seething cauldron—again.

8.

Sustainability

HABITAT SURVIVAL DRIVES SPECIES SURVIVAL DRIVES HABITAT SURVIVAL

IN THE STORY OF THE SCORPION AND THE FROG, A scorpion and a frog meet on the bank of a stream, and the scorpion asks the frog to carry him across on its back. The frog asks, "How do I know you won't sting me?" The scorpion says, "Because if I do, I will die too." The frog is satisfied, and they set out, but in midstream, the scorpion stings the frog. The frog feels the onset of paralysis and starts to sink. Knowing they both will drown, he has just enough time to gasp, "Why?" Replies the scorpion: "It's my nature."

No doubt most have heard this story. It is widely applicable but never more consequential than when applied to how our species is destroying the very habitats upon which its survival depends.

The overarching principle guiding evolutionary sustainability appears to be species survival. Survival and habitat sustainability are inseparable.

As we have seen, whenever evolution has given rise to life of any form or complexity, it has invariably first established a sustainable habitat with everything needed for survival; then, out of that habitat, life emerges. A sustainable habitat is not only a prerequisite for the evolution of life but also indispensable for the long-term survival of a species. And while it might appear from a human perspective that the Earth will last forever, Earth's sustainable-capable habitats often do not, and it would appear evolution has arranged things so nonintelligent species have a predisposition to maintain their habitats but left it up to intelligent species to figure out on their own that their survival will depend upon their doing so too. Failure to maintain a habitat, whether by neglect and/or deliberate destruction, degrades that habitat's ability to sustain life, and unless corrected, can result in complete habitat collapse and the cessation of its ability to sustain life.

Fig. 8-1 The Survival Sustainability Cycle

As The Survival Sustainability Cycle above shows, habitat, species and Homo sapiens' survival form an inter-related cycle. Species depend upon habitats, and vice versa, for survival, and humans depend upon both.

The evolutionary hand an emerging intelligent species appears to have been dealt is the responsibility to care for and preserve its habitat. In exchange, the habitat might last as long as it takes for that species to attain evolutionary maturity. If habitat collapse occurs nonetheless, an intelligent species must be free, ready, and able to migrate to other habitats to survive as our forebears did out of Africa, or go extinct.

Given that the Earth might last another 4 billion to 5 billion years, evolution appears to have built in more than enough time for our species to mature before Earth's warranty begins to expire. The only issue facing Homo sapiens then is: Can we take care of Earth's habitats so that they remain sustainable until we attain evolutionary maturity and become capable of interplanetary migration? If we don't, or won't, it's highly likely we will go extinct.

But it is not all doom and gloom. As we have seen, evolution is rooting for intelligent species' survival. So with evolution on our side, the real question then becomes how, to the extent it's up to us (and it may not always be up to us, as habitats can become unsustainable with or without our help as they did for our forebears), can our species ensure that the Earth remains a sustainable habitat long enough for us to attain evolutionary maturity and migrate, and thereby avoid extinction.

Homo sapiens is at that critical juncture when our species must become mature enough to avoid self-extinction and become capable of interplanetary, inter-solar and probably even intergalactic migration—possibly a not-so-well-hidden goal of evolution and the potential outcome for Homo sapiens if we begin to prioritize reaching that survival plateau above everything else.

But what will it take to ensure the Earth remains a sustainable habitat? What does sustainable mean in the context of a planet that has seen repeated extinctions, repeating cycles of global climate change driven in part by the Earth's dance around the sun? Is there anything a species can do in the face of these awesome forces to maintain its home world's habitats and survive? Well, believe it or not, evolution is counting on species doing just that and has even hinged survival to a species' ability to maintain its habitat even as that habitat undergoes global climate-change events.

This is somewhat difficult to accept until one realizes that an emergent intelligent species, regardless of climate-change events, has no choice but to survive on its home world until evolutionary maturity is attained and it becomes capable of interplanetary migration. Getting to maturity can take several million years. It was more than 5 million years ago (and several global climate-change events) that the first hominins began to emerge, and it could take a few million more before maturity is attained.

Nonetheless, built into a species' evolutionary maturity development path is climate change—a clear means for prodding a species to mature, migrate, survive, or go extinct.

But what exactly is "sustainability"? Sustainability in this context is evolutionary sustainability, the only form of sustainability that will ultimately matter to our survival. It is probably unlike anything that comes up when discussing sustainability today.

> *Evolutionary sustainability is the care and feeding of its habitat by an intelligent species to ensure mutual survival, even during climate-change cycles, and for as long as it takes to attain evolutionary maturity and migrate to survive.*

As noted earlier, it would seem that evolution went to a whole lot of trouble to make the Earth just right for complex animal and intelligent species (and we would like to think just for us Homo sapiens). The early Earth was anything but a sustainable-capable habitat. Any life evolving during the first 700 million years would have been repeatedly annihilated due to heavy bombardment by meteor showers. It took evolution nearly 4 billion years to establish sustainable-capable habitats for complex animal life that only took off about 550 million years ago in the Cambrian explosion. It was going to take yet another 500 million years or so before the Earth was finally ready for hominins and, eventually, Homo sapiens.

Yet even then, Earth was no utopia. Countless millions of species have evolved and gone extinct. Countless habitats have been obliterated, some by way of plate tectonics, super-volcanoes, earthquakes, ice ages, and glacial periods. But through it all, life found a way to survive. Dinosaurs were taken out as recently as 65 million years ago. Early hominins went extinct at various times during the last 5 million to 6 million years, amidst geological upheavals, climate change, environmental chaos, and habitat destruction. Our forebears fled Africa during yet another ice age, which engulfed them through no fault of their own.

So while the Earth has afforded many sustainable-capable habitats over extended periods measurable in tens of millions of years, there have also been periods when those habitats were from time to time destroyed independent of anything extant species might have done.

To survive, an emergent intelligent species must come to grips with the inevitable need to maintain a sustainable habitat, even on a planet like Earth that comes with all the habitability preconditions met, even if it is predisposed to sustain complex animal life, even if it has been doing so for almost a billion years, and even if it has repeatedly demonstrated its ability to jump-start life all over again after numerous near-total mass extinctions.

Today we are becoming increasingly aware that climate change has happened in the past, is currently happening, and will continue to happen. And we have noted that there have been times when certain species have contributed to climate change, i.e., cyanobacteria initiated early Earth's CO_2/O_2 atmospheric gas exchange and, as scientists believe, Homo sapiens are accelerating CO_2-driven warming of the Earth's atmosphere today. However, for the most part, long-term (geologically speaking) climate change appears to be driven primarily by the orbital elements of the Earth, the sun, and where the solar system might be in its journey through the galaxy.

Climate change, then, is an integral condition of Earth's habitats.

Scientists are well on the way toward understanding the connection between the Earth's orbit and climate change; this orbit-climate-change connection subjects Earth's climate to the fluctuating gravitational field of the solar system. The gravitational impact of the sun, moon, and planets appears to orchestrate large-scale climate fluctuations, and these form the backdrop for the evolution of species, including our own. This orbit-climate connection was first determined by a Serbian mathematician named Milutin Milankovitch, who theorized that the ice ages occurred when orbital variations caused the Northern Hemisphere around the latitude of the Hudson Bay and northern Europe to receive less sunshine in the summer. Milankovitch predicted that the ice ages would peak every 100,000 and 41,000 years, with additional "blips" every 19,000 to 23,000 years. The paleoclimate record shows peaks at exactly those intervals.

Sunspot activity has been associated with mini ice ages such as occurred across Europe between 1300 and 1850. Major ice ages (glacials) and warm periods in between (interglacials) are associated with a change in Earth's orbit around the sun (its eccentricity), which occurs once every 100,000 years; the angle of its tilt (its obliquity, which varies between 22 degrees and 24 degrees every 41,000 years); and its wobble on its axis as it spins toward and away from the sun over the span of 19,000 to 23,000 years (known as its precession). It is these variations in Earth-sun geometry that changes how much sunlight each hemisphere receives during the Earth's year-long trek around the sun, where in the orbit (the time of year) the seasons occur, and how extreme the seasonal changes are.

Climate change events are thus an integral part of the Earth's habitats, and we will just have to learn how to live with them, as early humans did, if we are going to survive. And while we today have no species memory of living through an ice age, the archaeological and genetic data, and our existence today, suggest that our forebears survived, if only just barely, the last glacial period and, hopefully, we will be able to survive climate change again.

> *The takeaway here is that cyclic, global climate change, rather than being an aberration, is an integral part of evolutionary sustainability. The good news is that Earth offers evolutionary sustainability despite global cyclic climate change.*

Orbit-driven climate change seems to be a cyclic solar phenomenon as well as a planetary phenomenon. Scientists believe the entire solar system might be heating up. New evidence suggests the solar system is moving into a new energy zone as it travels through the galaxy, which is altering the magnetic fields of the planets.

So, migrating to another planet won't necessarily mean escaping the effects of climate change, since every planet no doubt comes with its own set of orbit-driven climate-change events. Mars, like Earth, experiences ice ages due to wobbles in its orbit. An ice age that is believed to have begun 370,000 years ago is now coming to an end. NASA scientists say global dust storms might also be contributing to global warming on the planet. These have been known to blank out the sun, cover continent-sized areas of the planet's surface, and last for weeks at a time.

But unlike Mars, which has long since lost its atmosphere and oceans, and many other planets and moons thought to be potential Earth 2.0s but, as we have seen so far, have no ability to sustain complex animal life, Earth offers a sustainable-capable environment that has rebounded and restored its habitats after repeated glaciations and interglacial warm periods of climate change.

The challenge facing our species, then, will be learning how to adapt and survive not only during an interglacial warm period as we are doing today, but also in the long term during an ice age as our forebears did during the Last Glacial Maximum, when vast ice sheets 3 to 4 kilometers high (1.9 to 2.5 miles) covered much of North America, northern Europe, and Asia and profoundly affected Earth's climate, causing drought, desertification, and a global sea level drop of about 120 meters (390 feet).

More likely in the near to intermediate term, though, before the iceman comes again, we will have to adapt and survive in a world of increasing global temperatures; scorching temperatures that models show can reach 50°C-plus in already hot and dry regions and 40°C-plus in previously temperate zones; ocean temperatures too hot for the production of oxygen by phytoplankton, threatening complex animal and intelligent species extinction; and decreasing access to clean water. And we will have to adapt and survive in a world seized by potential global conflict due to scarcity of food, water, and climate change refugees.

What this all hints at is a repeat of a now-familiar set of evolutionary survival patterns. As early humans driven by climate change fled Africa during an earlier glacial, we will more than likely have to confront and survive climate change as well, or migrate to other planets to survive, or go extinct.

In the near to intermediate term, climate change appears to be taking the form of global temperatures approaching near Permian-Triassic extinction levels, rapid increase of CO_2 and other greenhouse gases, and declining O_2 levels. In the much, much longer term (perhaps so long as to be irrelevant), we can anticipate one or more of these cycles of ice ages, glacials, and interglacials.

If any of the above scenarios were to play out in sufficient intensity, our species would be in for quite a few extreme climate events, some of which not even the earliest hominins would have experienced, let alone anyone living today.

Some of these climate events may occur before we're capable of migrating to an Earth 2.0. Until then, our species had better come up with innovations and adaptations to enable survival during these extreme climate events and embark upon a World War II-like effort to reverse, if possible, the destruction done hitherto to the Earth's flora, fauna, ecosystems and habitats.

If short-term climate change driven by El Niños and La Niñas lasting no more than decades drove many early human civilizations to collapse, consider what pressures long-term global climate change such as the Last Glacial Maximum will exert on modern civilization and our species. In a warming world, the map of the world will be redrawn not by armies but by climate change, with major coastal population centers submerged by the oceans; in a colder world, long-submerged cities and the remains of long-lost civilizations will be laid bare as sea levels fall.

While most hominins and early humans have experienced and survived some of these events, no one living, or for that matter anyone who has lived within the last 12,000 years, would have any memory of such climate events.

Homo sapiens is a species with a short-term perspective in a long-term world. We evolved with our horizons set in a 24-hour day and a 365-day year, with an average individual life-span that has yet to exceed 100 years and with little generational memory that, when it exists, rarely goes back beyond great-great-grandparents. What we think of as recorded history is barely 5,000 years old, beginning with the Sumerians of Mesopotamia and the Early Dynastic Period of Egypt.

We have evolved in a vast habitat with no inborn knowledge or experience to tell us how to survive; a habitat whose climate seasons of hot and cold extremes span geologic intervals of time. We have no survival lessons learned and passed down by species memory across ice ages, such as when our forebears migrated out of Africa, or warm periods, such as the one we live in today. Yet it is in these geologic time scales and their cycles of global climate change that the drama of species evolution, adaptation, innovation, maturity, migration, survival and/or extinction is enacted on this vast stage, the habitats of Planet Earth.

But for those looking for clues, there is a message left in an "ice-age bottle," and it's not all bad news.

It was falling ocean levels due to a glaciation event at the time of early human migration that facilitated the migration across the Red Sea, and not only out of Africa but across land bridges, such as the Bering land bridge, or Beringia, which emerged between Siberia and Alaska during the last ice age, now long submerged by the interglacial that predictably (we now know) followed. And with geologic hindsight, as it were, it's hard to miss the message in the ice-age bottle that that long-term global climate event left us; it's hard not to see how these colossal climate events have choreographed the last great species survival migration that Homo sapiens embarked upon. It could only be to our ruin if we fail to heed the advance notice left to us as to what the onset of the next equally predictable (we now know) major climate change event implies for us.

Homo sapiens will probably survive if we take notice and straighten up before it's too late. Early humans did. But my guess is that if we do, we could come out the other end a very different, chastened, and hopefully more mature species. And that may well be the evolutionary point of this impending cycle of climate change. As noted earlier, climate change is evolution's tool to drive species on to greater maturity, to migrate and survive, or to extinction.

For those currently charged with planning and designing for sustainability, the first takeaway is that sustainability planning, in the context of global climate change, cannot be limited to national or regional boundaries but has to take on a global perspective. If one wants to get some sense of how climate change on that scale might rearrange national boundaries and the planet, short of getting into a time machine and heading back to the last ice age, the movie *The Day After Tomorrow* might not be a bad preview. Such a scenario would be quite ironic, because it won't be Mexicans heading north but Americans streaming south, hat in hand, climbing over their own border wall to escape the ice. Evolution, after all, does have a sense of humor.

But how exactly does evolutionary sustainability differ from what is commonly understood as sustainability? Evolutionary sustainability is global, but that's only where it begins.

Given how short our species memory and perspective is, it seems preposterous that a species that has lived during only a tiny fraction of an interglacial warming period, as Homo sapiens currently is doing, could imagine that maintaining a sustainable habitat includes doing so even during that habitat's climate change cycles and for as long as it might take to attain evolutionary maturity, migrate, survive, or go extinct. One only has to look at what we view as sustainable land use, sustainable cities, and sustainable living to realize how out of touch we are with the concept and requirements of evolutionary sustainability.

To begin to get a grip on the sustainability that evolution might have had in mind when it launched life on Spaceship Earth for an 8- to 10-billion-year journey through galactic space, consider what NASA and others had to say about sending members of our species to Proxima b, a planet 4.2 light-years away (a light-year is about 5.88 million, million miles, so about 25,000,000,000,000 miles away), orbiting Proxima Centauri, the closest star to our solar system and part of a triple star system called Alpha Centauri. Proxima b is about 5 million miles from its star, compared with Earth's 93 million miles from the sun, and is believed to orbit within the habitable zone, where liquid water might appear on the planet's surface. (However, NASA has concluded that it's unlikely it is capable of sustaining human life.)

Given current technology, such a journey would take from 80,000 to 100,000 years and last more than 2,250 generations—almost half again as long as the interval since early humans migrated from Africa 70,000 or so years ago. None of those who might get to Proxima b would have any more generational or species memory of their forebears who left Earth than we do of those who left Africa. And that's only for a journey of 4.2 light-years and less than 3,000 generations.

Now, in order to begin understanding what evolutionary sustainability means, consider what evolution's design requirements for Spaceship Earth might have been for a journey that will last 8 billion to 10 billion years.

Everything on Spaceship Earth would have to be 100% sustainable and recyclable, and its energy source would have to be powerful enough to last the entire journey, yet far enough away to avoid destroying its fragile occupants. It would also need variable, perpetual, and renewable sources of food and water and anything else needed to survive the voyage; an inner atmospheric system, temperature, pressure, and just enough gravity to keep it all from skipping off into space; an outer tier of larger ship-like planets surrounding the inner core and protecting life from incoming asteroids and debris; and renewable habitats teeming with just the right set of species to support life all the way and for as long as that journey might take. In other words, Earth would have had to be designed as a living spaceship. Evolution needed to design a living planet capable of caring for and feeding itself and all species on board for billions of years. Now let's see how NASA, SpaceX, or Beijing might approach such an assignment.

How does one begin to design a vessel that can sustain itself and its passengers over such vast geologic intervals? Evolution's focus on species survival is obviously a hint. The key to Earth's ability to survive is undoubtedly its biodiversity—its species. Earth's species are not just passengers but are in fact an integral part of the planet's survival mechanisms that together form a symbiotic, interdependent, and recurrent set of processes that drive Earth's ability to remain a living planet. Species survival ensures habitat survival, and habitat survival ensures species survival.

And indeed, some Earth and life scientists seem to agree. According to James Lovelock in his book *The Vanishing Face of Gaia*, in a statement called the Amsterdam Declaration made at a meeting of the European Geophysical Union in 2001, more than 1,000 scientists signed a statement that began with the words: "The Earth System behaves as a single, self-regulating system comprised of physical, chemical, biological and human components." As homeostasis in the human body enables maintenance of an internally balanced environment to support life, so too Earth's self-regulating systems enable the maintenance of habitability and species survival. "This is indeed a view of Earth as self-regulating with the community of living organisms, including Homo sapiens, in control and is totally inconsistent with prior views that Homo sapiens are somehow separate and in charge and empowered by its god to do with Earth's flora and fauna as it so pleases."

One need only compare Earth's topsoil with the soil samples astronauts found on the moon and those analyzed by Mars rovers to understand what separates a living planet like Earth from a dead planet like Mars. Two evolutionary mechanisms—a broad spectrum of biodiverse species performing the biological and chemical processes they are evolved to do, and a sustainable-capable habitat designed to interact with those processes—need to exist and function synergistically in a symbiotic, reciprocating manner, in sufficient numbers and variety for the planet to remain alive.

A planet without these evolutionary biodiversity mechanisms is like a spaceship without a life-support system. Without these, Earth would be orbiting the sun lifelessly, like Mars, Venus, and Mercury. Earth is alive and Homo sapiens exist today only because of the early emergence of photosynthetic life in the form of cyanobacteria and later plants that together enabled a buildup of free oxygen in the atmosphere. This oxygen buildup outpaced oxidation by rocks in the Earth's crust (through a process known as chemical weathering). This increasing net free oxygen reacted with hydrogen molecules that were skipping off into space due to the sun's UV rays splitting water molecules in the atmosphere and thereby stopped the resulting net loss of water.

These photosynthesizing species drove an oxygen buildup that not only enabled Earth to hold on to its water and consequently its oceans but also enabled eukaryotic cell and eventually complex animal and intelligent species evolution. In the absence of photosynthesizing life, Mars and Venus lost whatever atmospheres and oceans they might have had, and they are now quite dead. The day Earth begins to lose its species and habitats in sufficient numbers (as Mars and Venus might once have done), it, too, will die.

The Earth's abundant and innumerable species comprise a complex matrix of interwoven, redundant, and interdependent ecosystems and food webs that maintain Earth's habitats and form numerous intricate, overlapping, self-replicating food chains capable of sustaining the planet for billions of years. Taken altogether, it is a self-replicating, survival-driven design pattern even NASA might find useful.

Who would have guessed that the interactions between the very physical, biological, and chemical life processes of a ship's passengers, food sources, and waste, working symbiotically with the ship's systems, would be the best way to ensure the survival of both on protracted interplanetary or interstellar voyages? The living Earth presents Homo sapiens with orders of magnitude levels of complexity the comprehension of which will take millennia. Fiddling with it by destroying species, habitats, changing its atmosphere, and heating its oceans can and have already begun to show signs of dire consequences.

One would find it very difficult to imagine what could possibly motivate the crew to deliberately destroy their life-support systems (as aboard the spacecraft rerouted to salvage Matt Damon, who had been inadvertently abandoned on the surface of Mars in the movie *The Martian*). Nonetheless, we have been doing just that since our species boarded Spaceship Earth some 300,000 years ago. Like that scorpion we encountered at the beginning of this chapter, Homo sapiens—even if it results in its own extinction—has embarked on a mission of relentless and ever-intensifying destruction of the planet's flora and fauna.

Still regarding Earth's biodiversity primarily as just a source of food and materials to be exploited for economic gain, Homo sapiens is just beginning to realize it is much, much more, that it is in fact a carefully engineered and complex web of interlocking life forms, habitats, and ecosystems that together make up the life-support system of the planet, a critical component in Earth's complex set of survival mechanisms. What's more, we have no idea how and when Earth will respond to the continuing destruction.

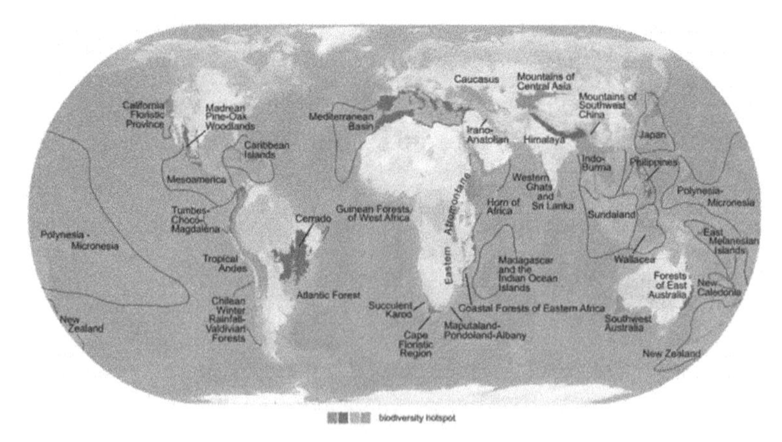

Fig. 8-2 Biodiversity hotspots

(Source: *https://www.sciencedaily.com/terms/biodiversity_hotspot.htm*)

Hopefully, this belated but increasing awareness hasn't come too late and has led to increased attention to regions of the planet being designated biodiversity hotspots, locations with significant levels of biodiversity that are threatened with destruction. There are thirty-five areas around the world that qualify as hotspots, and while they represent just 2.3% of the Earth's surface, they support more than half of the world's plant species that are found no place else. Nearly 43% of bird, mammal, reptile, and amphibian species also found no place else on the planet to live.

The list of hotspots currently includes the following and are shown on the map above: North and Central America: California Floristic Province, Madrean pine-oak woodlands, Mesoamerica; The Caribbean: Caribbean Islands; South America: Atlantic Forest, Cerrado, Chilean Winter Rainfall-Valdivian Forests, Tumbes-Chocó-Magdalena, Tropical Andes; Europe: Mediterranean Basin; Africa: Cape Floristic Region, Coastal Forests of Eastern Africa, Eastern Afromontane, Guinean Forests of West Africa, Horn of Africa, Madagascar and the Indian Ocean Islands, Maputaland-Pondoland-Albany, Succulent Karoo; Central Asia: Mountains of Central Asia; South Asia: Eastern Himalaya, Nepal, Indo-Burma, India and Myanmar, Western Ghats, India, Sri Lanka; South East Asia and Asia-Pacific: East Melanesian Islands, New Caledonia, New Zealand, Philippines, Polynesia-Micronesia, Southwest Australia, Sundaland, Wallacea; East Asia: Japan, Mountains of Southwest China; and West Asia: Caucasus Irano-Anatolian.

However, while a good start, attention to these designated hotspots is not a panacea and omits many areas of equal concern. In particular, these hotspots do not adequately represent other forms of species richness (e.g., total species richness or threatened species richness—count—of species); do not adequately represent taxa (a group or rank in a biological classification into which related organisms are classified) other than vascular plants (e.g., vertebrates or fungi); do not protect smaller-scale richness hotspots; do not make allowances for changing land-use patterns; do not protect ecosystem services (provisioning, such as the production of food and water; regulating, such as the control of climate and disease; supporting, such as nutrient cycles and crop pollination; and cultural, such as spiritual and recreational benefits); and do not consider phylogenetic diversity—a way to measure biodiversity at the level of features.

Designation of these hotspots notwithstanding, widespread ecocide continues, and with the ongoing destruction of Earth's biodiversity, ecosystems, and habitats, Homo sapiens are essentially destroying Earth's ability to remain a living planet and thereby endangering Homo sapiens' survival.

As alarming, scary, and terrifying as the threats are that climate change pose to Homo sapiens survival, the threat posed by continued ecocide—the rapid acceleration in the destruction and decline of the natural world—is much, much greater even if the effects are not nearly as well known. The clearing of forests, over-exploitation of seas, soils, and the pollution of air and water have driven the living world to the brink, says a huge three-year U.N.-backed landmark study published in May 2019.

According to this study the natural world appears to be in a state of continuous, rapid and uncontrollable decline, and the planet's support systems are so stretched that widespread species extinctions and mass human casualties and migration are inevitable unless urgent action is taken. That's the warning hundreds of scientists are preparing to give, and it's stark. According to a report by IPBES (Inter-Governmental Science-Policy Platform on Biodiversity and Ecosystem Services):

> *"The loss of species, ecosystems and genetic diversity is already a global and generational threat to human well-being. Protecting the invaluable contributions of nature to people will be the defining challenge of decades to come."*

Sir Robert Watson, overall chair of this study, is reported to have said: "We are at a crossroads. The historic and current degradation and destruction of nature undermine human well-being for current and countless future generations. . . . Land degradation, biodiversity loss and climate change are three different faces of the same central challenge: the increasingly dangerous impact of our choices on the health of our natural environment."

We are just starting to unravel how evolution has engineered human cells, organs, and body, and we are just beginning to comprehend the effects of tinkering with the planet's food chains through rampant destruction of flora and fauna, and we have no idea how close we might be to destroying one ecosystem or species too many. We are now only grudgingly beginning to acknowledge the potential consequences of disturbing the delicate balance of the planet's atmospheric and oceanic temperatures, and the impact this might potentially have on our survival.

Evolution, then, has laid out the gold standard for the fundamental characteristics of sustainability. These fundamentals continue during all or most global long- and short-term climate change events at different times and in different places across the Earth. The fact that early humans evolved, adapted, innovated, and were able to migrate to survive when habitats failed during some of the worst times since hominins emerged, and we are still around to talk about it, is testament to Earth's ability to serve as a sustainable-capable habitat for complex animal and human life on the one hand, and early Homo sapiens' willingness to partner with evolution to do whatever it took to survive on the other.

The only question, then, is: Can modern Homo sapiens be persuaded to care about species survival before it's too late? Does it want to survive as a species, and is it willing to go to the lengths it will take, and that evolution has shown it had been willing to go in the past to ensure early human survival?

> *It would seem that as a species, we have yet to make this fateful connection between our own survival and the preservation of Earth's species, ecosystems and habitats.*

I loved rotating the globe once mounted on a metal pedestal next to my grandfather's chair. I loved looking at the continents and oceans of the Earth and different geographical features. I would turn on and turn off the light within the globe and watch day and night on different continents as the globe rotated. And while with a smartphone and Google Earth or NASA's Earth Now one can look these up at any time, just having that physical globe now in my study and being able to touch and feel and rotate it reminds me of where I am and why it matters.

Can fixing in one's mind the image of the lonely Earth burrowing its way through the Milky Way remind one of the fragility of our existence?

Fig. 8-3 Views of Earth from space
Courtesy of NASA

The evolutionary pattern is clear. If the most important reason to maintain habitat sustainability, perhaps the only reason, is species survival, what does that tells us about Homo sapiens' treatment of Earth's species, ecosystems, and habitats?

Not since that giant asteroid collided with the Earth 65 million years ago, wiping out half the world's species in a geological instant, has anything arisen to become the greatest catastrophic threat. Now Homo sapiens has become the dominant species on Earth, and our impact has been equally devastating. If we continue to destroy the environment as we do today, half of the world's species will become extinct early in the twenty-first century.

Going by our recent use of Earth's habitats and abuse and destruction of our own and other species, it's hard not to conclude that for Homo sapiens, species survival has not been and still isn't a priority.

The Earth, its species, habitats, and ecosystems, once independent forces, have been reduced to an assortment of exploitable resources to be negotiated in the marketplace as mere commodities. The continents, oceans, atmosphere, electromagnetic spectrum, and now even the gene pool, are all being commercialized, their value measured almost exclusively in monetary terms. Instead, we have come up with and doubled down on multiple philosophies, religions, economic models, political systems, innovations, and technologies that have in each prior age exacerbated ecocide even as they failed to prevent the collapse of human civilization after human civilization.

Our most recent and prevailing beliefs today find morality in rational self-interest and virtue in selfishness. Unlike evolution, which seeded Earth's habitats for all, we have taken the view that man exists for his own sake; that, instead of pursuing species survival, man's pursuit of his own happiness is his highest moral purpose; that instead of freely sharing Earth's habitats with the many, he must not sacrifice himself to others, nor sacrifice others to himself. Altruism or any other nonmaterial "sappy motive," like the common good and species survival, is trampled in pursuit of self-interest.

We put our trust in an economic model that assumes economic actors behave rationally and market forces, unimpeded, are the best mechanism for the management and distribution of the Earth's resources, whereas evolution has shown that it is Earth's species, habitats, and ecosystems that have for billions of years done just that, and if given half a chance, could ensure the long-term survival of a nascent emergent intelligent species.

So, compared to evolution's example in providing for an entire planet of species survival for literally billions of years, where has Homo sapiens' self-centered use of the Earth's resources led? Two sets of statistics should more than suffice to answer this: world income inequality and world military spending.

According to the 2017 Oxfam International's global income inequality report, just eight men (updated to 26 in 2018) own as much wealth as half the world. That's right. Eight men, mostly Americans, now control as much wealth as the world's poorest 3.6 billion people. The men—Bill Gates, Warren Buffett, Carlos Slim, Jeff Bezos, Mark Zuckerberg, Amancio Ortega, Larry Ellison, and Michael Bloomberg—are collectively worth $426 billion. And what is true of the world in general is also true of some of its developed and emerging economies. In India, fifty-seven billionaires have the same amount of wealth as the bottom 70%. In the U.S., the richest 1% control 42% of the wealth.

In contrast, almost half the world—over 3 billion people—live on less than $2.50 a day. At least 80% of humanity lives on less than $10 a day. More than 80% of the world's population lives in countries where income differentials are widening. Using the World Bank definition of the global poverty line as $1.25 a day, as of September 2013, roughly 1.3 billion people remain in extreme poverty. Nearly half live in India and China, with more than 85% living in just twenty countries.

Meanwhile, 11.3% of the world's population is hungry. Poverty is the principal cause of hunger, and drought is one of the most common causes of food shortages in the world. According to the U.N., natural disasters such as floods, tropical storms, and long periods of drought are on the increase, with calamitous consequences for the hungry poor in developing countries.

Oxfam's report shows how our broken economies are funneling wealth to a rich elite at the expense of the poorest in society, the majority of whom are women. The richest are accumulating wealth at such an astonishing rate that the world could see its first trillionaire within twenty-five years. To put this figure in perspective, you would need to spend $1 million every day for 2,738 years to spend $1 trillion. "Such dramatic inequality is trapping millions in poverty, fracturing our societies, and poisoning our politics," said Paul O'Brien, Oxfam America's vice president for policy and campaigns.

Military spending is another area where Earth's resources are being diverted away from species survival. Among the things evolution thought necessary to support species sustainability, the use of Earth's resources for military purposes is conspicuously absent. Nonetheless, Homo sapiens has managed to develop a civilization that depends upon ever larger military budgets and mutually assured destruction in the "peace through strength" mantra.

Global defense spending has gone up for the first time since 2011. According to the World Military Expenditure Report published by the Stockholm International Peace Research Institute (SIPRI), countries around the world spent a total of $1.68 trillion (1.56 trillion euros) on arms in 2016 and increased slightly again in 2017. But how might our species benefit if only it could find a way to "beat its swords into plowshares" and make equitable use of Earth's resources and habitats? As shown in the UN's Sustainable Development Goals diagram below:

Allocating only around 10% of the world's military spending would be enough to achieve major progress on some key United Nations Sustainable Development Goals (SDGs), supposing that such funds could be effectively channeled to these causes despite major obstacles, such as corruption and conflict.

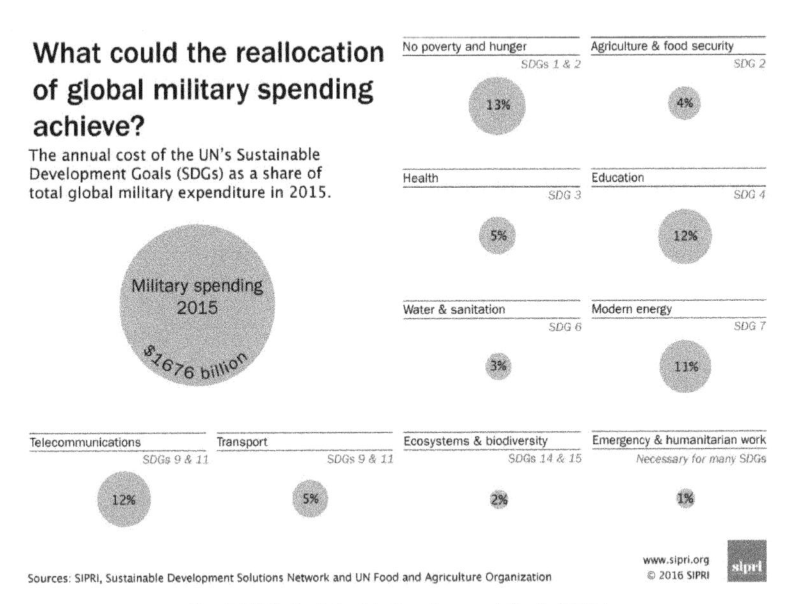

Fig. 8-4 UN's Sustainable Development Goals (SDGs)

Here are the ten countries with the largest militaries, by spending, according to the Stockholm International Peace Research Institute. The U.S. outpaces all other nations in military expenditures. World military spending totaled more than $1.6 trillion in 2015. The U.S. accounted for 37% of the total. U.S. military expenditures are roughly the size of the next seven largest military budgets around the world combined.

Table 8-1 Ten most powerful countries' 2015 military spending

(Source: SIPRI)

Country	Military Spending in 2015	Most Powerful Countries Rank
U.S.	$596 billion	1
China	$215 billion	3
Saudi Arabia	$87 billion	9
Russia	$66 billion	2
U.K.	$55 billion	5
India	$51 billion	14
France	$51 billion	6
Japan	$41 billion	7
Germany	$39 billion	4
South Korea	$36 billion	10

These inverted priorities, as the sociologist Immanuel Wallerstein notes, are not the neutral decisions of a market; they are the priorities of powerful people in powerful nations, mostly men, whose gender, race, and class interests drive the worldwide capitalist political economic system of accumulation and deprivation.

For as long as our species has existed, we have struggled with and reflected deeply on these issues, what we could be versus what we are. Endowed like no other species with the freedom to choose, we have been unable to change course. In the well-known lyrics of his song "Imagine," John Lennon invites all to imagine what the world could be like if we lived in brotherhood with one another, in peace, "all people sharing all the world." There would be no countries, religion, possessions, greed or hunger, "nothing to kill or die for." The "world will be as one."

The direction Homo sapiens appears to have chosen and is currently pursuing, whether deliberate or in spite of itself, is directly orthogonal to the apparent intended use and purpose of Earth as a habitat and is definitely not the one John Lennon imagined.

The difference between what appears to drive evolution in enabling Earth's habitats versus what drives Homo sapiens in the use of Earth's habitats couldn't be starker.

Evolution appears to be seeking intelligent species survival and maturity, self-extinction avoidance, and continuity on to interplanetary migration, and, it seems, even population of the stars might be a potential evolutionary goal. Failing to grasp this potentially bigger picture, Homo sapiens find themselves squandering and squabbling over Earth's resources, putting at risk the much greater prize.

By seeking, instead of species survival, corporate profits, personal wealth and power, and something we call "growth," even when doing so means wanton destruction of Earth's habitats and members of every species, might Homo sapiens be unknowingly missing the opportunity to be *the* resident intelligent species on Earth and, if evolution gets its way, possibly, one day the stars?

You might have heard the lyrics of the old Anglican hymn "All things bright and beautiful, all creatures great and small, all things bright and wonderful, the Lord God made them all." While some still debate whether all creatures great and small owe their existence to God or evolution, what is not in dispute is that Homo sapiens did not make any of them, though one would never guess that from the way our species has selfishly commandeered Earth's resources for our own use.

All the things we could not and still cannot do for ourselves—choosing the right planet orbiting the right star in the right neighborhood of the galaxy and providing a perpetual and comparatively safe renewable source of energy as well as clean air, clean water, lots of organic, natural, renewable food sources and natural renewable habitats filled with renewable and recyclable materials that can be used and reused from one generation to another—evolution selected or evolved and did not look for payment in return. All evolution appears to want out of this arrangement is for an emergent intelligent species like Homo sapiens to attain evolutionary maturity, avoid self-extinction, overcome the universe's existential threats, survive, and possibly migrate to the stars—a potential vision of the future our species, blinded by growth and profits, has yet to grasp.

Homo sapiens did not invent the Earth and its bounty, did not even need to terraform it to survive, as would be necessary anywhere else in the solar system, and perhaps even the galaxy. It was all here billions of years before we even arrived.

Homo sapiens did not invent food or water, yet so much of what is done to the soil in the name of agricultural science—biotechnology, genetically modified crops, herbicides, pesticides, insecticides, and endless varieties of chemicals—threatens to destroy not only the food supply, the rivers, streams, and fresh water sources evolution so freely gave, but also the fish, birds, animals, and insects that comprise the very food webs upon which our survival depends.

Homo sapiens did not invent energy; it came with the solar system. We call it the sun, and it has been burning brightly, providing fusion power, for well over 4 billion years, with at least another 4 billion estimated to still be in the tank. That energy enabled the first eukaryotic cell and drives photosynthesis in plants and plankton in the oceans to provide the oxygen we breathe today, enables the crops and livestock we eat, and serves as a giant heater, keeping the Earth's temperature in a range suitable for complex animal and human life. It has, one might note, been doing so for billions of years without once polluting the atmosphere, fouling the water supply, poisoning the food supply, destroying wildlife and natural habitats, or endangering the survival of millions of species. From an evolutionary survival perspective, using fossil fuels as an energy source is a clear nonstarter.

But since the energy industries haven't yet found a way to charge us for sunlight (I am sure they are working on it) the way the food industry does for water, or the pharmaceutical industry does for patented versions of natural medicine freely available in nature, they have come up with other forms of energy that they can exploit for profit. They pollute the atmosphere, poison millions, and scar landscapes by fracking for natural gas, mining coal, and drilling for petroleum products, in the process spilling endless barrels of fossil fuels in the oceans, streams, and vital coastal waters, destroying wildlife and natural habitats. They do all this while relentlessly mounting effort after effort to slow down or block clean renewable energy sources.

Whereas evolution appears to have left a basket of resources for the many, Homo sapiens appears to have appropriated and given those resources to the few. Whereas evolution appears to have laid out a path to life and survival at no cost whatsoever, Homo sapiens appears to prefer a path to profits at any cost, mostly for the few, even when the pursuit of profits and growth means species extinction.

Somehow, despite all the survival bread crumbs evolution has left over the eons, as a species we have yet to figure out why we are here on this planet; why, against all odds, we exist; and why we might now be trapped, at the mercy of our own devices, with no ability to migrate to a potential Earth 2.0 in the face of another global climate-change event, one that, for the first time, we may well have precipitated.

Somehow we manage to see our current path through the lens of growth, industrial progress, technology, innovation, and entrepreneurship; towering skyscrapers and resplendent and shining cities; and economic systems that drive investment and uplift a modern technological civilization, but are unable still to learn from the lessons written in the ruins of prior civilizations, long collapsed, strewn across the surface of the planet.

Somehow we believe it is this path that lifts the poor out of poverty, as if economic growth at any cost and riches regardless of how they're obtained are a prerequisite for survival. We have come up with innumerable rationales, justifications, and euphemisms for plain old-fashioned greed and, yes, innumerable systems and organizations that ensure the enrichment of "the few."

Under the mistaken belief that this is what makes up a modern technological civilization, we pursue this path even when doing so means ignoring the fundamental purpose for our species' existence—survival. Failure to identify and adopt species survival as the most important—if not the only—goal is where and why our species has gotten derailed. For evolution, habitats and habitat sustainability are meant to preserve life and enable species survival to evolutionary maturity. For Homo sapiens, Earth's habitats and their sustainability matter only when it's good for business and profits. Yet it wouldn't be the first time our species has gone down this dead-end path.

Can a species break free from this repeating snare? Unlike all others, Homo sapiens is the only surviving intelligent species evolution has given the freedom to choose its own path. Yet Homo sapiens seems to have become lost and entrapped by its own thinking, powerless to recognize and follow the evolutionary path to survival. It seems not to be able to reject the utter absurdity of the appropriation by "the few" of most of Earth's resources, which are intended for the survival of all of the planet's species.

Yet because evolution has endowed our species with unique abilities, including the ability to destroy Earth's habitats, other species, and even itself, can one dare to hope that evolution is not about to walk away from the table just yet but has more than a casual interest in our survival and probably, even hopefully, has a few more cards, if not dice, left to play?

We now know from the species survival evolutionary patterns what happens to species that, despite evolution's best efforts, fail to get it. They never reach evolutionary maturity and eventually go extinct.

Here is where global climate change appears to come in. Evolution erases the record and rejuvenates Earth's habitats. Given that another 4 billion years remain on Earth's warranty, there's still time to start all over with another—hopefully more intelligent—species. Yes, this too, is an evolutionary pattern—the extinction pattern, a pattern that has swept away millions of species and replenished the Earth and its habitats to sow it afresh with new species. It was the extinction of the great and mighty dinosaurs after a nearly 175-million-year run that made room for the emergence of the then humble and tiny mouse-like mammals. Sixty-five million years later, could it be that Homo sapiens, their very, very distant descendants, now themselves big and mighty, if only in their own eyes, are coming face-to-face with their own time of reckoning?

Human history shows that when a civilization gets caught up in its own machinations, as modern civilization appears to be at the moment, and completely loses track of the a priori evolutionary purpose for being; when species survival becomes less important than profits, or race, or religion, or creed, or whatever is in vogue at the time, or is not a consideration at all, it becomes almost impossible to liberate it from the intoxicating web of false philosophies, memes, cultures, and belief systems. It becomes almost impossible to shake it free from the stranglehold of kings, queens, presidents, dictators, military oppression, slavery, warlords, emperors, religions, political parties and systems, economic systems, military power, and, yes, even false trust in technology and innovation as the panacea for all of a species' ills. It is then that climate change blows through to reshuffle the species survival deck and start the game once again.

Where are the Sumerians, the Assyrian Empire, the great Egyptian Pharaohs, the Greek Empire, the Roman Empire, the Tang Dynasty, the Kublai Khans and Genghis Khans, the Maya, the Moche, the Incas, the ancient Pueblo people, the ancient Khmer kingdom of Angkor Wat fame, and, yes, the British Empire, and more? Most were swept away by droughts, famines, and other forms of climate-driven habitat change that managed to reshuffle the Homo sapiens deck, allowing for yet another fresh start, to try yet again to get it right. No one knows just how many more of those iterations remain—we might be running out of time.

Will we make it? How long will it take before we know? How many more governments, economic systems, philosophies, religions, military doctrines, adaptations and innovations, wars, and genocides are yet to play out? How many civilizations have yet to collapse? How many millions still must disappear into the dust before Homo sapiens gets it? Perhaps the more troubling question is, is Homo sapiens, one of the seven surviving great apes in the family Hominidae, the third chimpanzee, capable of getting it?

This last question can be answered only with a resounding "Yes we can!" Why would evolution take the risk of evolving a clutch of emergent intelligent hominin species, carefully winnowing that collection down to one, and endow it, unlike any other species, with free will, thus unshackling it from the constraints of biological evolution, if there were no chance of that species ever being able to choose to avoid self-extinction and grow to evolutionary maturity?

Einstein once famously said, "God does not play dice with the universe," to which Stephen Hawking responded, "Not only does God play dice, but he sometimes throws them where they cannot be seen." Thankfully we are only talking here about the survival of the lone, surviving, emergent intelligent species on Earth, potentially in the entire galaxy, and might be able to find a little hope in the observation that evolution tends to err on the side of species survival. If so, the dice might be loaded in Homo sapiens' favor.

In deeply contemplative moments such as these, one might find hope and solace, or frustration, in the decidedly unscientific yet relevant themes of the hauntingly melancholic lyrics of Bob Dylan's "Blowing in the Wind." Like him one sometimes wonders how many more roads to travel, and seas to cross; when will we realize that people are not free, and too many people have died. The answer we now know is not "blowing in the wind" but instead written in the fossil record of our hominin forebears, and that of prior collapsed Homo sapiens civilizations, but can end when our species chooses to follow the evolutionary patterns to survival, or, heedlessly continue on to extinction.

An intelligent species, then, must not only maintain its habitats, ecosystems, and sustainably use Earth's biodiversity; it must also use its intelligence in the face of climate change to adapt just to survive. In the next chapter we explore whether Homo sapiens has managed to figure out the evolutionary rules for survival-driven adaptation any better than it has those for survival-driven sustainability.

9.

Adaptation

MUST BE SURVIVAL-DRIVEN TO PRESERVE HABITATS AND BIODIVERSITY

SINGLE-CELLULAR LIFE MAY WELL PERVADE THE universe, some scientists think, and for billions of years it was the predominant life-form on Earth. Earth is the only planet we know of in the entire galaxy, probably even the universe, that has been terraformed for the evolution and survival of complex animal and intelligent species life. It took well over 3.5 billion years before complex animal life appeared, and almost another 500 million before intelligent species emerged. But one shouldn't take Earth's early, tiny inhabitants for granted, as they are believed to have engaged in planetary re-engineering at a scale far, far beyond the capabilities of modern Homo sapiens today, and perhaps for millennia to come.

During the first 3 billion years, single-cellular organisms (prokaryotic bacteria) were hard at work using sunlight, water, and carbon dioxide to terraform the Earth by generating atmospheric oxygen, thus paving the way for the emergence of complex animal life. As noted earlier, cyanobacteria, through photosynthesis, generated most of the early Earth's oxygen, triggering Earth's first mass extinction in the so-called Great Oxidation Event, and thereby may have transformed the planet forever. They didn't stop there either. These little critters may well have been responsible for the first snowball Earth by oxidizing atmospheric methane, resulting in the first ice age and much more. So you might want to thank these tiny organisms each time you take a deep breath. And given their work on Earth, one can't help but wonder what they might be up to on Earth-like planets across the galaxy.

Whether it is primordial Bacteria, Archaea, or Eucarya (branches on the tree of Earth life), all life-forms need to metabolize: to feed, process chemicals, and thereby obtain energy to reproduce, develop, evolve, and survive. Without energy, life as we know it would not exist, and while there are life-forms that survive on chemicals emanating from hydrothermal vents at the bottom of the oceans, the main source of energy for Earth life is the sun. It is sunlight that provides the energy to convert carbon dioxide and water into complex chemical compounds with many chemical bonds that store energy. By breaking these bonds, the energy that enables life is released. Evolution has chosen to make this energy source available to all Earth life through a single power main—photosynthesis, Earth life's only way to connect to its only source of energy.

A forest is not just a forest, then, but a power grid, and a tree is not just a tree but a discrete connection to the sun's energy grid through photosynthesis. And while in James Cameron's movie *Avatar*, the Tree of Souls is the Na'vi link to the spiritual and guiding force Eywa, Earth-life trees and forests are much, much more. Collectively, they comprise Earth's energy- and oxygen-supply grid, and Homo sapiens need to be as unrelenting as the Na'vi in preserving their only connections to the energy grid evolution provided. Once you're aware of this energy connection, you can never look at a tree the same way again. Have you hugged a tree recently?

In other words, Earth's trees, forests, phytoplankton afloat on its oceans, and cyanobacteria, together through photosynthesis, produce more than 99% of the oxygen in the atmosphere today. Without them not only would there not be enough oxygen to sustain aerobic complex animal and intelligent species life such as Homo sapiens, but equally important (and one doesn't get to pick one's poison in this case), Earth life would have no way to access the sun's energy.

And since there is not yet an alternate way known to capture the sun's energy, if photosynthesis stopped, because of widespread deforestation, for example, one would essentially be pulling the plug on all Earth life, including the life of Homo sapiens. Already, more than 50% of Earth's trees have been lost since humans first started cutting them down, and NASA predicts that if current deforestation rates continue, the world's rain forests will be completely wiped out in another century. Imagine evolution posting a pair of blindingly bright, flashing red Do Not Destroy! signs. They would without a doubt be posted around forests, on trees, and on cyanobacteria and phytoplankton communities.

From the loftiest species to the humblest, from Homo sapiens to dung beetles, most Earth life has obtained energy by feeding on plants or the herbivores that feed on those plants, food that is ultimately derived from the sun through photosynthesis. All Earth life must evolve in and adapt to a photosynthesis-capable habitat that enables it to access the energy and oxygen it must have to survive.

> *Evolutionary survival-driven adaptation for intelligent species, then, is the process of learning how life works on Planet Earth and using that knowledge to figure out how to sustain life in order to survive without destroying itself, the Earth's habitats, and other species.*

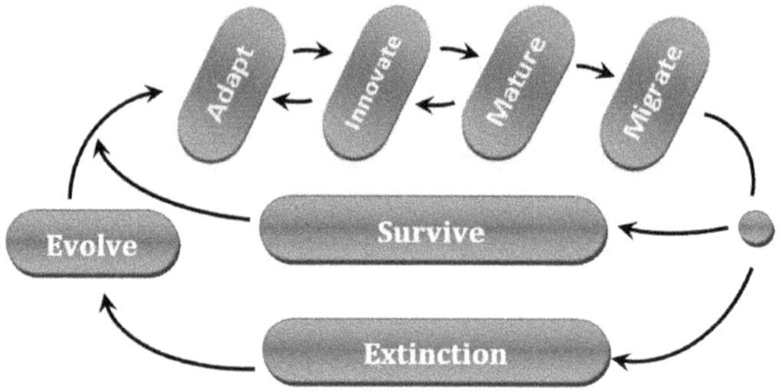

Fig. 9-1 Evolutionary survival patterns
(Source: Author)

As depicted in the diagram above, adaptation is the first in a set of inexorable evolutionary survival patterns (adapt, innovate, mature, migrate, survive, or go extinct) that all nonintelligent species automatically and instinctively follow (because they evolved in an a priori state of evolutionary maturity, already programmed that way), but that all intelligent species, like Homo sapiens, must learn just to survive. These patterns form an indispensable, repeating set all species follow to survive. When followed, these evolutionary species survival patterns will enable an intelligent species to attain evolutionary maturity—survival beyond the point of self-extinction.

To survive, then, an intelligent species must learn how to adapt, innovate, mature, and migrate, as many hominin species have, in a way that does not destroy Earth's habitats and species, or it will go extinct.

This chapter examines how our forebears and modern Homo sapiens have chosen to execute the first among the evolutionary survival patterns, adaptation; what it means to adapt from an evolutionary survival perspective; and how those adaptation choices can and have impacted their and our potential for survival or extinction.

In the prior chapter on sustainability, we noted that the key to Earth's ability to remain a living planet, that which differentiates it from dead planets such as Mars and Venus, is its biodiversity—its species. These are not just passengers on Spaceship Earth along for the ride but are in fact an integral part of the planet's life support and survival mechanisms. Together these form a symbiotic, interdependent, and recurrent set of processes that drive Earth's ability to remain a living planet. For Earth to survive, Earth's habitats must survive, and for Earth's habitats to survive, Earth's species must survive. And in yet another of those uncanny evolutionary yin and yang survival patterns,

> *species survival ensures habitat survival, and habitat survival ensures species survival.*

Evolution, then, appears to have given all Earth life, including Homo sapiens, the same equation to solve. A species must be able to obtain all it needs to survive from its habitat, and do it without rendering that habitat incapable of sustaining life. Built into this simple requirement is the implication that preserving it is not just for one's own survival but also for the survival of other species that share that habitat.

Embodied in the above are the first two rules of evolutionary, survival-driven adaptation, and they differ markedly from what Homo sapiens have come to regard as rules of adaptation. Essentially, these rules say do nothing to destroy one's habitat and do nothing to destroy the habitats of other species. These two rules are inseparable, as there is no way in an Earth ecosystem to destroy one's own habitat without simultaneously destroying the habitat of other species. These survival-driven adaptation rules imply a deep sense of species interdependence, interrelatedness, and coexistence with all other species and with the habitat itself, and hence a need for harmony and balance. Adapting to an Earth ecosystem therefore requires a species to learn to live and let live. This often means intervening in natural ecosystems, but Earth's sustainable-capable habitats are engineered to endure and recover from some amount of damage.

Should a species' habitat become incapable of sustaining life (Earth's habitats, as we have seen, are not inexhaustible), whether due to its own actions or to non-species-driven causes, the species must be able to migrate to a more sustainable-capable habitat to survive, or it will go extinct. And for billions of years, all other life-forms, all nonintelligent species that evolved in a state of evolutionary maturity, with the possible exception of cyanobacteria (in the Great Oxygenation Event) and perhaps a few others, have managed to follow these rules, adapting without destroying Earth's habitats and driving themselves or other species to extinction.

It follows that to be a good Earth life species, each must see itself as being codependent, a part of a community of species sharing common habitats. This was the prevailing behavior of most, if not all, of Earth life, including hunter-gatherers, whose way of life was practiced for all of Homo sapiens' existence until 10,000 years ago, when agriculture became the dominant method of subsistence and settled communities replaced the mobile ones. Hunter-gatherers lived in close harmony with the species that made up Earth's ecosystems and are believed to have seen themselves as belonging to and being one with the biodiversity of Earth's habitats, not above or apart from them.

Modern Homo sapiens, however, do not share this view but rather see themselves as separate, apart from, and above all other species on the planet, flora and fauna both. Indeed, they see themselves as their lord and master, so endowed by God and/or science, enjoined to use and abuse the Earth's biodiversity as it sees fit—a misguided perspective that over the last 10,000 years has led to the devastation of Earth's flora, fauna, and habitats to such an extent that the survival of the Homo sapiens species itself might now be at risk.

Here, then, is where modern Homo sapiens got off the evolutionary, survival-driven adaptation path and, departing from the sustainable subsistence lifestyle of its forebears, began to make its own rules for adapting to life on Planet Earth. This was a fundamental shift and has governed most interactions between humans and all other Earth life; a perspective and way of thinking that may have begun with the transition to agriculture and settled communities but did not stop there; a transition that sent Homo sapiens down the path of living apart from and outside of Earth's ecosystems; a way of thinking and living that became entrenched and greatly accelerated in the last two centuries.

From an evolutionary survival perspective, then, modern Homo sapiens' view of itself as being separate, apart, and above all other species on the planet is antithetical to survival-driven adaptation.

The problem intelligent species like Homo sapiens face is how to take from the environment what they need to survive in a way that does not render the environment incapable of continuing to sustain life; how to balance their needs against the carrying capacity of the planet's ecosystems—a problem hunter-gatherers were able to solve for 99% of our species' existence. Only after the shift to agriculture, settled communities, and subsequent civilizations did sustainable living within Earth's habitats become the challenge it still is today.

Once Homo sapiens adapted agriculture as its primary way to make a living, it began to live outside of Earth's ecosystems. The challenge since has been knowing when—and finding the economic, political, and social will—to stop before Earth's habitats, subject to agricultural adaptations, become irreparably damaged. But the more complex Homo sapiens' civilizations and accompanying adaptation strategies grew, the less capable it was of stopping, or finding the will to restrain itself, especially in the last two centuries, when it turned that decision over to the blind and mindless forces of the market economy, embodied most purely in multinational corporations—legal persons, one might note, that need neither trees nor oxygen to survive.

From the time the first human ancestors appeared, between 5 million and 7 million years ago, until about 10,000 years ago, they survived in hunter-gatherer communities and succeeded in finding that balance, but since then, most successor civilizations have failed to do so. Civilizations that emerged since the adoption of agriculture have been unable to strike a balance between the need for food and the ability of the environment to sustain the intensive agriculture this required. Invariably, the demands of these increasingly complex societies began to exceed the capacity of the agricultural base. The results were failing food production, leading to social, economic, and political chaos, wars, and collapse.

After the slow and eventual rise of agriculture and settled communities, and the accompanying attrition of hunter-gatherer communities, Homo sapiens' struggle to provide enough food became for the first time a constant and central feature of nearly all civilizations and remains acute for the majority of people in the world today. Rapid advances in agricultural technologies in the last two centuries have lifted much of this pressure, so much so that less than 5% of the world's 7.8 billion are engaged in agricultural food production. With this much surplus food, the struggle has been transformed from a search to provide food for the many to a search to find ways to use that surplus to amass and conceal wealth, profits, and endless growth for "the few." But the agricultural innovations of the last two centuries may yet fall short of meeting the needs of a world population set to be approaching 10 billion by the end of the twenty-first century.

It has not always been this way. The specter of hunger, starvation, endemic, and epidemic diseases, mass death, and widespread inequality that became the scourge of Homo sapiens civilizations arrived only after the transition from hunting and gathering to agriculture as the dominant way of life.

It may come as somewhat of a surprise to many to learn that the "Greatest Generation" was not those born between 1930 and 1946, as Tom Brokaw claimed in his book *The Greatest Generation*, nor was it the baby boomers that followed, nor any generation since, for that matter—iPhones notwithstanding. The "Greatest Generation" would have been members of hunter-gatherer communities, the first and possibly the last great Homo sapiens civilizations. That generation was part of Homo sapiens' best attempt to live as an integral part of Earth's ecosystems, in harmony with its flora and fauna. Their hashtag might have said something like #Survival. We did so without destroying the Earth.

Most of what we might consider indispensable for twenty-first-century life our early predecessors managed to do just fine without. Recent studies confirm that for all but the last 10,000 years of our species' existence, Homo sapiens obtained their subsistence by hunting and gathering and lived in small mobile groups. Their simpler way of life was the most successful and flexible adopted by our species. Theirs was a way of life that caused the least damage to the Earth's natural ecosystems and sustained our species' migration across the globe to survive. It was hunter-gatherer Homo sapiens who became the first Earth-life species to dominate and exploit every terrestrial ecosystem. And they did so in a comparatively sustainable manner, and for the most part without destroying too much of the Earth's habitats and biodiversity.

Hunting and gathering was a highly stable and very long-lasting way of life, and for hundreds of thousands of years, until about 10,000 years ago, when this subsistence method began to change, it was the only way Homo sapiens were able to extract the necessary sustenance from the environment. Theirs was a nutritionally adequate diet selected from a wide range of food sources, and they survived with only a handful of goods. Obtaining food and the goods they needed occupied only a small portion of their day, with less than half of the group engaged in these activities, leaving large amounts of time for leisure and ceremonial activities. Hunting might have contributed less than a third of their diet; some meat might have come from scavenging other predators' kills. Among some groups, women are believed to have brought in twice as much food as men, working less than three hours a day.

Unlike those who developed subsequent agriculture-based settled communities and the civilizations that slowly replaced them, hunter-gatherers formed egalitarian communities with shared responsibilities. They were dependent upon neither king nor queen nor tribal chief for leadership. These communities might comprise fifteen, twenty, even as many as fifty individuals who would move several times a year, fifteen kilometers or more each time. They did not live in constant fear of starvation, hoping for a good harvest, nor were they dependent upon the one or two crop varieties of agriculture-based civilizations; rather, they were able to select their food from a wide variety of food sources and locations.

Hunter-gatherers' approach to food and possessions was also quite different from subsequent civilizations. Everything they owned had to be carried; hence they kept very few possessions. Food was neither owned nor stored and was available to all as needed. Thus, they avoided that whole set of problems that came with what we now call "civilization": surplus food from good harvests gave rise to rulers to manage and distribute it, which led to religions to convince the peasants of the divine authority of the rulers, which led to a class of elites and bureaucrats that extorted tithes and taxes, along with militias to enforce the laws of the rulers. This all led to increasing inequality for the farmers and peasants, who became the new poor, and the imposition of numerous vexations that became the foundations of modern Homo sapiens civilizations.

Hunter-gatherers were also much healthier than most of the settled agriculture communities and civilizations that followed them, even well into the early nineteenth century. Malnutrition and deficiency diseases are believed to have been rare due to the wide variety of food sources. Their small communities were not conducive to the spread of infectious diseases. The diseases they were believed to have had came from intestinal parasites due to improper cooking and unclean water, and even these are believed to have occurred at low rates. However, infant mortality was high, believed to be about a fifth of births, and life expectancy is believed to have been about the same as in India in the nineteenth century (twenty to twenty-five years) with some of the population exceeding age sixty.

But because they were mobile and did not live in settled agricultural communities, they avoided many of the diseases that became endemic due to poor sanitation and waste disposal, lack of clean drinking water, and densely packed living situations that facilitated epidemics that would become common in villages, towns, and, eventually, cities.

Hunter-gatherer communities also avoided most if not all of the diseases that accompanied the domestication of animals that often shared the tiny, windowless huts peasants and farmers are thought to have lived in. Most common diseases are variations of those found in domestic animals. Homo sapiens now share sixty-five diseases with dogs, fifty with cattle, forty-six with sheep and goats, forty-two with pigs, thirty-five with horses, and twenty-six with poultry. The common cold is believed to have come from horses; smallpox might have come from cowpox; measles from a cattle disease called rinderpest, and leprosy from water buffalos. Tuberculosis and diphtheria are believed to be due to cattle, and influenza is found in pigs and birds. It is clear that the adoption of agriculture and the establishment of settled communities were the kinds of adaptations that exacted a heavy price on early Homo sapiens, a price they continue to pay today.

From an energy-use perspective, all the energy hunter-gatherers needed was entirely renewable. They used mostly their own human energy and that afforded by the occasional use of animals. Human energy and animal energy were soon complemented by fire, wood, water, and wind. It was not until around the late nineteenth century that Homo sapiens began to turn to nonrenewable energy sources, particularly fossil fuels, which are today responsible for 80% of the world's energy.

Just as agriculture and the establishment of communities and civilizations that subjected populations to innumerable diseases and a life spent in constant fear of starvation well into the late nineteenth century, and which still persist in many undeveloped parts of the world in the twenty-first century, the use of fossil fuels (coal, oil, and natural gas) has turned out to be another adaptation that has brought in its wake seemingly irreparable harm to our species and the environment. Fossil fuels brought the global warming that now threatens not merely the spread of more animal-originated pandemics, as prior adaptations did, but, unless a solution can be found in time, something much worse—potential collapse of world civilization, if not extinction of our species.

Finally, as seen in the chapter on climate change, our hunter-gatherer forebears evolved and survived a time of many glacial and interglacial cooling and warming periods. Many earlier hominin species in the long line of the genus *Homo* evolved, added their distinctiveness, and went extinct, some, like Neanderthals, as recently as 30,000 years ago. Their way of life enabled them to adapt, endure, and survive some of Earth's worse climate-change events for hundreds of thousands of years and may well have positioned them to deal with and survive climate change better than twenty-first century Homo sapiens appear to be today.

The hunter-gatherer way they pioneered began to come to an end some 10,000 years ago and with it perhaps the last best hope for our species to survive and coexist with Earth's habitats and biodiversity.

In hindsight, hunter-gatherers may well have been pushed into farming, as more and more groups began to settle and more and more land became dedicated to agriculture. Today, the hunter-gatherer way of life is severely restricted. Just a handful of groups persist, including the Bushmen in southwest Africa, the Hadza in east Africa, a few groups in India and Southeast Asia, some aborigines in Australia, some Inuit in the Arctic, and native inhabitants of the tropical forests in South America.

If hunting and gathering can be regarded as Homo sapiens' first adaptation to surviving on Planet Earth, then the shift to agriculture must be seen as the second. The shift to agriculture as the predominant form of subsistence became the foundation for a number of relatively rapid follow-on adaptations that led to twenty-first-century Homo sapiens civilization. It was at first slow and gradual and may have begun as long ago as 18,000 BCE. Agriculture was independently adopted in different parts of the globe and at different times, with southwest Asia becoming the first region where Homo sapiens began to derive its subsistence entirely from farming, followed by China and parts of Central and South America. It is believed that most major crops and farm animals across the world had been domesticated by 2000 BCE.

No one knows for sure exactly how and precisely when mobile hunter-gatherer communities became settled communities. They may have gradually become settled, partially at first, then slowly over thousands of years. As more and more people were able to find and grow enough food in small areas, they became increasingly less mobile. With settled communities came increased population numbers, and as their numbers grew, it became harder and harder over time to eke out a living in their regions. Pressure to split up and spread apart and/or adopt more intensive farming techniques increased. With time, better tools, including bone sickles with flint blades, grinding stones, and mortars and pestles gradually became available and may have facilitated the shift to cultivation.

As population grew, our forebears migrated across the Earth into all major ecosystems within a span of 30,000 years: first, about 100,000 years ago to Eurasia, then around 50,000 to 60,000 years ago to Australia and New Guinea, around 13,000 years ago to Siberia and North and South America, and most recently, only 4,000 years ago, to the Pacific Ocean Islands. At the same time, humans underwent a massive demographic boom, expanding from a few million people 50,000 years ago to around 150 million in 2000 BCE. As more and more people became settled, fewer and fewer retained the hunter-gatherer skills and ways. And while population growth became a driving force, in the longer term, climate change may well have been the main driver of this transition. Around 10,000 BCE, the last ice age was ending and a warming interglacial was beginning to take hold, driving changes in the variety and availability of food sources that made the ability to obtain more food from a smaller area, as was possible with farming, seem more attractive.

> *But the shift to agriculture was more than a shift to a new form of subsistence. It was a complete change from the way Homo sapiens had survived, thrived, and successfully adapted within the Earth's ecosystems for more than 99% of our species' existence. It was the beginning of Homo sapiens' assault on Earth's habitats, ecosystems, and biodiversity that 10,000 years later has resulted in a level and rate of extinction of Earth's flora and fauna the planet had hitherto not seen, mass extinctions notwithstanding.*

The impact on the environment was small at first but visible in regions with less-robust ecosystems and areas with large populations. The need to cook, build houses, and heat them drove the demand for wood and the beginning of deforestation. Large tracks of land were constantly being cleared for cultivation, and falling productivity due to soil erosion required even more land. Soon slash-and-burn agriculture developed to obtain more and more fertile land as productivity in prior tracts fell.

Sustaining their lives from outside of Earth's ecosystems may not even have been a choice; it might have been necessary for survival. If it was optional, like moving to another region had been, it still might not have been a conscious choice at first, and more than likely wasn't seen as such. Even if it was a choice, it's likely they made it because they had assumed they could continue to live in harmony with Earth's habitats and biodiversity as their ancestors had for millions of years. How could they have imagined where this path would take them?

Nonetheless, it did happen. It may well have started Homo sapiens down the road to endless adaptations. Each subsequent adaptation fixed the problems created by prior adaptations, thereby putting us in the position to potentially adapt ourselves all the way to the sixth mass extinction. This is hardly the legacy one would envision passing down to subsequent generations.

This is a perspective that surviving members of Brokaw's Greatest Generation, baby boomers and all generations since might want to reflect hard upon given the adaptations we are choosing today. Ten thousand years ago, our forebears could not have foreseen the consequences of the choices they were making, but with hindsight we might want to stop and ponder the kind of legacy we might be leaving all future Homo sapiens.

But here is the thing: today one doesn't even get to choose; our world appears to be set on autopilot.

Unlike our hunter-gatherer forbears, whether or not you or I want to live sustainably and in harmony with nature is mostly immaterial. That decision is a priori made for us by the many multinational corporations that supply us with food choices and almost everything but babies (not yet, anyway). International institutions, financial institutions, and nation-states, including our own governments, are all competing in global markets with little or no thought given to preserving the planet's biodiversity or understanding their role as the life-support system that enables Earth to remain a living planet.

With hindsight we can see how our species' adaptation of agriculture was naïve and ill-advised, a blunder into a way to survive outside of Earth's ecosystems that became deliberate with time, the preferred way of Homo sapiens subsistence. It was at that fork on the geologic road where Homo sapiens began to go down a path that makes one wonder how different it might have been if another way forward had been found. That in turn makes one wonder how much of an accident of history it was in light of the role climate change has played in this particular transition and in shaping the destiny of our species—a topic we will explore later.

Whereas all other species remained faithful to the "rules" of survival-driven adaptation, content to adapt and live within Earth's ecosystems, Homo sapiens developed an artificial habitat that involved clearing land, severing natural nutrient cycles and thereby cannibalizing the ecosystem and replacing its natural variety and permanent ground cover with a small number of crops and animals. This left much of the soil unused and exposed to increased heat, rain, and wind that resulted in increased soil erosion, which in turn led to two other adaptations, the use of fertilizers to maintain soil productivity and irrigation to water the soil.

Once again, seemingly complementary (some might even say commonsense) adaptations needed to fix unforeseen failings in prior adaptations were to become the beginning of yet another journey, one that would help for a while but would in the end hurt for far longer. Intense irrigation soon led to increasingly higher water tables, causing waterlogging of the soil, altering the mineral content and increasing the salt content; in lower altitudes, this can render the soil completely unusable for agricultural purposes. Such fields would have needed to be abandoned for extended periods before they could be returned to agricultural production, driving the need for more and more arable land and the clearing of more and more forests. In some regions, barely a thousand years after the widespread emergence of settled farming communities, it became necessary to abandon whole villages due to soil erosion caused by deforestation.

Today, much of the food chain has been contaminated with chemical fertilizers, insecticides, and herbicides, resulting in, among other things, as much as 75% of the planet's pollinators being wiped out by insecticides. Nitrogen and phosphorus-rich agricultural runoff is causing dead zones in the planet's marine ecosystems, areas where the oxygen in the water has been depleted by accelerated algae growth, which then blocks the sunlight, making it impossible for aquatic life to survive. The largest dead zone in the United States—about 7,829 square miles—is in the Gulf of Mexico and occurs every summer as a result of nutrient pollution from the Mississippi River Basin. Today, manure, excess fertilizer applied to crops and fields, and soil erosion make agriculture one of the largest sources of nitrogen and phosphorus pollution.

Today too, scientists report that irrigation is used across 20% of global cropland producing about 40% of global food production while rain watered areas serve 80% and produce nearly 60% of global food. But researchers warn that the more irrigation is used to increase food production, the more saline soils become. According to a report by the U.S. Department of Agriculture, "Concern is mounting about the sustainability of irrigated agriculture," and suggests crop yield reductions due to salinity occur on an estimated 30% of all 56 million acres of U.S. irrigated land.

In many countries irrigated agriculture has caused waterlogging, salinization, and depletion and pollution of water supplies. The UN warns that the 70% food production increases needed to feed the world by 2050 is unlikely given the increasing cropland salinity due to poor irrigation practices and sea level rise. The great challenge of the century may well be to find ways to use less water, to manage it better and to grow food without turning the soil into salt.

Thus, from their humble beginnings, fertilizers, which began as simple manure, and irrigation have become the scourge of almost every ecosystem on the planet. The creation of artificial environments to continuously grow more food for a continuously increasing population has created a death spiral that now seems almost impossible to alter.

Some in agribusinesses may not want to, but it's not too late to remember that these adaptations became necessary only when Homo sapiens began to live outside of the planet's nutrient-rich and naturally irrigated ecosystems. Hunter-gatherers who lived within the planet's ecosystems for millions of years apparently had no need for either.

The shift to agriculture engendered numerous adaptations. In fact, this change in the method of subsistence completely rearranged the lifestyle of hunter-gatherers and laid the foundation for what would afterward be called "civilization."

For the first time, a once-proud, economically self-sufficient, egalitarian, comparatively healthy, free, and independent mobile community, leading an open lifestyle with allegiance to no one, would discover that the transition they were undergoing required them to adapt to a new and very different reality. They would need to adapt to living in a settled community, in increasingly crowded quarters shared with domesticated animals and exposed to their diseases. They would need to adapt to economic insecurity due to a reduced number of food sources and the risk of crop failure.

And, eventually, they would need to surrender even their freedom and independence in exchange for allegiance to a ruler of some sort—a chief or a distant king or queen and later priests and an ever-increasing retinue of bureaucrats and craftspeople who lived off the surplus of their harvest and in time would take this surplus by force.

In the words of Clive Ponting in his book *A New Green History of the World*, "Human history over the last 8000 years has been about the acquisition and distribution of this surplus, and the uses to which it has been put became the foundation of all later social and political change. The ability to support non-farmers made possible all subsequent cultural and scientific advances; the emergence of religions and the ability for rulers to forcibly compel compliance and slavery and engage in warfare."

Acquisition and distribution of surplus became the foundation of some of the earliest civilizations that arose in Mesopotamia, Egypt, China, and India, to mention a few. According to Ponting, "In these early civilizations, most people became peasants, landless laborers and/or slaves subject to extensive expropriation of their food, forced labor and warfare, living constantly on the margin of subsistence, suffering hunger and the ever-present threat of famine." This became and remained the way of life for thousands of years.

Hunter-gatherers could not have imagined what they were getting themselves into. During the long transition to widespread adoption, as they began to realize all the extra work and effort that came with farming, as well as the loss of freedom and economic independence, why did agriculture take hold? What would have driven an otherwise free people to take this path?

Anthropologist Mark Cohen, in his book *The Food Crisis in Prehistory*, claimed it might have been due to a food crisis resulting from the extermination of megafauna such as the woolly mammoth and the American bison. In most places around the planet, megafauna extinctions occurred shortly after the arrival of our forebears. On finding fresh hunting grounds, our ancestors encountered animals with no experience of human predators. Like the ultimate invasive species, we quickly obliterated species that didn't know how to stay out of our way. Cohen believes mass extinction of these large animals, a primary subsistence source, may have forced our early forbears to change their social organization wherever they could find fertile soil, water, suitable climate, and animals that were easily domesticated.

If indeed climate change and a rapid increase in population created a food crisis that led to the extinction of much of Earth's megafauna as our forebears migrated across the planet, was their extinction inevitable? Was this behavior a one-off, the aberrant result of a desperate species driven by the need for food? Far from being a one-off, Homo sapiens' history shows a penchant for hunting species to near extinction, stopping only when there are so few left it is no longer worth the effort or cost of pursuing them. It was neither a food crisis nor climate change but greed and a search for profits that drove the fur trade and the whaling and sealing industries; it was not out of need that the passenger pigeon and numerous other species have been hunted to extinction or near extinction.

Even with an awareness of the impact this massacre of species has had across the centuries, there is still no sign that our species is prepared to change, or that it is even capable of changing its ways. The takeaway here is that we seem to be able and willing to stop only when the resource being consumed is exhausted. If this turns out to be the way Homo sapiens respond to the atmospheric rise in CO_2 (with April 2018 monthly average exceeding 410 ppm), due primarily to the burning of fossil fuels, we won't have much longer to worry about our inability to course-correct, as global warming may well do that for us. The words of Emily Dickinson come to mind as one contemplates our predicament: "Because I could not stop for Death, he kindly stopped for me."

If Homo sapiens are going to learn from hindsight, they must find the social, political, and economic will to choose sustainable paths that preserve the Earth's species and its habitats. But alas! Homo sapiens may have already turned over its ability to choose to nonhumans—multinational corporations and a free market system driven by a capitalist ideology seemingly designed for a limitless universe, that see the Earth's habitats and biodiversity simply as capital, items to be traded for profit. And now we are getting ready to further relegate our ability and freedom to choose to an emerging artificial intelligence.

Why did hunter-gatherer Homo sapiens choose to adapt as they did in response to shrinking food sources, increasing population growth, and climate change? One might similarly ask, why have twenty-first-century Homo sapiens surrendered their freedom to choose to impersonal market mechanisms, multinational corporations, global institutions and nation-states to which economic development and market growth are all that matter? Why have they become mere consumers whose existence is important only in terms of the labor and skills they provide and the goods they purchase that help perpetuate this endless cycle of plundering the planet?

Homo sapiens' adaptations and innovations were not only material in form but also included ways of thinking, perceiving, and imagining the world. It was that power that engendered the changes we see around us today. If you ever wondered why modern civilization is the way it is and how it got to this place, how we ended up with the belief systems that now largely dominate the world—the religions, forms of governments, and forms of social, scientific, and economic thinking that became dominant—I will briefly share some of the mental adaptations that might have led us here.

Historians claim that the way of thinking about the world that has become dominant in the last two centuries—particularly about how Homo sapiens should relate to the planet's habitats and biodiversity and how resources should be used and distributed—originated primarily in Europe—its classical, religious, economic and scientific thought. In fact, since the world is largely led (for now) by Western and, often, so-called Christian nations, it is probably not surprising that their view that God has put human beings in charge of all of nature and they can do with the planet's habitats and biodiversity as they choose is based on classical economic thinking, and earlier Christian and Jewish religious beliefs. One must keep in mind as one explores where this now-dominant anthropocentric way of thinking has led our species that it is only as old as classical and Judeo-Christian thought and traditions—less than 5,000 years old.

Our species followed neither the current economic ideology and scientific principles nor religious beliefs and practices for 99.9% of its existence. In fact, there were and are other traditions and ways of thinking and perceiving this relationship that are diametrically opposed to the classical and Judeo-Christian view of the world. Hunter-gatherers, aboriginal peoples, and many Eastern religions and philosophies, as old as our species in some cases, saw humans as an integral part of the natural world, just another species among many, albeit an intelligent one. So, without further ado, let's explore a few of the mental adaptations that got us here.

If one starts off believing, as Ponting notes in *A New Green History of the World*, what the Christian Bible claims in Psalm 115—"The Heavens are the Lord's but the earth he has given to the sons of men"—or in Psalm 8—"Thou has given man dominion over the work of thy hands," it should come as no surprise that early and medieval Christians accepted that their god had given humans the right to exploit plants, animals, and the whole world for their benefit. In this view, Ponting writes, "Nature is not seen as sacred, but open to exploitation by humans without any moral qualms—indeed humans have the right to use nature in whatever way they think best. . . . Interventions and modifications to the natural world such as by extending cultivated areas, deforestation, and driving species to extinction and appropriating the planet's resources were readily interpreted as taking part in God's plan to improve upon creation."

Ponting notes that this belief system "produced a highly anthropocentric view of the world and had an enduring impact on later European thought. From such views it was easy to conclude that humans, the only creatures with a soul and a life after death (according to Christian belief), enjoyed a wholly different status from other animals and it was a small and logical step therefore to welcome greater control over the natural world and to believe that this would be pleasing to god."

On top of this religious ideology, modern secular writers added their views: René Descartes claimed that "he did not recognize any difference between machines . . . and animals, and animals were therefore machines"; Francis Bacon wrote: "Let humans recover that right over nature to restore the dominion over the world that had been lost with the fall of Adam and Eve in the Garden of Eden"; and Sir Isaac Newton saw the world as a "giant clockwork wound up by God, where the entrepreneur, merchant, industrialist and scientist become God's counterpart, skilled technicians who use the mechanical laws and principles that operate the universe to assemble the stuff of nature." Ponting notes that "in all of these views, nature is seen as a dead and mechanized world—a view that permits people to think of ecosystems and biodiversity as mere resources for human use."

> *However, while these religious and philosophical views continue to affect how modern Western Homo sapiens view the world, it was economic thought and cultural and social reorganization, in the form of modern societies and competing capitalist economies, that dominated the last two centuries that would change the lives of modern Homo sapiens every whit as much as the transition to agriculture and settled communities changed the lives of our hunter-gatherer forebears.*

This transformation, though, was going to hijack our personal freedom and the ability to make informed cultural, social, economic, political, and environmental choices. It appears to have set a trajectory we seem unable to extricate ourselves from, unable to rescind this license to rape the planet set in motion by religious and philosophical thought, seemingly unable to halt the mad rush to what might well turn out to be a sixth mass extinction.

If the food crisis, climate change, and population growth in prehistory drove the change to agriculture that began the destruction of Earth's ecosystems, so the change to an anthropocentric view of the world, combined with the social, cultural, and economic thought, reorganized the day-to-day life of modern Homo sapiens, leaving them bereft of freedom to choose and locking them into a global economic machine obsessed with growth, capable of stopping only when all the planet's natural resources have been exhausted, regardless of the cost to flora, fauna, or human life. An alien takeover of the planet could not be more complete.

Only in the last two centuries have societies developed that are controlled by the free operation of economic markets; in the last century, unrestrained free-market capitalism has subordinated all other considerations to profits, growth, and personal wealth accumulation.

For 99% of our species' existence, until 10,000 BCE, hunter-gatherers were mobile and healthy, ate abundantly and nutritionally from available food sources, yet owned neither food nor land and had few possessions. They were free, egalitarian, and owed allegiance to neither king nor ruler until the change to agriculture and settled communities that over time led to the early city-states and eventually to the first civilizations in Sumer, Egypt, India, and China. These early civilizations developed systems for food redistribution. People lived in small, self-contained units and engaged in semi-subsistence farming, using any surplus that was not seized to buy or barter for local craft-produced items. They had no contact with what we now regard as a market economy.

From the fifth to the fourteenth century, social and economic relations in Europe were based on agriculture and feudalism, a caste system based on land ownership, a way of structuring society around relationships derived from the holding of land in exchange for service or labor. Agricultural labor was performed by serfs, who were bound to the land and the local landowner; these small landowners, in turn, were vassals to overlords who were members of the nobility. Feudalism began to decline during the Middle Ages due to new opportunities for trade that triggered a shift from a land-based economy to a money-based economy, particularly in England. Feudalism was followed by capitalism.

Historians describe feudalism as an open form of political and economic domination; the king and his vassals held authority over serfs, and the extraction of wealth was direct and visible. When capitalism emerged, a market-based form of authority replaced this politically based form of authority. Serfs, now called "at-will employees," were free to work for any employer.

The clear and visible feudal relationships of vassal, serf, and land was replaced under capitalism by the relationships among investor, shareholder, company, employer, and employee, now veiled with an appearance of choice on the part of employees. They can work for whom they wish but can now be fired and left without a means of existence—a much less likely risk under the feudal system. Feudalism benefited everyone, creating an obligation for the powerful to help and protect the weak, giving all a chance to survive. Capitalism tends to benefit investors, shareholders, and upper and senior management, rewarding the latter with contracts and lavish compensation and remuneration packages, including golden parachutes. Employees are lucky to get a living wage.

The serf had a guaranteed job, a place to live regardless of harvest output, and a community to rely on. Under capitalism, an unemployed person without savings to tide him or her over, or relatives and friends to help, often falls into abject poverty, or, becomes reliant upon government social programs (which did not exist during the late 18th century, early 19th century when the UK, for example, was in transition from an agrarian to an industrial economy and as Charles Dickens documents so well in his novel *Hard Times*). Without these government programs, people can and did become homeless and destitute. Under capitalism, the worth of Homo sapiens is measured solely by the labor and talent each can offer. When that talent or labor is no longer needed—due to falling revenues, new technologies and equipment, or relocation of a factory or service to lower-cost countries like China, India, or Mexico—entire communities can be devastated economically, socially, culturally, and politically, as is now visible in many Rust Belt cities across the USA.

Over the last two centuries, then, Homo sapiens have come up with a capitalism energized by modern liberal economic thought, an adaptation that is at once a form of subsistence, a way to organize and distribute scarce resources, and a new way to organize the social, cultural, and political relationships that constitute a society. It is an adaptation that has the potential to render Homo sapiens as optional extras, a distinct possibility as more and more jobs are eliminated by technology or are soon taken over by AI-enabled robots, without the social and cultural umbrellas such as a living wage or non-work-based forms of compensation.

> *Capitalism has emerged as the new alpha species. Homo sapiens have demoted themselves to being mere consumers, beings that exist only for the purpose of acquiring, accumulating, and consuming, like machines on an economic production line, existing only to consume the output of an endless flow of products and services that capitalism extracts from the planet's habitats and biodiversity to keep its profits growing.*

From the cradle to the grave, at every stage of life (from birth, kindergarten, elementary school, middle school, high school, college, graduate school, work, marriage, raising a family, sickness, death and burial), like a cog in capitalism's liberal economic free market wheel, our species has reduced itself to being a dispensable and minor robot-like component, ceding control over most, if not all, meaningful social, cultural, economic, and political choices that hunter-gatherers, our forebears, once proudly held.

Under this twenty-first-century capitalist, global, free-market adaptation, Homo sapiens has allowed itself—one has to assume unintentionally—as an employee, a human slave, to become part of the output, mere products to be bought and sold on the open market for a profit.

With capitalism as another adaptation, our species has not only reduced itself to a potential optional extra at the individual level, but with current economic practices has also weakened the political arm of the government that once controlled the levers of power and sustenance. Instead, nationalized industries have been privatized, weakening the ability of employees to negotiate, and most of the control of the economy has been ceded to the "invisible hand" of the market, central banks, multinational corporations, global financial institutions, international institutions, and free trade agreements between competing nation-states, blind to biodiversity and environmental issues, that together comprise the bastions of economic power in this new system.

Under the prevailing economic system, the governments of nation-states have become mostly powerless against global economic institutions and must compete against one another to curry favor with them to attract investments. Indeed, under this not-so-new liberal economic free-market system, the balance of power has clearly shifted to the side of these multinational corporations, leaving governments weakened and people powerless.

> *This modern free-market system can essentially be viewed as an economic version of an AI: a global economic AI, or "eAI."*

Like an artificial intelligence, it is almost autonomous, mostly if not completely run by computers and completely unaware of and insensitive to the impact on humans, ecosystems, and biodiversity that result from its financial and trade inputs and outputs. It knows no way to account for the ecosystem services provided by Earth's habitats. It is choreographed by market movements, thinks in dollars, and communicates in market growth and profits. It has no capacity to care about the harm this growth is doing to our species, the environment, Earth's habitats, biodiversity, or the buildup of CO_2 in the atmosphere and consequent global warming; it never eats, sleeps, breathes, or drinks, so it couldn't care less about the integrity of the human food chain, clean drinking water or clean air, or the health systems that are under siege by the chemical, biochemical, biotechnology, and agricultural businesses that are high on profits and low on species survival and the health of Homo sapiens.

It is unaware of the disappearing pollinators and what's happening to the oceans: dead zones created by agricultural runoffs; dying coral reefs and the aquatic ecosystems they enable; islands of plastic forming from the 18 million tons dumped into them each year. It wouldn't know that plastics emit methane and ethylene, greenhouse gases, when they are exposed to sunlight and degrade; nor could it care that microplastics are toxic to phytoplankton which supplies 50% to 70% of the oxygen we breathe. It wouldn't know, too, that phytoplankton take carbon dioxide from the atmosphere and ocean to produce carbohydrates via photosynthesis, and zooplankton; neither does it care that microplastics eaten by plankton may impair oceans' ability to trap and transport CO_2 to the seafloor, enabling it instead to escape back into the atmosphere, thereby creating a positive feedback loop that can accelerate global warming and warming seas, that in turn further threaten our atmospheric oxygen supply with a potential phytoplankton die-off.

Its algorithms do not enable it to recognize whether that dollar comes from the sale of an endangered species, an illegally harvested human organ, humans sold for sex, or proceeds from an auction for human slaves. And finally, it has no ability to stop until and unless the market-driving natural resources are exhausted, or until humans stop buying what it produces, and stop feeding it their skills, labor, innovations, and technologies.

In this insatiable free-market system, Earth's habitats, biodiversity, and Homo sapiens are reduced to mere commodities to be bought and sold on the open market. It would seem that the only way out is to slay this capitalist-driven free-market economic monster, to starve it of resources, to stop consuming its products and services; and deny it access to our technologies, innovations, and human resources—and therein lies the problem.

Somehow the Homo sapiens species always seems to put itself in an unenviable position where it must be willing to kill itself in order to save itself, in this case to starve itself in order to starve this dragon, to live or die by the fortunes of its adaptations. Despite the example of Jesus Christ of Nazareth and Mahatma Gandhi, self-sacrifice does not appear to be something members of our species have been particularly good at. So we are stuck. And this phenomenon of becoming stuck in our own adaptation schemes has become a repeating pattern since we opted to become farmers.

And, just in case one happens to be among those worried about what newer adaptations like AI will eventually do to our species, take a good look at what its ugly economic twin, capitalism embodied in free-market systems, has already done. Witness how regulatory efforts to rein in unrestrained capitalism were brushed aside. Recall how attempts at maintaining human control of this species- , ecosystem- , and Homo sapiens-destroying economic engine now on autopilot have been abandoned. Why would it be any different with AI? We should be very afraid.

Homo sapiens, like that old lady who swallowed a fly in the Burl Ives ditty, has always been incapable of abandoning its adaptations before it's too late. When population increase and changing climate drove a prehistoric food crisis, we ate our way clear across the planet and drove megafauna to extinction. We couldn't stop ourselves from adopting agriculture, abandoning the clearly superior lifestyle our hunter-gatherer forebears led in harmony with the environment for millions of years. Agriculture soon became mechanized monoculture farming that, until the first half of the twentieth century, gave rise to food shortages and a perennial fear of starvation that stalked civilizations for millennia, and drove some to collapse.

Agriculture eventually led to genetically modified crops. Simple manure fertilizer morphed into deadly chemical fertilizers accompanied by equally deadly herbicides, insecticides, and fungicides. And we have seen where all that is taking us. Over time, irrigation raised the water table, waterlogging and changing the chemical composition and salt content of the soil, which led to salinization, abandoned fields, and more and more deforestation, and more and more soil erosion. Have we not seen the ravages that came with settled communities; the diseases that arose from the animals we domesticated; the resulting rigid social, cultural, religious, economic, political, and military schemes with their institutionalized inequality and racism, now exacerbated by capitalism, that we celebrate as "civilization"?

What's the point? you ask.

Here is the point. Burl Ives said it best in that ditty "I know an old lady who swallowed a fly." Burl who? I know. Burl Ives is not exactly up there with Tupac, Puff Daddy, and Justin Timberlake, and his old lady wasn't Beyoncé either; moreover, most Millennials and Gen-Xers will not have heard of this particular old lady, so let me share her story as Homo sapiens' pattern of adaptations is not unlike hers. Problem is, it didn't end so well for her. Having swallowed a fly to begin with, this old lady swallowed a spider to catch the fly, then a bird to catch the spider, and she continued swallowing progressively larger animals, each time with the hope that swallowing just one more would solve her problem. She went on to swallow a cat, a dog, a goat, a cow, and finally died when she swallowed a horse.

Homo sapiens' history, like the old lady's, has been a never-ending succession of ever more complex and environmentally more damaging adaptations to meet the same basic needs. Indeed, we might be just a few adaptations away from swallowing that horse.

What a far cry, then, are Homo sapiens' adaptations from the first two simple evolutionary survival adaptation rules: live within the Earth's ecosystems and live without destroying them and the other species they sustain. Live and let live.

But here is the rub, we once knew how to do that and had done just that for millions of years.

Just as population growth and changing climate drove a food crisis that may have forced an adaptation to agriculture and settled communities, with this twentieth-century adaptation—global capitalism and liberal economic free-trade markets—Homo sapiens are once more confronting an existential crisis, this time one of our own making. And we may well have finally come up with a subsistence adaptation that, if left unchecked, has the capability, the means, and the intent to drive our species and Earth's biodiversity to extinction. But don't worry if economic AI (eAI) fails; its technology twin, technology AI ("tAI"), is waiting in the wings.

Yes, Homo sapiens in the last 10,000 years have come up with numerous adaptations, but this is not an attempt to explore them all. However, there are a few more salient ones that can help decide whether the adaptations we have and are making will increase our chances for survival or drive us to extinction.

Organized religion turns out to be another significant adaptation that has had an outsized impact on our species. According to Wikipedia, organized religions emerged as a means to provide social and economic stability to large populations; to justify a central authority that possessed the right to collect taxes in return for providing social and security services to the state. Ancient empires such as India and Mesopotamia were theocracies with chiefs, kings, and emperors playing dual roles of political and spiritual leaders. Virtually all state societies and chiefdoms around the world have similar political structures, where political authority is justified by divine sanction.

But it is important to recognize that the period of religious history that has had the greatest impact on our species only began with the invention of writing, about 5,200 years ago. In contrast, our species has existed for millions of years.

Like many other Homo sapiens' adaptations, religion may once have served a good purpose—if serving as an arm of the state or religious hierarchy to subjugate the masses of humanity is a good purpose. As Karl Marx so aptly put it, "Religion is the opium of the poor." As Germany celebrates the 500th anniversary of the Reformation, it's worth remembering that Martin

Luther's Ninety-Five Theses were, among other things, a protest against the Catholic Church turning the path to salvation into a business opportunity. And now, 500 years later, the entire world seems to be joining Martin Luther in protest against the Church—this time for turning the saying of Christ to "suffer the little children to come onto me and forbid them not" into an opportunity to turn the church into a brothel for pedophile priests.

Today it's hard to justify the continued existence of religions purely on the basis of their charitable work, when NGOs and nonreligious philanthropic organizations manage to do that work just as well. But Homo sapiens' adaptations have a way of lingering on long past any usefulness, real or imagined, that they might once have had.

Among the not-so-great things about this adaptation are the endless series of wars it engendered and the suppression of women. In modern times, religious teachings have been hijacked to fan culture wars and weaponized as sermons of terrorism. Some have said that organized religion emerged as a means of maintaining peace between unrelated individuals, but events of the last few decades strongly suggest the contrary.

How has Islam and the teachings of the Prophet become the inspiration of multiple terrorist groups and organizations? And how does one account for what has been considered one of the most peaceful, if not *the* most peaceful, of religions, Buddhism, getting all huffy in Myanmar about a poor, despised Muslim minority, the Rohingya, sufficiently so as to want them gone?

These events are not unique to the twenty-first century but have persisted throughout history. Remember the Crusades. Indeed, political Islam has more than met its match in the atrocities of political Christianity. So, from an intelligent species survival perspective, the topic of this book, it's hard to see how any form of religion has really helped our species move closer to the survival side of the ledger.

But religion's greatest harm to Homo sapiens and to the planet might be the anthropocentric view of the world it inculcated in the minds of

Western thinkers, which led to the unprecedented current level of species extinction and plunder of the planet's ecosystems.

Today, many people no longer follow any form of religion; they do not believe in any kind of god, expect to go to heaven or hell, think humans have an immortal soul, or expect to participate in a resurrection or be reincarnated. Indeed, with the emergence of modern scientific thinking, many now accept Homo sapiens are made of the same stuff as the rest of the universe, particles and fields that obey the ironclad laws of physics, that we are a collection of biological cells passing electricity and chemicals back and forth as we metabolize free energy from the environment.

But alas! Five thousand years later, despite this new scientific understanding of our nature, the damage is already done. From crusaders marching into the Middle East, to missionaries taking diseases and death to the Maya, to justifying slavery and racism, to an epidemic of pedophile priests, to the rise of classical and modern economics and capitalism, religion stands out as one of the doozies our species has tried in its efforts to eke out a living on Planet Earth.

And what about science, innovations, and technologies? In some respects, science has become the new religion; its many research universities and organizations, the new churches; its professors and academicians, the new priesthood; its research papers, the new Bible. Science commands as much allegiance, devotion, and worship as any of the great religions that have captivated the mind of our species. Unlike traditional religions, though, Homo sapiens' achievements due to scientific breakthroughs have been remarkable, and science is rightly lauded for its many achievements and contributions to the advancement of knowledge, our understanding of the universe and nature, and to the way of life we have built. Yet when viewed from the perspective of the survival of our species, few if any of these wonderful achievements seem to have helped us return to an environmentally sustainable way of living within Earth's ecosystems and habitats without destroying them and the species they support, and have put our very survival at risk.

> *It was science and innovation that led to the chemical weapons of WWI and the dreaded WMDs of WWII that marked the beginning of a new era in Homo sapiens evolution—our ability to drive our own extinction.*

It's not clear why we tend to think of science and technology as our ace in the hole, our go-to place for all that ails our species. From its inception, science has been associated with war. Famous inventors such as Archimedes and Leonardo da Vinci focused on war-related problems, as well as many individual scientists. In their book, *Accessory to War: The Unspoken Alliance Between Astrophysics and the Military*, Neil deGrasse Tyson and Avis Lang tell how the methods and tools of astrophysics have been enlisted in the service of war. The overlap is strong and flows in both directions they claim. They observed that many of the same things that science cares about, the military does too, including multi-spectral detection, ranging, tracking, imaging, high ground, nuclear fusion, and access to space.

This science-war relationship was at first intermittent. Over time, however, it grew into an unholy alliance and intensified. Soon science became an integral part of the military, particularly during WWII, and has become anything but unspoken. According to Encyclopedia.com it became the first war in history in which the weapons deployed at the end of the war were significantly different from those that opened it. "Many of the new developments—radar, jet propulsion, ballistic missiles, the atomic bomb—were developed largely or entirely in the course of the war."

It wasn't long before more than half the research and development done in the United States—government, corporate, and university-based—was military. The military services, or defense-related agencies such as the Atomic Energy Commission, became the principal supporters of research in nuclear physics, computers, microelectronics, space, and other scientific and technical fields. Then in 1958 President Dwight D. Eisenhower signed into law the National Defense Education Act, which funded graduate study in science and technology for thousands of American students. More broadly, military funding supported a significant percentage of university research during the Cold War and helped to shape these institutions.

This alliance had grown so intense that in his farewell address in 1961, Eisenhower referred to it as the "military-industrial complex" and warned, "Public policy could itself become the captive of a scientific-technological elite." Given the role of "Big Tech" today, his fears were not unfounded.

Indeed, the scale of environmental problems that have been created as a consequence of all this scientific and technological progress is unprecedented, and of a complexity that defies solution. Futurists like to point to the exponential increase in knowledge and the innumerable creature comforts that became possible in the last century but fail to note how many of these same achievements have begun to put our existence at risk. Of course, their hope is that we will innovate our way out of such threats, yet no one knows for sure whether that will be possible.

It would seem that each successive adaptation Homo sapiens have tried has dug us deeper into a state from which it has become extremely difficult, if not impossible, to extricate ourselves. Yet all is not lost. Some events and changes taking place may well be an indication that at least some Homo sapiens have begun to slowly walk back some of those brash adaptations we rushed to make. Efforts to curtail the spread of WMDs and current attempts to return to renewable energy sources as our forebears used for millions of years must count among these efforts. But will they be enough? No one would dispute that much, much more is needed.

Now consider that for more than 99% of our species' existence, and with none of the science and technologies modern Homo sapiens so highly esteem, our forebears managed to live sustainably and in harmony within the planet's ecosystems and shared its habitats with all its flora and fauna.

Scientists are beginning to note that the hunter-gatherer lifestyle may well have been the healthiest, most successful, politically free, and economically independent and secure lifestyle our species has ever known.

For all but the last 10,000 years, our forebears had no need of any of the seemingly indispensable adaptations—products and services, technologies, innovations, rulers (kings, queens, presidents, and governments), religions, medical science and healthcare systems, formal education systems, writing, economic and monetary systems and their market mechanisms—that comprise modern Homo sapiens' civilization in the twenty-first century.

And if the goal is survival through attaining evolutionary maturity, has anything in the list above that took our species away from a life within the planet's ecosystems increased our ability to survive, or increased the chances of our extinction? Build yourself a simple three-column Survival, Extinction, and Maybe table. In the rows, assign each Homo sapiens' adaptation to one of these categories to see where you come out on this issue.

Consider, too, that hominins originated from the Hominidae family, which includes the great apes. There were six extant species in four genera until a seventh, Pongo tapanuliensis, or the Tapanuli orangutan, was recently discovered in Indonesia's island of Sumatra. The extant species of the great apes include orangutans (now three species in genus *Pongo*), gorillas (two species in genus *Gorilla*), and chimpanzees (two species in genus *Pan*). There are multiple hominin species, Homo sapiens (humans) being the last, in genus *Homo*, according to the Encyclopedia of Life. Yet none of these other great apes encountered anything remotely resembling the survival challenges described herein, except for those Homo sapiens inflicted on them through habitat destruction and wildlife poaching and trading.

So, the 5-million–6-million-year-long question is why Homo sapiens encountered such adaptation and survival challenges while the other great apes didn't? How is it that they managed to continue to survive in accordance with the evolutionary survival adaptation rules we noted at the beginning of this chapter, while Homo sapiens seem so ready to abandon them?

The perspective taken in this book is that the genus *Homo* and the hominins that followed comprised a nascent breed of intelligent species that seemed to have lacked the preprogramming of how to instinctively live within that evolutionary survival framework and needed to discover these rules on their own just to survive and to learn how to sustain themselves within Earth's ecosystems without destroying themselves and 4 billion years of evolution. A successful outcome would imply that an intelligent species would have managed to attain to that level of evolutionary maturity (i.e., survival past the point of self-extinction) that all nonintelligent species seem to have attained by default.

The difference being, of course, that if an intelligent species did manage to get there, it would have gotten there by choice in spite of all the adaptation setbacks we observed above, caused in no small part by that very intelligence that made it different and guided its choices. Here, then, is a reason to hope that, despite all the struggles our species has encountered, there is still a chance, if it is not already too late and if we so choose, to get to that level of maturity.

This perspective reveals how and why Homo sapiens seem to have gotten it so wrong. Why was it we got off the evolutionary survival-driven adaptability trail while most every other species remained? This pattern of evolution reflects the growing pains of a human infant as it passes through the different stages of life to maturity and adulthood, and some, as we know, never make it all the way there. Today, Homo sapiens are becoming increasingly aware that our species might be in some kind of evolutionary growth phase and might well be in an adolescent, if not still in an infant, stage.

Homo sapiens' challenge, then, is to discover before it's too late what it wants to be when it grows up; what it now needs to do to make it all the way to evolutionary maturity. Simply put, does it want to survive, or does it want to go extinct like the dinosaurs?

One final note to ponder before we leave this topic is the role climate change might have played in shunting our species down the path that led to agriculture and potentially all that followed.

As noted in a prior chapter, climate change seemed over geologic time to have choreographed the options from which our species had to choose and thereby the outcome. So how much did a food crisis driven by climate change and aggravated by overpopulation nudge early humans down the road to agriculture? It seems to make sense, doesn't it, that an emergent intelligent species absent the built-in preprogramming that would have made it stick with the rules of adaptation, rules it might not have doped out as yet, would come up with its own solution, perhaps Homo sapiens' very first adaptation of making a living outside of Earth's ecosystems—hence, agriculture.

The key issue here is that Homo sapiens were able to use that nascent intelligence to choose a solution, even if, with hindsight, we can see it's not the right or best solution. From this perspective, agriculture may not be too shabby a solution after all, in spite of all the blind alleyways it has taken us down so far.

So, if agriculture was the result of an untutored nascent intelligent species' attempt to solve a seemingly existential food crisis, is there a way to do agriculture that largely respects the adaptation rules now that Homo sapiens know them? Some scientists say there is—a return to sustainable agriculture is possible. Yet, in pursuit of economic growth and profits, that is not how most agriculture is done. It is this choice, this preference, then, that is at the root of all Homo sapiens' adaptation failures.

If our species were to fail to survive, it would not necessarily be because we have made bad choices; after all, isn't making poor choices a necessary part of an intelligent species' maturing process? Perhaps it is, instead, our inability to abandon those choices once we detect our mistake. Unfortunately, instead of correcting them, we seem to take a strange delight in doubling down on these apparent missteps.

Why have Homo sapiens' adaptations gone so horribly wrong? Whether they were social, cultural, religious, economic, political, or scientific, it's been hard to identify a set of these adaptations since the move to agriculture that has facilitated a way of life that does not result in the destruction of the planet's ecosystems and biodiversity—a non-negotiable evolutionary prerequisite for Earth-life survival. Instead, these adaptations have etched a swath of death and destruction across time and across the planet, its habitats, and its species, and they now threaten the Homo sapiens species itself. Unless and until we change course and prioritize survival, then we already know how AI, genetic engineering, and many brighter and shinier innovations in the works may well end—just like plastic has today. As *National Geographic* observed: "We made plastic. We depend on it. Now we are drowning in it." And so it has been with all we have made.

If you ever watched the TV series about air crash investigations titled *Mayday*, you would be familiar with the excruciating attention to detail and enormous effort of the NTSB and others to ferret out the root cause of a particular crash. They are thorough because hundreds of lives are at stake. Yet when it comes to taking action to investigate what's going wrong with Homo sapiens' adaptations, not many—not the NTSB, and not even NASA—seem to be too worried that Spaceship Earth is showing signs of crashing and burning, even though not just hundreds but billions of lives are at stake.

Fortunately, there is no need for the NTSB or NASA in this case, as evolution has already identified the problem. The reasons are plain and open for all who care to read nature's evolutionary survival patterns. Earth's species and habitats have been splayed and scattered across the Earth like the parts of that jumbo jet strewn across the Alps after being deliberately flown into the mountainside by its pilot. That crash sadly was no accident. It was not even pilot error. That crash was caused by a pilot who knew all the rules yet broke them. And Homo sapiens, like that pilot, mentally deranged or not, has broken all the evolutionary rules for Earth-life adaptation and survival.

> *Homo sapiens' adaptations are failing because they are not survival-driven, but driven instead by profits, greed, and pursuit of economic growth. Indeed, Homo sapiens' adaptations will continue to fail unless and until they become survival-driven.*

In the next chapter we begin to look at whether Homo sapiens' ability to innovate fared any better than its ability to adapt, and whether that ability can enable our species to avoid civilizational collapse and survive. But be warned, the innovation and adaptations of earlier civilizations were not enough to save them and may not be enough to save us.

10.

Innovation

FROM APE-MAN TO SPACEMAN IN TWO MILLION YEARS—TO . . . ?

"THERE IS ONLY ONE CORNER OF THE UNIVERSE where we know for sure that the laws of nature have conspired to produce a species capable of transcending the physical bounds of a single life and developing a library of knowledge beyond the capacity of a million individual brains which contains a precise description of our location in space and time. Two million years ago we were ape-men. Now we are spacemen," writes Brian Cox in his book *Human Universe*.

From ape-men to spacemen in 2 million years, and by the look of things, Homo sapiens is just getting started. According to futurists, the twenty-first century will see our species break into an exponentially accelerating innovative sprint that will make prior rates of innovation seem eternally slow. Indeed, it's hard not to be impressed by the remarkable achievements of our species given where it all started. Our early forebears departing Africa had no idea what they were starting. The chart below underscores how the rate of innovation accelerated over time.

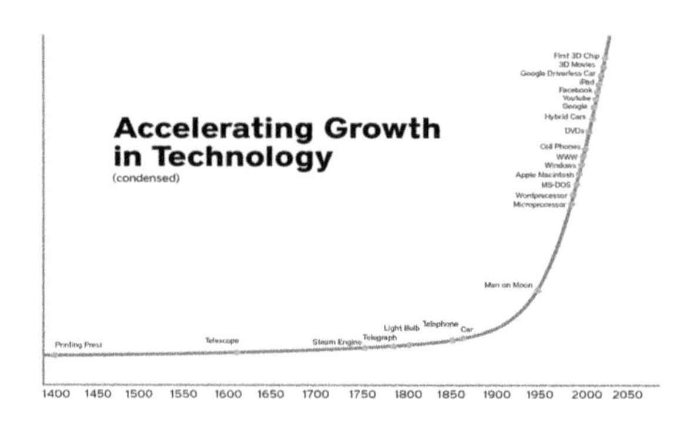

Fig. 10-1 Technology growth
(Source: http://renewableplus.blogspot.com/2018/09/does-man-drive-technology-or-does.html?m=1)

For the first 66,500 years, writes Chris Herd in an article titled "A Brief History of Humanity and the Future of Technology," published in *Peripheral Foresight* in 2016, Homo sapiens initially relied on verbal transfer of knowledge, which put significant limitations on bandwidth. One could know only as much as one had the ability to remember. All the information one possessed was shared history and conventional wisdom inherited from predecessors. As soon as anything was forgotten, it was lost forever. Homo sapiens nonetheless evolved from hunter-gatherers to farmers to city dwellers between about 10,000 BCE and AD 500, including all the requisite technological advances, which is tremendous progress, and a massive transition in a relatively short period of time. However, the real progress that led to modern technology gathered pace only with the ability to record knowledge. Today, futurists predict Homo sapiens are set to accomplish in the next twenty years what would previously have taken 200.

A decade ago, smartphones didn't exist. Three decades earlier, before the first personal computers arrived, almost nobody owned a computer and the commercial internet barely existed. Today, advanced technologies such as computers, genetics, nanotechnology, robotics, and AI are progressing at an accelerating pace.

But are we so immersed in today's technology that we have forgotten what has been accomplished in a single lifetime? We have gone from the Wright brothers' four brief flights on December 17, 1903, to landing men on the moon on July 20, 1969, and from Ernest Rutherford's splitting of the atom in 1911 to the bombing of Hiroshima in 1945 and the Cuban Missile Crisis in 1962, when Nikita Khrushchev and John F. Kennedy came close to potentially ending our entire species and the fruits of almost all 4 billion years of evolution. Now it's Donald Trump and Kim Jong Un's turn to rattle the world.

Is our view of our inventiveness so myopic that we fail to take in the wider context of what else we've done in this amazingly short interval? The charts below show the results of some of the innovations we tend not to want to talk about as much. They speak for themselves. Were we so intent on our technological innovations that we neglected to notice that we were simultaneously digging ourselves into a potential extinction ditch so deep that getting out of it might be beyond our innovative capabilities, or did we just not give a hoot?

Today, within limits, we can forecast the weather, link climate change on both sides of the Atlantic, anticipate the short-term climate change impacts of El Niños and La Niñas, and warn of impending tornadoes and, to a lesser degree, earthquakes and tsunamis. We have also mostly acknowl-

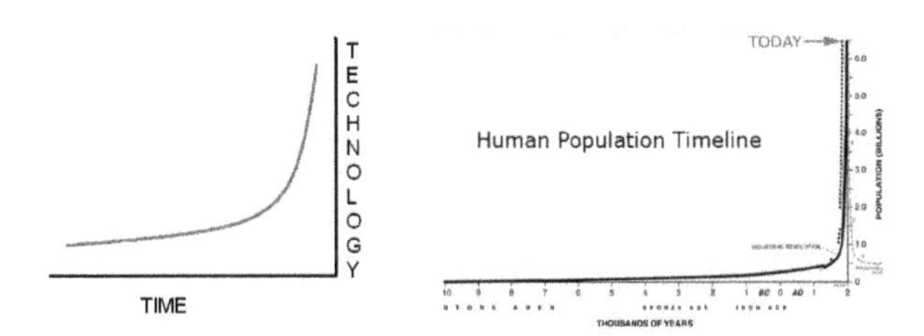

Fig. 10-2 Technology growth vs. time vs. population
(Source: The Bulletin of the Atomic Scientists - http://thebulletin.org)

edged that man-made global warming is in progress and we still can, time permitting, take corrective action to mitigate some of its potential devastating effects if we so choose. AI innovations can diagnose potential heart issues better than trained physicians, and cars are starting to drive themselves.

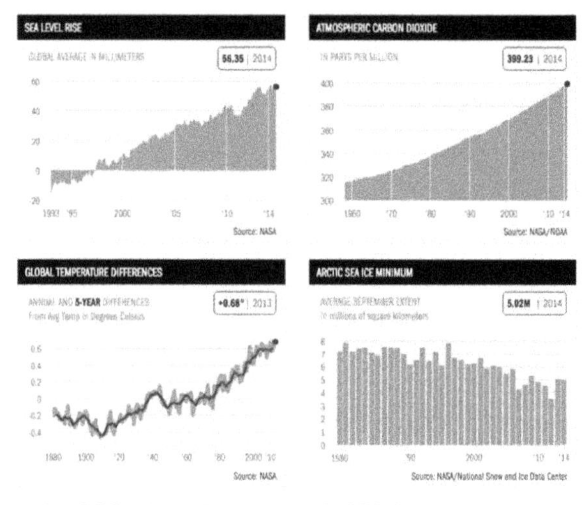

Fig. 10-3 Sea Level Rise, Atmospheric CO2, Global Temperature-Differences and Arctic Sea Ice Minimum

(Source: The Bulletin of the Atomic Scientists - http://thebulletin.org)

Futurists like Ray Kurzweil are predicting that by the 2020s self-driving cars will begin to take over the roads, and people won't be allowed to drive on highways; most diseases will go away as nanobots become smarter than current medical technology; and normal human eating can be replaced by nanosystems. By the 2030s, virtual reality will begin to feel 100% real, and we will be able to upload our mind/consciousness by the end of the decade. By the 2040s, nonbiological intelligence will be a billion times more capable than biological intelligence (aka us). Nanotech foglets will be able to make food out of thin air and create any object in the physical world on a whim. And by 2045, we will multiply our intelligence a billionfold by linking wirelessly from our neocortex to a synthetic neocortex in the cloud.

Aaaaaamazing! Great! Brilliant! Fantastic! Except that none of the above addresses the root cause of Homo sapiens' problems (the inability we saw in the last two chapters to live in a way that doesn't destroy Earth's ecosystems, habitats, species, and Homo sapiens themselves). And until relatively recently there has been little to no research, and for sure no innovations and/or technologies, that target this root cause.

While most AI experts worry less about machines rising up to overthrow their creators than about their making a fatal mistake, philosopher Nick Bostrom is among those who believe it's entirely possible that artificial intelligence could lead to the extinction of Homo sapiens. He also believes that much more research has been done on dung beetles and *Star Trek* than on the risks of human extinction.

The study of existential risks is really just getting off the ground. Research centers that claim to be doing research in this area all seem focused on identifying which technological advances pose risks that could wipe out humanity—or at least end civilization as we know it—and what could be done to stop them. They seem to forget that Victor Frankenstein, creator of the Creature, did not survive. What they appear not to be focused on is the root cause of all Homo sapiens-driven self-extinction-capable innovations and adaptations—the seeming inability or unwillingness to live within the planet's ecosystems and mature as an emergent intelligent species.

And yet the idea of Homo sapiens driving themselves to extinction by their own innovations is hardly new; it goes all the way back to Mary Shelley's novel *Frankenstein.* The real takeaway there was not how to control the Creature; it was that one should not invent technologies that can kill you. The point is that an intelligent species that has matured past the point of self-extinction would have been able to do so only by never inventing a technology that could have driven its own extinction to begin with. Not allowing its ability to innovate to outpace its ability to mature is a prerequisite—a topic we will explore in the chapter on evolutionary maturity. Received wisdom considers existential risk problems waiting to be solved with new technology, like diseases with new vaccines. None of this research appears to be looking at the real problem—Homo sapiens.

That many of Homo sapiens' innovations bring benefits, no one can deny. At issue here is when they are held up as the panacea for all that ails our species, or as indication that the world is getting better. It has never been about the world getting better; it's about Homo sapiens maturing beyond the point of self-extinction. And far from getting better, the world has continued to get worse.

While futurists are busy churning out the next iteration of these wonderful technologies, others are equally busy devising ways to put the current and previous generations to devastatingly destructive use. Like the orchestra on the *Titanic*, the innovators and their industries keep playing on, perhaps hoping bigger, better gizmos will distract us, keep us from noticing the sea level is creeping up, the oceans are heating up, the atmosphere is becoming increasingly hotter and more polluted, and the water and food chains are slowly being laced with carcinogenic agents and other elements just as harmful.

Our best minds talk about how innovations in technology extend our reach and push back the frontiers of scarcity. Seen through the lens of technology, they'd say, few resources are truly scarce; they are just temporarily inaccessible. In his book *Abundance: The Future Is Better Than You Think*, Peter H. Diamandis invites our species to "imagine a world of nine billion people with clean water, nutritious food, affordable housing, personalized education, top-tier medical care, and nonpolluting, ubiquitous energy. Building this better world is humanity's grandest challenge."

But there is also another side to this future, a dark side that some Techno sapiens unfailingly ignore. That side includes terrorists with miniaturized WMDs and drones to deliver them instead of yesteryear's crude IEDs (improvised explosive devices); rogue states aided by militarily strong nations that destroy their own people, driving regional and global instability; refugees and migration, with one in every seven people on the move; threats of nuclear war and the apparent inability or unwillingness of nations to settle differences peacefully; shameful income and global wealth inequality, with eight men owning as much as half of the world's population; and emerging worldwide anxieties and threats to peace driven by any number of the preceding. The history of our species shows that notwithstanding the best technologies of their time, human civilization after human civilization collapsed.

Dr. Diamandis tells the story of why aluminum disappeared soon after it was discovered more than 2,000 years ago. Though it resurfaced, its use became commonplace only with the invention of electrolysis in 1886. His takeaway is that aluminum was scarce only until a cheap and relatively inexpensive innovation came along to extract it from bauxite. Once again technology had pushed back the frontiers of scarcity.

However, there are a few other takeaways from this story as told by Pliny the Elder in book 35 in his *Naturalis Historia*. It's also about human greed, which technology does not solve. It's about how a powerful Roman emperor (Tiberius) sought to monopolize precious metals and control their value and how the poor lost their livelihood once again, and even their lives, as in this case, the poor man was beheaded. The tale is not unlike the way rich bankers are alleged to have fiddled with the LIBOR (London Interbank Offered Rate) rate in the UK, when many midlevel bankers lost their livelihood. The moral, if there is one, is that, as the history of innovation shows, it is always the rich and powerful that benefit most, and the poor are considered prosperous because it takes only six months earnings to be able to afford a prior generation iPhone to google the world's trove of knowledge, unaware their privacy is being mined and sold by Big Tech even as they connect on Facebook. Yes, that washes down well with stale, leftover coffee, doesn't it?

What if there were an XPRIZE for better ways to govern to ensure economic fairness and an even distribution of a country's resources? — XPRIZE designs prizes in the domain areas of Space, Oceans, Learning, Health, Energy, Environment, Transportation, Safety and Robotics and has awarded seventeen last count. What if there were an XPRIZE to find ways to eliminate inequality? What if there were an XPRIZE to address the tensions in the Middle East? What if there were an XPRIZE to come up with solutions to the problems driving migrant crises worldwide? What if some of the superwealthy Techno sapiens were to use their money to address the inequalities and economic neglect that breed unrest and terrorism among minority populations? What if there were an XPRIZE to motivate Homo sapiens to solve its biggest problem—how to mature beyond the point of using all these brilliant innovations and technologies to destroy one another, Earth's species, habitats, and environments?

Today we can launch an intercontinental ballistic missile from a submerged submarine that could destroy an entire nation, ignite a global war, and probably end our species. By 2050, climate change will push European cities toward a breaking point. A third of the world's crops depend on bees and other pollinators, and by 2050, bees might be among the first species we lose. By 2050, too, according to the World Bank, climate change could force more than 140 million people from their homes. And depending on the amount of climate-warming greenhouse gases are emitted during the next few decades, between 2047 and 2069 mean annual climate on Earth will slip past the most extreme conditions experienced in the past 150 years. Indeed, it is estimated that up to 5 billion people will face an "entirely new climate." Now, who is working to find innovative solutions to these problems?

Regrettably, inherent in each of our amazing innovations is the power to do good and the power to do evil, to build up and to tear down, to give life and to end it, to nurture and protect our species or drive it to extinction. And running through that brief period of history in which we accomplished all those marvelous things is a trail of death and destruction of other species, Earth's habitats, and members of our own species. In the slightly paraphrased words of Carl Sagan, "We are an interesting species. We're capable of such beautiful dreams, and such horrible nightmares."

This seeming inability to stop our species from committing harikari underscores the fact that while our amazing ability to innovate has turbocharged the advance of our species, and might yet turn out to be essential to our survival, it will never be sufficient by itself to ensure Homo sapiens' survival. It is this sad prospect that continues to cast deep doubts on any hope one might have for a technologically-driven survival strategy, whether here on Earth, or anywhere else in the galaxy.

Cassius once said to Brutus, "The fault, dear Brutus, is not in our stars, but in ourselves," and, indeed, the problem, dear Techno sapiens, is not in our ability to innovate, but in our inability to mature; not in our technologies, but in us. It's precisely here that innovation is desperately needed and remains conspicuously lacking.

Unlike any other species so far, Homo sapiens evolved the capabilities to change the outcome of nature, both for better and for worse. Some uses of this ability have been downright scary. The problem our species must confront is not an inability to innovate but an unwillingness to mature past the point of self-extinction, addressed in a later chapter.

However, those who continue to pin their hopes on our species' ability to adapt and innovate itself out of any existential threat using technology must accept that our ability to survive as a species will have to come from an entirely different quarter—an as-yet-poorly-explored innate capability. It is a capability that evolution has planted deep within our species along with our ability to adapt and innovate. It's the next in that set of survival patterns (adapt, innovate, mature, migrate, survive, or go extinct) we have been examining—the requirement to mature, and the freedom to choose survival over extinction. Almost like an inactive gene waiting to be turned on, Homo sapiens' ability to choose and reach for evolutionary maturity must be activated.

But like the evolution of most things, Homo sapiens' innovations, like batteries, are not included, nor are they needed to access and use this ability. All that is needed is for Homo sapiens to prioritize survival of Earth's species, ecosystems, and themselves over extinction, choices we have so far been reluctant to make as a species, though less so as individuals. The real message in the emergence of AI and other existential threats is, perhaps, that our species is running out of time to make those choices.

Long before Homo sapiens developed its ability to adapt and innovate, evolution had been busy doing some innovating of its own: innovating a planet with sustainable-capable habitats that can last for billions of years in order to ensure the evolution and sustainability of complex animal life and emergent intelligent species like Homo sapiens.

Like the ultimate long-term investor, billions of years ahead of everyone and everything else, evolution began preparing by plowing resources into the galaxy. First it selected the right type of star that can last the billions of years it would take for life to go from slime to ape-man to spaceman, and then, hopefully, to evolutionary maturity; ensured it was a mineral-rich, rocky planet with a moon to stabilize its orbit and with outer-orbit planets to shield it from incoming threats; placed it in a stable orbit in the habitable zone of its star; supplied it with running water on its surface and an oxygen-rich atmosphere with an ozone shield to block DNA-destroying UV and gamma rays; and provided a temperature range suitable for complex animal and human life. Some scientists think finding another planet with a significant fraction of these Earth attributes is unlikely.

As if that were not enough, evolution went on to ensure that the chosen planet would have access to a perpetual and comparatively safe source of renewable energy, clean air, clean water, lots of organic, natural, renewable food sources, and natural habitats filled with renewable and recyclable materials that can be used and reused from one generation to another. Then it went on to populate it with a menagerie of species, enabling one, Homo sapiens, unlike all others, to evolve the ability to determine its own evolutionary destiny by choosing to avoid potential self-extinction, grow to evolutionary maturity, and go on to populate the stars, or choosing to go extinct, taking 4 billion years of evolution along with it.

Sounds like a nice yarn, but most of the above has already occurred, except for the bits about extinction or growing to evolutionary maturity and populating the stars. What a track record—one that even futurist Ray Kurzweil can envy. If I were a futurist or a big-time market investor like Warren Buffett, my money would be on the rest also happening. Which futurist are you betting on?

But haven't you noticed? It is so easy to miss this most important of points. The focus of evolutionary innovation appears to have been on one thing and one thing only: species evolution and survival, including our own. Evolution's innovations targeted species survival. The motivation and drivers of evolutionary innovation were also exclusively species evolution and survival, and as such evolution came from a place that elevated habitat sustainability and species survival above all else. The outcome evolution appears to be seeking is for emergent intelligent species, such as Homo sapiens, to avoid self-extinction, reach evolutionary maturity, and go on to populate the stars. Evolution seemed even willing to "go all in" and bet 4 billion years of evolution that a lone emergent intelligent species endowed with freedom to choose its own destiny would ultimately choose evolutionary maturity over self-extinction.

However, like any smart investor, evolution appears to have a few moves left should things go south with Homo sapiens. With at least another 4 billion years left before Earth's warranty expires, evolution can start all over with another, hopefully more intelligent, species should there be a Homo-sapiens-induced global extinction. I seriously doubt any of our wiliest investors, given a similar choice, would have bet on Homo sapiens as evolution appears to have done.

Unlike evolution's, Homo sapiens' innovations are ultimately driven by only one outcome—profits. Whereas evolution's innovations are motivated, driven, and directed to ensure habitat sustainability and species survival, Homo sapiens' innovations are ultimately motivated, driven, and directed to ensure ever-increasing economic growth and a lopsided market share. Whereas evolution ultimately seeks species maturity and the avoidance of self-extinction, Homo sapiens are ultimately seeking to amass personal, corporate, and global wealth, economic growth, and, wherever possible, monopolistic control of scarce resources. And since we are willing to stop at nothing to attain these outcomes, even if that entails devastating Earth's habitats, heating its oceans, killing off its wildlife, and razing its forests, we are destroying any potential shot we might have at avoiding self-extinction and attaining evolutionary maturity.

The contrast here couldn't be any starker. For evolution, drilling for fossil fuels for energy and pumping CO_2 into the atmosphere would have been a nonstarter. For evolution, polluting the atmosphere, heating the oceans, scarring the Earth in search of mineral resources and fossil fuels, and destroying habitats and the species they support in order to do so must be an anathema. For evolution, wantonly destroying one's own species, which evolution has gone to such lengths and cost to evolve and sustain, is probably enough to write Homo sapiens off and start all over again—if evolution's and Homo sapiens' roles were reversed. Yet the fact that our species is still around leads one to suspect there might be more to this galactic intrigue than meets the eye. It doesn't require evolutionary intelligence to have expected Homo sapiens to be an evolutionary accident waiting to happen, that such a potentially disastrous outcome was highly likely if not inevitable.

If the outcome had been so predictable, why, one might ask, would evolution evolve a species and essentially give it the keys to the planet, knowing that there was a high probability of it all ending in self-extinction? Why would evolution have turned species' survival and habitat sustainability over to an alpha species like Homo sapiens, with no a priori guarantee of a successful outcome?

Might it not be precisely because hominins are the only species, as far as is known, with whom evolution appears to have decided to share control of their evolutionary outcome, knowing that outcome can potentially bring an end to more than 4 billion years of evolution on Earth; that, if successful, evolution will be jump-starting an even more exciting evolutionary phase, wherein for the first time evolution would be partnering with a species to chart the next phase of its evolutionary journey? What next phase? you're wondering.

In Chapter 5 we called this new potential evolutionary phase "species-driven evolution," and we wondered aloud whether species-driven evolution might not be a prerequisite for intelligent species survival and extraterrestrial intelligent species evolution. So, discussing each of these concepts in turn, let's look first at what species-driven evolution might mean.

Can one have an intelligent species if it is unable to choose its own path and destiny? Clearly not, as it is this "intelligence" that makes it different from all other species not so endowed. Once one entertains the notion of an intelligent species that is able to go its own way, one whose brains are no longer stuck running the same instinctive base routines that execute and control the behaviors, responses, and outcomes of nonintelligent species, then one would expect this intelligent species to also be capable of making choices that result in evolutionary maturity and avoidance of self-extinction. One has to believe that because such a species has evolved, its very existence must be a prerequisite to something else, and that something else may well be species-driven evolution.

Species-driven evolution might be an easier concept to grasp today than just a couple of years ago. Indeed, it's not only evolution that is experimenting with turning over some control to species it has evolved; Homo sapiens is turning over control to some of its own inventions and appears to be as apprehensive as evolution is, if one can characterize evolution as being apprehensive. Passengers routinely risk their lives in airplanes flown on autopilot (and some literally did recently in two Boeing 737 Max 8s), and we are getting ready to hand the steering wheel over to self-driving cars (even though some people have lost their lives there too). Somehow the fear that these could go horribly wrong hasn't tempered our enthusiasm for pressing on. Clearly, at least some of us have concluded that the benefits outweigh the risks, just as evolution might have concluded that accepting the risk is indispensable to getting to the next phase of intelligent species evolution.

So here is what might be driving this entire process. If a potential evolutionary outcome is the spread of intelligent life across the galaxy, and if the peculiar circumstances that allowed complex animal and intelligent species life to evolve on Earth is unlikely to exist on most other planets, as some scientists are beginning to suspect, then one possible way, if not the only way, to achieve that goal would be to export fully evolved and mature intelligent Earth life-forms to suitable planets.

And if that were true, it would make sense that evolution would be interested in evolving emergent intelligent species like Homo sapiens on Earth first, then using these to spread intelligent life to other planets. If one of these intelligent species managed to make it past self-extinction and reach evolutionary maturity, evolution would then have the means to begin populating the galaxy with mature emergent intelligent life, rather than starting from scratch and waiting 4 billion years all over again—a handicap, perhaps, even evolution could use.

But notice the emphasis on evolutionary maturity. We defined evolutionary maturity in Chapter 6 as a survival plateau, a state that indicates an intelligent species has matured beyond the likelihood of self-extinction. Evolution appears to be uninterested in the indefinite survival of species that seem unable or unwilling to get to evolutionary maturity. The reason for this is not so hard to understand, is it? Few if any Homo sapiens mothers are likely to entrust the care of their human babies to an unproven robonanny any more than evolution would be willing to entrust the universe to immature species, even if intelligent.

There is no point in spending 4 billion years preparing a planet to support complex animal life and evolve and groom an emergent intelligent species only to see it descend into self-extinction, potentially taking those 4 billion years of evolution with it. So it's not surprising, then, that evolution would not only limit expending resources on species that seemed unwilling or unable to evolve to maturity, but also would have a few, and one can think of at least three, built-in fail-safes to ensure that failing species remain trapped on their home worlds, unable to migrate to the stars and spread their immaturity across the galaxy. These fail-safes are (1) such species will more than likely drive themselves to self-extinction; (2) potential global climate change will drive them to extinction; and (3) unbelievable galactic distances separate Earth from any potential Earth 2.0.

Clearly, then, evolution has little interest in enabling the galactic battles imagined in *Star Wars*, or the ceaseless skirmishes between the Federation, the Cardassians, the Klingons, and the Vulcans of *Star Trek* fame.

Armed with the knowledge that evolution might well be engaged in nurturing species on Earth that could become the seeds from which intelligent life is sown across the stars, the meaning of extraterrestrial evolution becomes, perhaps, less obscure. Evolution may be planning to "partner" with intelligent species nurtured to maturity on Earth to continue the intelligent species evolutionary process across the galaxy in a way that is perhaps not unlike the way Homo sapiens are beginning to partner with AI products. Indeed, what bees have been to flowers, Homo sapiens might well become to exoplanets.

So why would evolution need to partner with an intelligent species to launch intelligent-life-based extraterrestrial evolution, potentially the next phase of Homo sapiens' evolution? It did quite well on its own on Earth, didn't it? Well, yes, but as noted above, while many scientists believe that microbial life might be abundant in the universe, jump-starting intelligent life on another planet or in another solar system using migrating mature intelligent species, such as Homo sapiens, has the potential to be much easier and faster than starting all over again from scratch on each planet. But that can hardly be the whole story, can it? To explore this question more completely, one must revisit and expand upon the concepts of evolution-inspired species-driven evolution and species-driven innovation touched on in Chapter 5.

If Homo sapiens were to avoid self-extinction, attain evolutionary maturity, and set out to migrate to other planets and solar systems, among the many issues would be how to get there. Homo sapiens probably won't have such capabilities for at least another 100 to 200 years, and getting to that level of technology might be where evolution-inspired species-driven evolution and species-driven innovation come in. Assuming identification of a suitable Earth 2.0 capable of sustaining Homo sapiens, migration issues to be overcome are many, but two in particular pose significant challenges—interplanetary and interstellar distances, and Homo sapiens' biology, sustainability, survivability, and suitability for undertaking such migrations.

These areas, were they to become a species priority, would be examples of, on the one hand, evolution-inspired survival-driven innovation to attain the technologies needed for interstellar travel, and on the other, species-driven evolution to enable potential biological enhancements that might be necessary to survive the rigors and distances of interstellar travel. But there just might be another way to migrate to the stars that one would be unwise to rule out.

Homo sapiens' ability to flee out of Africa to survive did not depend upon their innovations and technologies, and if modern Homo sapiens were to attain evolutionary maturity, their ability to flee to survive might not either. Falling sea levels due to an ice age that facilitated early Homo sapiens' migration out of Africa was probably evolution's way of enabling migration when seafaring and ocean-crossing innovations and technology were multiple millennia away.

One can't help wondering if that migration out of Africa would have occurred how and when it did, or whether it would have occurred at all, but for the sea-lowering effects of that ice age, and the earlier creation of a land bridge between Africa and Arabia by plate tectonics. As noted earlier, timing was everything. One also can't help wondering if another seemingly natural phenomenon might not serendipitously facilitate the next survival-driven species-wide migration, this time no doubt an interplanetary migration, with or without the advanced technologies and innovations we think it would require.

What would be the outcome if all or most of Homo sapiens' innovations and adaptations were to be directed away from profits and toward species survival and habitat sustainability? We have seen where most profit-driven innovations end up—enabling and potentially hastening the extinction of our species. Not that there haven't been outcomes that temporarily benefit our species. The problem is that many of these innovations, however marvelous, when wielded by an evolutionarily immature species, like a four-year-old playing with a hand grenade, can end up contributing to its demise.

Such were the concerns of Stephen Hawking, Elon Musk, and a few of the top Techno sapiens, and they are all sounding the alarm about AI's potential to end life as we know it. Hawking said it well when he wrote, "In the short term, AI's impact depends on who controls it; in the long term, it depends on whether it can be controlled at all." Elon Musk warns that "without oversight, AI could be an existential threat: we are summoning the demon." Bill Gates has also expressed concerns about AI. During a Q&A on Reddit in 2015, Gates said, "I am in the camp that is concerned about super-intelligence. First the machines will do a lot of jobs for us and not be super-intelligent. That should be positive if we manage it well. A few decades after that, though, the intelligence is strong enough to be a concern. I agree with Elon Musk and some others on this and don't understand why some people are not concerned."

Their AI fears are probably not unfounded. Evolution was daring enough to give Homo sapiens the keys to the planet, and look at what we've done to it. Now Homo sapiens is beginning to be worried about its technological innovations becoming the source of its own extinction. Will AI choose to cooperate with Homo sapiens, or will AI choose to destroy all biological species and take the planet for its own? Given the ability to choose, would its choice be Homo sapiens survival or Homo sapiens extinction? Evolution, as we have seen, has left itself an out with its three ways (at least) to stop an immature species from spreading that immaturity across the stars. What would be Homo sapiens' out?

One can't help but notice the irony of this dilemma. AI may well be evolution's way of sharing with Homo sapiens a teachable moment: if Homo sapiens want AI to make the right choices, perhaps Homo sapiens should consider making the right choices itself. It is highly unlikely that a species that has evolved past the point of self-extinction and attained evolutionary maturity will ever devise a technology with the potential to ensure its own extinction.

So, unlike many of today's futurists, I invite you, as John Lennon did, to imagine a time when Homo sapiens has reached that survival plateau, a time when the very notion of self-extinction would seem like a horror movie out of the past. Now imagine what innovations leading up to this new phase of Homo sapiens evolution might look like. One might think of it as survival-focused innovation. Wouldn't it be wonderful if today's Techno sapiens and Moneyed sapiens begin to prioritize, refocus, and see their innovations and investments through the lens of species survival over profits, and species extinction avoidance over personal and corporate wealth accumulation?

How might our world change when the focus, motivations, drivers, and outcomes for all of Homo sapiens' innovations and investments were to transition from profits to species survival and habitat sustainability, like evolution's? How might it look when the single mantra pervading every aspect of individual, family, civic, national, and international life is not how to grow up to become an entrepreneur, CEO, big business tycoon, imam, pastor, president, dictator, terrorist, or any other goals we have defined for ourselves, but how to reach evolutionary maturity and enable our species to survive?

Imagine a time when the aim of governments is not to roll back environmental laws and regulations that protect species, habitats, clean air and clean water, like Presidents Trump and Bolsonaro, but rather, how to enable species survival and habitat sustainability instead of bigger military budgets, exclusionary economic zones and military spheres of influence, and support for states that lay waste to their own people; a time when all nations gladly open their borders to refugees and migrants, as few do today; a time when leaders seek the health, welfare, and care of their people and put in place measures to help attain individual, community, city, state, and national evolutionary maturity; a time when the need for law enforcement, penal codes, and prisons is significantly diminished.

Imagine a time when food becomes organic again; when the air and water supply are no longer being polluted; and when the oilmen have pulled their rigs and frackers have dismantled their gear for the last time and fossil fuels no longer power our economies and pollute our world.

Futurists talk about a future of abundance, a time when technology will eliminate all scarcity, but it does not take a Techno sapiens to see there is no lack of abundance today—just artificial scarcity created by man-made greed. As Peter Diamandis says in his book on abundance, if one can find a way to push back the frontiers of scarcity, just look at the abundance that will become available to all. I think he envisaged technology driving this abundance. But the problem of getting to abundance was never due to a lack of technology but rather a matter of scarcity created by greed. There was abundance before there was any man-made technology.

> *The future of our species can never be about abundance—evolution already gave us that. The future of our species is about survival.*

How different might the world be if it were no longer possible for eight men to hoard half of an entire species' resources? How different would the world be if those eight men identified by Oxfam were to distribute their billions across the half of humanity struggling on $10 a day and in some cases less than $1.25? How different the world would be if all those offshore tax havens were emptied out and the money repatriated to the countries and peoples to whom they probably rightfully belong.

How different a world it could be, too, if all those corrupt governments, politicians, obscenely overpaid business executives, and organized crime bosses and drug traffickers could no longer be rewarded with or steal the resources of the people. How different a world it could be if all those military budgets were instead reallocated to species survival and habitat sustainability. And, alas, how different a world it would be if the trillions hoarded in corporate treasuries, private investment funds, private equity firms, and by market speculators were reallocated to meet humanity's needs. And wouldn't it be wonderful to have corporations that work for all citizens instead of just for themselves and their shareholders?

Now that would be a step toward the kind of world evolution might have envisaged 4 billion years ago; that's a world one can begin to get excited about, a world that is definitely worth investing in. That's a world worth innovating to make happen, and worth fighting for. Imagine the abundance these few changes, having little to do with innovations and technology, would release. Imagine that!

How might such a future take form in your imagination? Not unlike the adventures of Dorothy in *The Wizard* of *Oz*: "If you don't know where you're going, any road will take you there." It's up to each member of our species to choose. Evolution went to 4 billion years' worth of effort to equip each member of our species with the freedom to choose. If we can't imagine it, we can't dream it, and if we don't dream it, we won't create it, and if we don't create this future, all we will have left is the future many envision for us now—more electronics, more technologies, bigger, better, and potentially deadlier stuff, and, yes, more death, and probably eventually extinction. But we have a choice. Evolution made sure of that and is counting on each of us to make the choice to attain evolutionary maturity and survive.

Noam Chomsky once asked in a talk at MIT: Is it better to be smart than to be stupid? It's worth remembering that species that have survived multiple mass extinctions, some still with us today, wouldn't be viewed by us as intelligent, and never came up with anything we would recognize as innovations—yet they survived. They survived because species survival does not depend upon species innovations but on a species attaining evolutionary maturity. I believe Chomsky concluded that it was better to be stupid than smart. I also have a question: Is it better to be a dinosaur or a beetle? Imagine that!

If from an evolutionary perspective, innovations are ultimately about getting to maturity and thereby survival, then what innovations, and how should these be pursued to ensure our survival—questions each entrepreneur, investor, consumer and concerned entity must ask and answer in the affirmative if our species is going to survive.

There is an alternative way to ask these questions and it is: How is it that most other species, including microbes, managed to adapt and innovate for billions of years, and our hunter-gatherer forebears for millions of years, without driving to near-extinction Earth's ecosystems, habitats, other species, and themselves? Let's see whether we can find out.

It was Steve Jobs, at that memorable commencement address he gave in 2005 at Stanford University, who said, "You can't connect the dots looking forward; you can only connect them looking backward. So, you have to trust that the dots will somehow connect in your future." And throughout this book we have gone back to our species' past to better understand its future.

When we did, we found evolutionary patterns, rules, and repeating events, such as climate change, and how these drove our forebears to adapt and innovate to survive—including migrating out of Africa. But Homo sapiens weren't the first Earth-life species to make adaptations and innovations or embarked on migrations to survive. Indeed, early animals changed their habitat by crawling out of the oceans for the first time to live on dry land—and that led to the adoption of land as a new habitat. And we marveled at the kinds of adaptations and innovations they had to make to survive, including transitioning from fins to limbs; from gills to lungs; from seeing through water to seeing through dry air; and more.

What was common to all those adaptations and innovations seen in early animals' ocean-to-land migration was that (1) they were survival-driven; (2) enabled by biology—genetic adaptations; (3) they functioned within a common framework of chemical and biological laws—in the local ecosystem; and (4) all outputs were inputs to another process—they were no waste products—no garbage.

For example, early animals did not emerge from the oceans in a skintight water-suit the way humans must wear a diving suit or scuba gear to breathe under water, or a space suit to breathe in space—they evolved lungs, and their gills were not discarded as waste. Neither did they crawl out of the oceans wearing some mechanical prosthesis; their fins evolved into limbs. All adaptations and innovations had to inter-operate and/or fit within the common framework afforded by their biology and the ecosystems within habitats.

By limiting to biological changes the kinds of adaptations and innovations nonintelligent species could make, evolution ensured they stayed within that live-and-let-live guardrail, and that in turn ensured co-existence and mutual survival among all species. Unlike humans, such species then were unlikely to come up with new, novel and non-obvious innovations like synthetic chemicals such as PFAS —Per- and polyfluoroalkyl substances, so-called forever chemicals— or new elements that can wipe out an entire species, poison the food chain, or dump non-biodegradables, like plastics, and/or toxic waste into the environment, but rather, ensured that all mutual needs were accommodated within the local ecosystem.

Homo sapiens, however, changed the evolutionary way to adapt and innovate: from biological to non-biological; from doing so inside the ecosystem to doing so outside; from ensuring the output and waste of its products became input to another; from following the live-and-let-live survival constraints ecosystems imposed, to abandoning the ecosystem altogether and thereby its constraints. These changes not only differentiated how intelligent versus nonintelligent species adapt and innovate but are also the reason Homo sapiens' adaptations and innovations fail while all other Earth-life species succeed.

What was once enabled only by genetic code (DNA) changes, and may have taken generations to become standard in an emerging land animal population, became a simple knowledge transfer and/or a technology upgrade for emerging intelligent species like Homo sapiens, and often in the same or succeeding generation. Not that there was something wrong with this change from genes to memes, but it has resulted in several orders of magnitude increase in the human innovation adoption rate, a pace so frenzied that it threatens our ability to keep up.

Hence, with enhanced abilities came the need to attain evolutionary maturity to safeguard against inadvertently adapting and innovating in ways that result in extermination of species, habitats and potentially ourselves—a path down which we seem well on the way.

However, while the rate and the how of innovation changed from nonintelligent to intelligent species, the why of innovation—survival—did not. The pace and method changed from genes to memes, from biological to non-biological and from instinct-driven to freedom-to-choose one's own evolutionary outcome, but survival, the evolutionary reason to adapt and innovate, did not. In other words, for any Earth-life species to succeed, its adaptations and innovations must be survival-driven. Survival remains the only evolutionary-given motivation to adapt and innovate. And it has been our ability to change the motivation, the reason to adapt and innovate, from survival to profits, greed, growth and gain that makes modern Homo sapiens' adaptations and innovations so different from that of all other Earth-life species and that has led to the risks and threats these now pose to our survival.

Whereas all species, including our hunter-gatherer forebears, for millions of years adapted and innovated to survive and did so within Earth's ecosystems, early humans abandoned them when we transitioned to agriculture. In addition, we abandoned the need to ensure that waste from our products became inputs to subsequent products. We also stepped beyond the evolutionary live-and-let-live reason to adapt and innovate and its intended outcome—survival—and chose instead to do so for personal gain, profit, economic growth and the pursuit of science and innovation. And, moreover, we did so at the risk and often at the expense of ecosystems, habitats and species survival, including "our own"—well not exactly—as the millions of Homo sapiens who, as African slaves, bore so much of this early scientific research cost wouldn't have been regarded by those scientists and innovators as "our own."

Indeed, so anxious and driven we were to adapt and innovate that not even the enslavement and death of more than 12 million African humans taken from their homes between the fifteenth and nineteenth centuries were allowed to get in the way of scientific and economic progress.

In a Sciencemag.org article published April 4, 2019, Sam Kean claimed that "scientific research not only depended on colonial slavery but enabled it and helped expand its reach." He wrote: "To gain access to Africa and the Americas, scientists hitched rides on slave ships on which men and women were chained up for weeks in hot, filthy holds, where diseases ran rampant and punishment for disobedience was harsh. Upon arrival, naturalists also relied on slavers for food, shelter, mail, equipment, and local transport. Indeed, major U.S. universities such as Yale, Georgetown, and Brown have acknowledged how they benefited from slavery," while others are just coming to grips with the link between early science and slavery and the role their institutions might have played.

"It was slave labor that built the first major observatory in the Southern Hemisphere, in Cape Town, South Africa. Astronomers such as Edmond Halley solicited observations of the moon and stars from slave ports, and geologists collected rocks and minerals there. Even a field as rarefied as celestial mechanics benefited from slavery.

> *Modern science and the transatlantic slave trade were two of the most important factors in the shaping of the modern world and historians are finally recognizing that they shaped each other as well."*

And it was not just scientific research that had its foundations laid in the blood of African slaves and slaves of numerous peoples, but most of the modern civilizations Homo sapiens has built. According to History.com, "Since the beginning of civilization (post-hunter-gatherers), slavery has played a huge role in society from the building of the pyramids in Egypt to indentured servitude in England. In fact, at the turn of the 19th century, an estimated 3/4 of the world was trapped in bondage against their will (some form of slavery or serfdom)."

But at what cost? What has been the outcome of ignoring evolution's live-and-let-live survival-driven reason to adapt and innovate? The consequences we now see are imminent, numerous, extreme and dire: extreme heat, extreme weather events, sea-level rise, glacier retreat, accelerating water shortages and droughts, ocean acidification and marine life die-off, ecosystem disruptions, numerous extinctions, and even worse, a seeming accelerating collapse of the natural world. In less than 200 years our species has managed to adapt and innovate its way to the edge of a self-extinction precipice—and, potentially, can end up taking with it 4 billion years of evolution as well.

It is not the planet our innovations have put at risk—Earth has survived numerous extinctions. Nor is it the dinosaurs—unlike megafauna that came later, they at least never had to deal with us. But it's us humans, the wise ones, Homo sapiens—the ones with the freedom to choose survival over extinction—evolution's last surviving Earth-based intelligent species.

To answer the "how" part of the above question then, how Homo sapiens should adapt and innovate—clearly, we should do so in such a way as to ensure habitat and species, including Homo sapiens', survival. If we can see our way clear to do that, the "what" part of the question becomes obvious—we must only adapt and innovate the kinds of solutions that enhance and ensure the chances for survival. In the end we would find that doing so would have been the best long-term return on our species' investments—our survival.

In view of the above, what innovations should be focused on immediately, and in the short-, intermediate-, and long-term? If one were to build an "Innovations for Survival" dashboard to lend focus and track progress, as well as suggest areas for inventors, innovators, entrepreneurs, humanitarians, philanthropists, and anyone else concerned about our species' survival to concentrate their efforts on, what would be the top ten one might want to begin with, and how would one classify them from immediate to long-term?

Here are some to consider, along with a rating assignment.

Homo sapiens—the number one threat to its own survival.

Rating: A historic, immediate, present, and dangerous near-term and long-term threat to survival.

Homo sapiens has been, is, and always will be the biggest threat to its own survival until it attains evolutionary maturity. There are no aliens attacking the Earth. We are the source and creator of everything that now threatens our very survival. These so-called threats are the symptoms of the real issue that evolution wants us to recognize and deal with before it's too late. None would exist if we could focus and solve this "mother of all threats." But be warned, you won't find this threat on any government or corporate research agenda, nor will you find any VCs (venture capitalists) looking to fund, or start-ups wanting to innovate, a solution to this problem.

And while philanthropists and their foundations pour billions into treating symptoms, like healthcare and education, none seem able to recognize that we are the root cause of all the problems they are trying to solve.

There probably isn't time left to kick this can of threats further down the road yet again. Today, nuclear missiles, WMDs, cyber warfare, pandemics, climate change, rising ocean temperatures, a growing water shortage and drought in many regions, and AI are among the top threats on any list. Yesterday, it was knives, pickaxes, swords, javelins, arrows and bows, and spears; tomorrow it will be even worse nightmares. Yet we can't even identify ourselves as the root cause of all these woes. Like an ostrich, we bury our heads in the sand; we choose to treat the symptoms rather than the cause. Is it really so hard? As a species, we are not even trying to address this most existential of threats. We can't, won't, or maybe don't even see ourselves in this way, or name ourselves as the threat we pose to ourselves.

It's not the Russians, the Chinese, the North Koreans, the Americans, or the Indians, it's us. Don't you remember? Homo sapiens. It's not the East, the West, the North, or the South. It's not communism, democracy, fascism, or capitalism. Nor is it Christianity, Islam, or Buddhism. We are not Africans, Americans, British, Canadians, Chinese, Europeans, Indians, Japanese, Israelis, Russians, Turks, Egyptians, Syrians, Taiwanese, or any of the other nationalities we label ourselves today—we are Homo sapiens. Anything else is just a label and a persona we assume. But we are none of these.

We are Homo sapiens, one species, the descendants of hunter-gatherers, our forebears who left Africa 70,000 years ago toward the end of the last ice age. Our mitochondrial DNA confirms that we all came from a single African female, our great- , great- . . . great-grandmother. We are, regardless of our labels, skin color, nationality, race, religion, or sexual orientation, one very, very big family, the Homo sapiens family, and it's way past time we start acting like a family. Homo sapiens are long overdue for a worldwide family reunion.

We are Homo sapiens first and foremost, so shouldn't preserving our species to evolutionary maturity be our highest priority? Nothing else we can do as a species will matter if we drive ourselves to extinction. If any of our species' Techno sapiens or Wise sapiens are reading this, there can be no more rewarding, profitable, or challenging venture to embark on than to save the world. Sending rockets into space is fine. Getting a team to Mars is great. Ridding the world of malaria or stopping the spread of Ebola, AIDS and/or COVID-19 would be fantastic. But nothing will top saving Earth's species, habitats, and ourselves from self-extinction.

It is particularly important to bear in mind that evolution prioritizes our efforts to mature as a species way above our efforts to adapt and innovate. And few things we can innovate will signal attainment of that level of maturity as achieving the ability to harmoniously collaborate to solve the existential challenges confronting our species. Our accomplishments would accelerate orders of magnitude more in any field compared to what's possible in our current separate, redundant, competing and increasingly adversarial world order. We did establish several organizations with worldwide purview over numerous areas with varying degrees of authority and success such as the United Nations and numerous affiliated organizations, the World Bank, the International Monetary Fund and more, as well as numerous arms control agreements and treaties, only to see these begin to be slowly pulled apart by moves towards nationalism.

We have managed as a species, with some degree of success, to achieve near-worldwide collaboration on issues like terminating the massive hole CFCs (chlorofluorocarbons) had punched into the ozone layer; we have also finally mounted a credible international effort to confront the existential threat global warming poses. The problem with this on-again, off-again approach is that it takes years to get to commonly accepted solutions even as the tempo of threats keeps accelerating. The current approach of sliding in for a meeting for a few days at international forums on these issues is crying out for a dedicated, coordinated and continuous solution approach.

These are problems searching for solutions our current world order seems ill-prepared or organized to provide. Might this be a potential role Techno-sapiens, billionaire Moneyed-sapiens and capable interested others can collaborate on and lend their time, talents and resources? Hint. This definitely beats running for president.

If we can't get it together here on Earth, why would anyone assume it will be different in a galaxy or universe populated by immature, warring Homo sapiens? Hence, the wisdom in making attainment of evolutionary maturity a prerequisite for survival.

Here are six additional areas of increasing risk to Homo sapiens that a group of atomic scientists thinks of as the most urgent. So concerned were members of the board of the Bulletin of the Atomic Scientists that in 1947 they established the Doomsday Clock as a response to nuclear threats. The Bulletin takes several factors into consideration when calculating the time on the clock and can be added to the list of threats to Homo sapiens: (2) nuclear threats, (3) climate change, (4) biosecurity, (5) bioterrorism, (6) miscellaneous threats including cyber warfare, and (7) artificial intelligence.

Their concept is simple—the closer the minute hand is to midnight, the closer the Board believes the world is to disaster. According to a 2017 article in *WIRED* magazine, "It's intended to be a symbol of global threats, and a way to inform the public about threats to the survival and development of humanity."

Factors that may have influenced the recent change in time include a darkening global-security landscape, U.S. President Donald Trump's comments on nuclear arms, his views on climate change, the rise of strident nationalism worldwide, and the extraordinary danger of the current moment.

The Doomsday Clock's hands had until recently stood at 3 minutes to midnight, where it had been because of unchecked climate change. *In 2020 the hand has been moved to 100 seconds to midnight*. Of those six we will discuss nuclear threats, climate change, and AI along with others not being monitored by these scientists.

Nuclear threats

Rating: An immediate, present, dangerous near-, intermediate-, and long-term threat to survival.

Fig. 10-4 The Doomsday Clock
(Source – Bulletin of the Atomic Scientists)

Nuclear threat is not going away anytime soon but rather has just increased a notch. With both the USA and Russia abandoning in quick succession in 2019 the Intermediate-Range Nuclear Forces Treaty (INF), and might well be aiming to abandon the much-broader New Strategic Arms Reduction Treaty (New START) due to expire on February 5, 2020. Should the latter occur, for the first time since 1972 there would be no limit on how many warheads either nation can build and deploy. In addition, Russia has or is planning to deploy a strategic intercontinental ballistic missile system, the Avangard, equipped with a gliding hypersonic maneuvering warhead that threatens the USA missile defense system and prompting the USA to embark on a new round of nuclear weapons. With both Russia and America looking to update their nuclear weapons, China, with its own arsenal, meanwhile, is declining any invitation to join a three-party nuclear agreement. Even Europe is beginning to contemplate the potential need for its own cache.

> *It is clear then that like the Sword of Damocles this nuclear threat will continue to hang over the heads of our species until we have grown past the point of self-extinction.*

It's unlikely, too, that America and North Korea will have come to a disarmament agreement anytime soon. In addition, America, the West, and Russia appear to be headed into another Cold War, igniting fears of another event like the Cuban missile crisis as Russian military advisers show up in a troubled Venezuela. Similarly, cyber warfare is fast becoming an emerging global threat that has the potential to disrupt nations as seen during the 2016 American and 2017 French presidential elections, potentially leading to conflict between nations.

Climate change

Rating: An immediate, present, dangerous, near-term and long-term threat to survival.

While it does matter whether one thinks climate change is man-made or not, from a survival perspective, increasingly warm temperatures threaten habitat changes our species is hardly prepared to deal with, regardless of the cause. An IPCC Climate Change 2014 Synthesis Report Summary for Policymakers claimed that human influence on the climate system is clear, and recent anthropogenic emissions of greenhouse gases are the highest in history. Now, four years later, the IPCC 2018 report claims that the 2020s could be one of humanity's last chances to avert devastating impacts and warns that averting a climate crisis will require a wholesale reinvention of the global economy.

By 2040, the report predicts, there could be global food shortages, the inundation of coastal cities, and a refugee crisis unlike any the world has

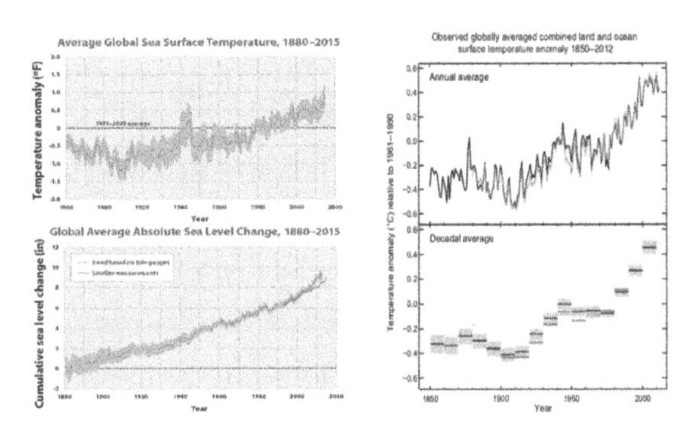

Fig. 10-5 Global trends in land and ocean temperatures 1850 - 2000

(Source: EPA)

ever seen. And as alarming as these predictions are, some scientists argue they understate the full extent of the threat and warn that the economic costs of severe storms and displacement of people due to drought and deadly heat waves would be much worse.

Recent climate change has had widespread impacts on human and natural systems. Warming of the climate is unequivocal, and since the 1950s, many of the observed changes are unprecedented. The atmosphere and ocean have warmed, the amount of snow and ice has diminished, and sea level has risen.

Each of the last three decades has been warmer at the Earth's surface than any preceding decade since 1850. The period from 1983 to 2012 was *likely* the warmest thirty-year period of the last 1,400 years in the Northern Hemisphere.

A *Climate Central* article by Andrew Freedman based on a study titled "The projected timing of climate departure from recent variability," says it is estimated that up to 5 billion people will face "entirely new climate" by 2050. Freedman writes: "The mean annual climate of the average location on Earth will slip past the most extreme conditions experienced during the past 150 years and into new territory by between 2047 and 2069, depending on the amount of climate-warming greenhouse gases that are emitted during the next few decades."

A recent study published in the journal *Nature* used a new index to show for the first time when the climate—which has been warming during the past century in response to man-made pollution and natural variability—will be radically different from average conditions that existed during the period from 1860 to 2005. The study shows that tropical areas, which contain the richest diversity of species on the planet as well as some of the poorest countries, will be among the first to see the climate exceed historical limits—in as little as a decade from now—which spells trouble for rain forest ecosystems and nations with a limited capacity to adapt to rapid climate change.

According to CO2.Earth's Homepage, recent carbon dioxide levels are hovering around 416.25 parts per million. If by the end of the century carbon dioxide in the atmosphere more than doubles, much of Africa, South America, and India will endure average daily maximum temperatures of more than 45°C. Jerusalem, New York, Los Angeles, and Mumbai could see summer temperatures reaching these levels, too. London will experience temperatures in the mid-20s, and Paris could see its July temperatures reaching the low 30s. Glaciers in the Everest region of the Himalayas could be almost completely eradicated by 2100 due to greenhouse gas emissions, scientists have warned, endangering a source hundreds of millions of people rely upon for the fresh water provided every summer by rivers fed by the Himalayan glaciers.

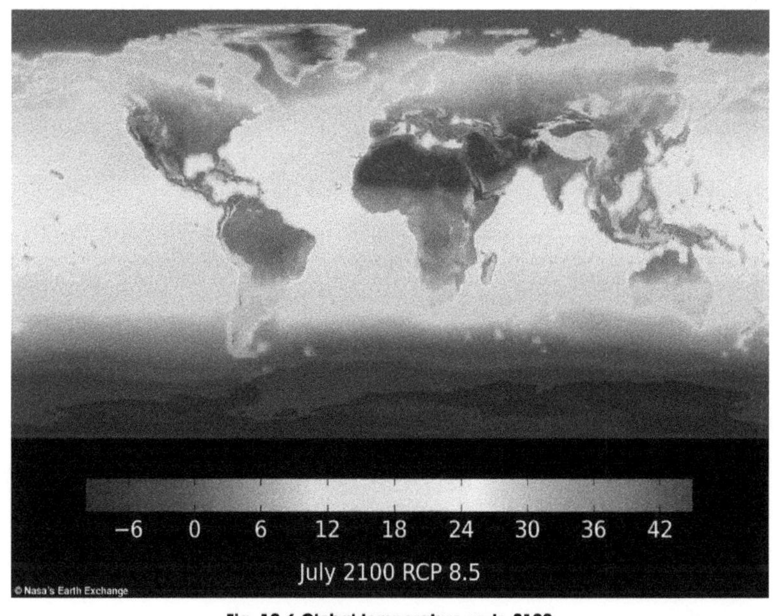

Fig. 10-6 Global temperature up to 2100

The new NASA global data set combines historical measurements with data from climate simulations using the best available computer models to provide forecasts of how global temperature (shown here) might change up to 2100 under different greenhouse gas emissions scenarios. (Source: NASA)

Rising ocean temperatures can lead to oxygen depletion, but not nearly enough people seem concerned. Low atmospheric oxygen was a primary driver during the Great Dying (aka the Permian-Triassic extinction), when nearly all life (an estimated 95%) in any form went extinct. When that increase in oceanic temperature gets close to or exceeds 6°C, phytoplankton can no longer carry out photosynthesis. About 50% to two-thirds of the planet's total atmospheric oxygen is produced by ocean phytoplankton, and therefore cessation would result in the depletion of atmospheric oxygen on a global scale. This would likely result in the mass mortality of animals and humans. Not only would the oxygen be threatened but also the entire ocean-based food chain.

Innovations needed: Many innovations are already at work enabling the transition from fossil fuel to wind, wave, solar, nuclear, and numerous bio-based technologies. Work on fusion technologies is progressing but perhaps decades away from becoming commercially available.

The biggest problem, however, is not a lack of nonpolluting alternative energy technologies, and maybe not even a lack of time to act to replace fossil fuels. The real problem is the unwillingness of "the few" to forgo the profits that will be lost by leaving fossil fuels in the ground and the determination to go against the world, as in withdrawing from the Paris Climate Accord, just to uphold ecocidal political and economic ideologies.

According to the Sierra Club, the Trump administration has been unrelenting in its attacks on public lands, clean air and water, threatened wildlife, and climate-change policies. A quick perusal of the tables below more than underscores the issues Homo sapiens will have to confront if we are to remain capable of choosing survival over extinction. At a time when the U.S. has begun rolling back environmental regulations, it is refreshing to see that Germany, the land of the automobile, has begun making it easier for cities to begin banning certain diesel-power vehicles.

Also, to no one's surprise, the current U.S. administration is thinking of shrinking four national monuments and opening six others to mining, drilling, and logging. Not that America has a monopoly on committing ecocide, but as the largest economy in the world, it does stand out.

Table 10-1 Twenty-nine rules overturned

(Source: NY Times)

29 RULES OVERTURNED

- Flood building standards
- Proposed ban on a potentially harmful pesticide
- Freeze on new coal leases on public lands
- Methane reporting requirement
- Anti-dumping rule for coal companies
- Decision on Keystone XL pipeline
- Decision on Dakota Access pipeline
- Third-party settlement funds
- Offshore drilling ban in the Atlantic and Arctic
- Ban on seismic air gun testing in the Atlantic
- Northern Bering Sea climate resilience plan
- Royalty regulations for oil, gas and coal
- Inclusion of greenhouse gas emissions in environmental reviews
- Permit-issuing process for new infrastructure projects
- Green Climate Fund contributions
- Mining restrictions in Bristol Bay, Alaska
- Endangered species listings
- Hunting ban on wolves and grizzly bears in Alaska
- Protections for whales and sea turtles
- Reusable water bottles rule for national parks
- National parks climate order
- Environmental mitigation for federal projects
- Calculation for "social cost" of carbon
- Planning rule for public lands
- Copper filter cake listing as hazardous waste
- Mine cleanup rule
- Sewage treatment pollution regulations
- Ban on use of lead ammunition on federal lands
- Restrictions on fishing

Table 10-2 Twenty-four rollbacks in progress

(Source: NY Times)

24 ROLLBACKS IN PROGRESS

- Clean Power Plan
- Paris climate agreement
- Wetland and tributary protections
- Car and truck fuel-efficiency standards
- Status of 10 national monuments
- Status of 12 marine areas
- Limits on toxic discharge from power plants
- Coal ash discharge regulations
- Emissions standards for new, modified and reconstructed power plants
- Emissions rules for power plant start-up and shutdown
- Sage grouse habitat protections
- Fracking regulations on public lands
- Regulations on oil and gas drilling in some national parks
- Oil rig safety regulations
- Regulations for offshore oil and gas exploration by floating vessels
- Drilling in the Arctic Wildlife Refuge
- Hunting method regulations in Alaska
- Requirement for tracking emissions on federal highways
- Emissions standards for trailers and glider kits
- Limits on methane emissions on public lands
- Permitting process for air-polluting plants
- Offshore oil and gas leasing
- Use of birds in subsistence handicrafts
- Coal dust rule

Among the reasons that inventions and technologies, even if available, are unlikely to be the basis of Homo sapiens' survival is the continuing ability of "the few" to superimpose their choices on the rest. Not until enough Homo sapiens can assert their will over that of "these few" and prioritize species and habitat survival over profits and endless economic growth can such solutions have a chance to turn things around.

(AI) Artificial intelligence

Rating: Some versions of AI will become dangerous and an intermediate- and long-term threat to Homo sapiens' survival.

If Homo sapiens designs an AI species, as in an intelligent hominin species, that potentially could succeed humans (if not on Earth, potentially on other planets), as humans succeeded Homo erectus, that may well turn out to be a potential evolutionary outcome as explored in the chapter on Species-Driven Evolution, an outcome humans may have no more choice or say in than Homo erectus had in what might have succeeded them—us.

If, however, Homo sapiens inadvertently designs an AI that can potentially destroy and replace humans and take over, then that problem is not very dissimilar to the creation of the atomic bomb or any other WMD—or a Frankenstein-like creature. In both AI scenarios humans end up being replaced—but the latter is not species succession but, rather, self-extinction. It is this latter AI scenario humans fear most.

> *The survival question here is which AI are we developing and how will we know?*

Evolution was daring enough to give intelligent species, and hence Homo sapiens, the keys to the planet and look what we've done and are still doing to it. So, fears about AI and Homo sapiens survival are not unwarranted, and the concerns expressed by Hawking, Gates and Musk (assuming they mean the Frankenstein version of AI) are also relevant in this context.

If indeed as the late Steven Hawking thought, AI can potentially replace Homo sapiens, it will be the first partial, or complete, non-biological species to have supplanted or succeeded a hominin species, making Homo sapiens the last of the hominins—and that says plenty about why Homo sapiens is unlikely to be able to develop an AI that would respect human survival constraints.

Recall that evolution ensured every species evolved from within a biological parent, and carried within itself the joint parental DNA and all the genetic nanomachines that comprise the components of a eukaryotic cell. In making each of us, our parents did not get a handful of clay, some water and some chemicals and fashion an offspring. Instead, we were all born with 4 billion-plus years of microbial programming that informs what it means to be human, Homo sapiens and descendants of the line of intelligent species—hominins.

We learned what it means to be human during those formative years in the care of parents, and in the crucible of the home, the school, family, friends, the village and other humans from childhood to adulthood. Our brains are programmed not by binary ones and zeroes, or computer software code, but by the innumerable interactions, experiences, discoveries, anxieties, joys as well as fears across a network of billions of neurons such that long before we become adults, we intuitively know what it is and how to be human. Something an AI's neural network, aided by deep learning and by copying us may one day hope to imitate—but that's exactly where our AI problems begin. How so?

Because even after all of that biological programming, all of that nature and nurture, not even humans behave the way humans expect humans should. One is often staggered and left flummoxed by the behavior we humans are capable of: indeed, the selfishness, violence, hate, cruelty, pain and destruction we inflict on one another, other species and on Earth's ecosystems and habitats upon which our survival depends, as well as the care, love, empathy, and selflessness of which we are equally capable.

So, if after all that, humans have a hard time behaving like humans, what can one hope to expect from what is essentially a machine, a cyborg, that is built in a lab or factory by engineers and scientists and fed unenumerable petabytes of human private information and experiences by profit-seeking corporations and suppressive governments who have already begun to steal our personal data so their AIs can learn and be "taught" how to look like and act human while thinking like a multi-national corporation?

If, unfortunately, what is driving these next technological advances is not Homo sapiens' survival, but rather, what has always driven all prior Homo sapiens' innovations—profits, greed, gain, economic growth and world dominance—can one really expect scientists, engineers, entrepreneurs, companies, industries and governments that have historically pursued and financed innovations for profit, growth, gain and war would all of a sudden have a change of heart and begin to develop an AI to save human lives above all else?

Even when new innovations intrinsically have the potential to be used for good, and many do, it is seldom long before these are preempted and repurposed (often with destructive consequences to humans, other species, ecosystems, and habitats) to serve the profit, gain, greed and growth motives that drive our species. So why would AI, Asimov's laws, neural networks, deep learning technologies or any we might conjure up, notwithstanding, be any different? As Harry Chapin so melancholily lamented in "Cat's in the Cradle" how his boy had grown up to be just like him, so humans will soon realize their AIs can only be just like them.

Recall how the emergence of social media engendered such hope and promise for society until it ran into Zuckerberg, who figured out how to turn people's privacy into profits, and into some of us, who figured out how to use it to vent our spleen across the globe spewing hate, prejudice and division and thereby shining a light upon our immaturity as a species. Similarly, CRISPR-Cas9 was going to help Homo sapiens revolutionize and speed up evolution, until a Chinese scientist, Dr. He Jiankui, and his colleagues went rogue and edited the DNA of twins in the womb of a pregnant mother, hoping to eliminate the possibility of these kids contracting AIDs, and simultaneously bewraying how such innovations are likely to be misused. There is probably still hope these technologies can yet become forces for good, but this is the way it has always been, and probably will continue to be, until Homo sapiens' adaptations and innovations become survival-driven rather than profit-driven.

The real issue then is not AI taking over—it already has—but rather how to break free of its ugly economic twin, eAI, the profit motive that, not unlike Boeing's alleged faulty 737 autopilot mechanism, insists on driving us all to extinction despite our best intentions. So yes, if AI turns out to be all it's cracked up to be we should be very afraid—not of AI, though, but of Homo sapiens—AI will have become like us.

> *Because we have failed to adapt the evolutionary survival-driven patterns explored earlier: i.e., how to make a living without driving to extinction other species and ourselves and destroying in the process Earth's ecosystems and habitats, it is highly unlikely human innovation can give rise to a species, biological or robotic, that will turn out to be any different from itself—remember Frankenstein.*

Based on these evolutionary survival patterns, it's clear that any AI is going to need much more than Isaac Asimov's et al's four Laws of Robotics, or any other human technology for that matter. According to Asimov et al a robot:

(1) may not harm *humanity*, or by inaction, allow humanity to come to harm;

(2) may not injure a human being or, through inaction, allow a human being to come to harm, except when required to do so in order to prevent greater harm to humanity itself;

(3) must obey any orders given to it by human beings, except where

such orders would conflict with the First Law or cause greater harm to humanity itself; and a so-called fourth law,

(4) must protect its own existence as long as such protection does not conflict with the First or Second Law or cause greater harm to humanity itself.

While these somewhat repetitious laws might work in the science-fiction world Asimov created, many see them as being no more than a crude starting point from which to begin thinking about potential AIs and human relationships in a world according to evolution.

Asimov et al envisioned robots that must be principally concerned with human survival. In contrast, evolution was and is concerned about the survival of all species, including humans, as well as the survival of Earth's ecosystems and habitats. Evolution "knows" that species survival depends on habitat survival and habitat survival depends on species survival. This evolutionary pattern, The Suvival Sustainability Cycle, is the foundation for all Earth-life survival, and Homo sapiens, seemingly due to climate change, is only just becoming aware of it.

Innovations needed? So here are a few takeaways for would-be AI architects:

(1) To design an AI (not an industrial robot) that could merely obey Asimov's four laws, one intended to exist and interact independently in a world with humans, one will have to start from redesigning humanity—humans routinely fail all of those laws.

(2) If the goal is to design an AI to ensure survival, then forget about Asimov and deep learning—only evolution's survival-driven rules can accomplish that, and humans, from whom AIs will learn, practice very little of that.

(3) Survival must mean *evolutionary-driven survival*—i.e., survival of all Earth-life species, humans, ecosystems and habitats.

(4) To accomplish evolutionary-driven survival, teach your AI to evaluate sources of all inputs, processes, outputs and the final outcomes by applying the *Survivability and Habitat Sustainability Test*.

To pass this test, AIs must be guided by this evolutionary truism—species survival equals habitat survival and habitat survival equals species survival.

Every activity an AI engages in must, at all stages, ensure the survival of *all species* and the sustainability of Earth's ecosystems and habitats. Otherwise, that activity or its outcome not only fails this test but will not ensure survival.

Only after Homo sapiens have learnt how to live by evolution's survival-driven rules can we hope to develop an AI that will not destroy us. Until then, any AI will be nothing more than a superhuman and will become orders of magnitude better at doing what our species have done best so far—drive Earth's species, ecosystems, habitats and potentially itself to extinction.

The problem of an AI-driven human extinction then is not an AI problem but a Homo sapiens problem.

In addition to the threats the Bulletin of the Atomic Scientists are monitoring one might want to add the following.

The food chains crisis

Rating: An immediate, present, dangerous, near-term and long-term threat to survival.

According to the FAO (Food and Agriculture Organization of the United Nations), the human food chain is under continued threat from an alarming increase in the number of outbreaks of transboundary animal and plant pests and diseases (including aquatic and forest pests and diseases), as well as food safety and radiation events. Avian influenza; peste des petits ruminants; locust infestations; wheat, cassava, maize, and banana diseases; armyworm; fruit flies; food-borne pathogens; and micotoxins are just some examples of threats to the human food chain that can have detrimental effects on food security, human health, livelihoods, national economies, and global markets.

An EPA report on persistent organic pollutants (POPs), toxic chemicals that adversely affect human health and the environment around the world, claims that POPs persist for long periods of time in the environment, accumulating in the body fat of living organisms and becoming more concentrated as they move from one creature to another up the food chain. Adverse health effects linked to POPs have been seen in human reproductive, developmental, behavioral, neurologic, endocrine, and immunologic systems. People are exposed to POPs mainly through contaminated foods, drinking contaminated water, and direct contact with the chemicals, which can be transferred through the placenta and breast milk to developing offspring.

The food chain is also threatened by the use of pesticides and fertilizers. It wasn't so long ago that all food was organically grown in a way that supported the ecosystem and environment as a whole. This all changed in the 1940s when the Green Revolution took hold and industrial, chemical-dependent farming techniques became the norm. Use of pesticides and fertilizers on farms has increased twenty-six-fold over the past fifty years, fueling increases in crop production globally, but there have been serious environmental consequences. Indiscriminate pesticide and fertilizer application pollutes land and water, and chemicals often wash into streams, waterways, and groundwater when it rains. Pesticides can kill nontarget organisms, including beneficial insects, soil bacteria, and fish. Fertilizers are not directly toxic, but their presence can alter the nutrient system in freshwater and marine areas, and excess nutrients can result in an explosive growth of algae. As a result, the water is depleted of dissolved oxygen, and fish and other aquatic life may die.

One particular kind of pesticide, neonicotinoids, or neonics—a water-soluble systemic class of neuro-active insecticides chemically similar to nicotine—is turning up seemingly everywhere—from deer, salmon, wild birds of prey, bees, and fish to lizards, frogs, and rodents. They are toxic to many species, and are getting into the broader food chain. A 2018 review in *Environmental Science and Toxicity* found that more than half of the fruits and vegetables served in the U.S. Congress's cafeteria contained neonicotinoids. And a recent *Environmental Health News* study found neonicotinoids even in some certified organic fruits and vegetables.

The industry, of course, blames today's environmental testing methods for being more sensitive and that the levels of the pesticide being detected are not harmful, and the EPA agrees. Scientists point out, however, that even at those low EPA levels, neonics are 5,000 to 10,000 times more toxic to honeybees than DDT was—a pesticide the same EPA banned in the 1970s. Is anyone surprised? Indeed, neonicotinoids have become the DDT of the 21st century the EPA still won't ban. Meanwhile, some governments, including the EU, have already begun restricting their use.

So why the fuss? These water-soluble systemic class of neuro-active insecticides is used to control "harmful" insects. Unlike the contact class of pesticides, which remain on the surface of the treated leaves, systemic pesticides are taken up by the plant and transported throughout the plant (leaves, flowers, roots and stems), as well as pollen and nectar. And the fuss is because (1) neonicotinoids are deadly to bees, which are indispensable to plant pollination, and (2) only 2% to 20% of the neonicotinoids applied to seeds actually make it into the plant. The rest—80% to 98%—are beginning to show up in other areas of the environment including surface waters and in untreated plants, and studies are pointing to toxic impacts on non-pest species from bees to deer, and to transmission beyond farm fields.

In addition, neonicotinoids are EDCs, endocrine disrupting chemicals. It's not just that birds are unable to migrate, bees can't find their nests, and deer are showing up with extra or misplaced organs and a serious overbite. As bad as these might be, endocrine disruptors interfere with human endocrine, or hormone, systems, and hormone disruptors can derail any system in the body controlled by hormones. They mimic naturally occurring hormones in the body like estrogens (the female sex hormone), androgens (the male sex hormone), and thyroid hormones, potentially leading to overstimulation. They bind to a receptor within a cell and block the endogenous hormone from binding. These disruptions can cause cancerous tumors, birth defects, and other developmental disorders.

Many manufactured chemicals mimic natural hormones and send false messages. Phthalates, for example, are a group of chemicals used in hundreds of products, including toys, vinyl flooring and wall covering, detergents, lubricating oils, food packaging, pharmaceuticals, blood bags and tubing, and personal care products such as nail polish, hair sprays, aftershave lotions, soaps, shampoos, and perfumes.

One potential consequence of EDCs, phthalates in particular, threatens our extinction as a species. You read that right—Homo sapiens extinction. This may sound dramatic and alarmist until one catches up with the research by "a team of epidemiologists, clinicians, and researchers from The Hebrew University of Jerusalem that culled data from 185 studies that examined semen from almost 43,000 men. It showed that the human race is apparently on a trend line toward becoming unable to reproduce itself. Sperm counts went from 99 million sperm per milliliter of semen in 1973 to 47 million per milliliter in 2011, and the decline has been accelerating." And while there are existing (in vitro fertilization [IVF]) and emerging (in vitro gametogenesis [IVG]) reproductive technologies that might prevent total extinction, the concern is whether another forty years or less might bring us all the way to zero.

In a September 2018 *GQ* article by Daniel Noah Halpern titled "Sperm Count Zero," Halpern writes: "A strange thing has happened to men over the past few decades: We've become increasingly infertile, so much so that within a generation we may lose the ability to reproduce entirely. What's causing this mysterious drop in sperm counts—and is there any way to reverse it before it's too late?" While there may be many causes, consensus appears to be gathering around the belief that this might well be a direct result of the Industrial Revolution, the move to fossil fuels, and the emergence of the twentieth-century chemical industry. "In short, humans started ingesting a whole host of compounds that affected our hormones—including, most crucially, estrogen and testosterone," says Halpern.

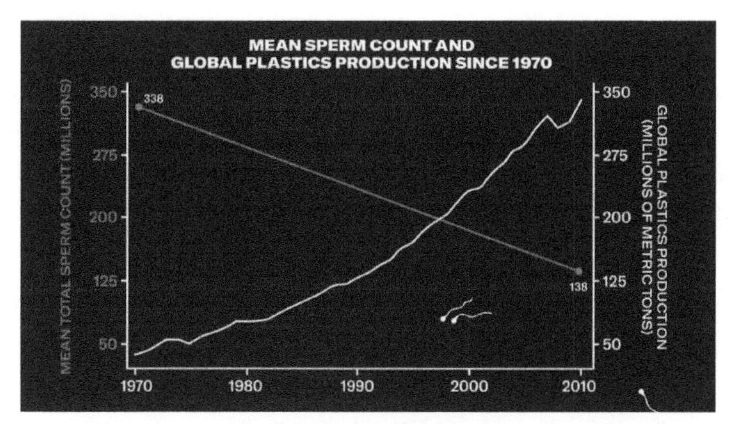

Fig. 10-7 Mean sperm count and global plastics production since 1970
(Source: Hebrew University of Jerusalem)

Consequently, huge concerns are beginning to mount about the quality of the food chain. People are starting to realize that modern agriculture is driving hunger, obesity, and disease, as well as threatening the food chain and, in some regions, worsening the water crisis. Virtually every growing environmental and health problem can be traced back to modern food production. This includes but is not limited to food insecurity and malnutrition amid mounting food waste; rising obesity and chronic disease rates, despite growing healthcare outlays; diminishing fresh water supplies; and toxic agricultural chemicals polluting air, soil, and waterways, threatening the entire food chain from top to bottom.

These problems are aggravated by an agriculture and food industry that is driven by profits and growth and couldn't care less about the health of their customers or the nutrition content of the produce and food they are serving up, abetted by politicians who care more about being re-elected to office and donations from the food industry. This has resulted in a growing consumer demand for organically grown produce and free-range farm animals.

As with most Homo sapiens' adaptations and innovations, the problem has been the pursuit of profits and endless economic growth, and in this case chemical WMDs led by the chemical industry and perhaps inadvertently by others like the agriculture and food industries that use their chemically derived products.

Some Homo sapiens demonstrated a willingness to use chemicals to poison 6 million of our species in Nazi gas chambers during the Second World War, and the poisoning with chemical weapons continues, as in the Syrian civil war. By the use of glyphosates (the primary active ingredient in the widely popular Roundup herbicide manufactured by the Monsanto Company) and plastics, the chemical industry and others seem prepared to poison, pollute, and destroy the food chain and drive disease and illness across the world. The recent civil case against Monsanto in California (juries in two key trials against Monsanto have unanimously agreed that Roundup—trade name for glyphosate—caused the plaintiffs' non-Hodgkin lymphoma cancers) as well as the EU's approach to the use of glyphosates and genetically modified organisms are small indications that our species is just beginning to wake up to the dangers to survival these industries have been able to perpetrate hitherto with impunity.

Indeed, with EDCs the chemical industry may well have found peacetime applications for its chemical products but this time with the potential to drive not just 6 million but our entire species to extinction.

And just when one may have thought these attacks on the food chain can't possibly get any worse, scientists are starting to report yet another threat—the reduction of nutritional content (protein, vitamins) in key staples, like rice, flour, potatoes, corn, and others, as increasing CO_2 in the atmosphere changes their basic chemistry. Since staples like rice are the main food for billions, the reduction in nutritional content could exacerbate an already emergent food chain crisis.

What innovations do you think will solve the food chain and chemical industry's EDC problem? Would you want to trust Homo sapiens' survival to yet another invention from these industries?

The growing extinction of bees and other pollinators

Rating: An immediate, present, dangerous, near-term and long-term threat to survival.

Many species of wild bees and other pollinators are in danger of extinction, warns a U.N. study, as reported in a 2016 Associated Press article.

Bees pollinate the largest number of plant species. More than $15 billion a year in U.S. crops are pollinated by bees, including apples, berries, cantaloupes, cucumbers, alfalfa, and almonds. U.S. honeybees also produce about $150 million in honey annually.

The vast majority of animal pollinators are insects such as bees, beetles, ants, wasps, butterflies, and moths; about 1,000 species of pollinators are hummingbirds, bats, and other small mammals.

At least one-third of the world's agricultural crops depend upon pollination by insects and other animals. Only fertilized plants can make fruit and seeds; without them, the plants cannot reproduce. "Everything falls apart if you take pollinators out of the game," said one bee expert. "If we want to say we can feed the world in 2050, pollinators are going to be part of that."

Innovations needed: It's unlikely that human pollinators can replace bees, so a return to organically grown foods in a way that supports the ecosystem and environment as a whole is needed as well as the elimination of all pesticides and GMO products from the food chain.

Drought and growing water scarcity

Rating: An immediate, present, dangerous, near-term and long-term threat to survival.

Cape Town, South Africa, just barely missed its countdown to "Day Zero" when its taps were expected to run dry. With great efforts this day has been pushed back but only for a while as experts think even if Day Zero doesn't happen in 2019, it is inevitable. What's happening to Cape Town is already happening in many areas across the planet and here is why. Here are some quick facts on the World's Dry Areas according to Dryland Systems:

- Cover 41% of the Earth's surface
- Inhabited by 30% of the world's population (2.5 billion people)
- Support 50% of the world's livestock
- Grow 44% of the world's food
- Account for the majority of the world's poor, with around 16% living in chronic poverty
- Most of the world's poor live in dry areas—with 400 million living on less than $1.25 per day.
- Drylands lose 23 hectares per minute to drought and desertification—a loss of 20 million tons of potential grain production every year.

So here is the problem with droughts and increasing water scarcity that should keep everyone up at night:

(1) Drought has been the major catalyst in the collapse of most prior Homo sapiens civilizations. Innovations and technologies notwithstanding from the Sumerians of Babylon, the Assyrian Empire, the great Egyptian Pharaohs, the Greek Empire, the Roman Empire, the Tang Dynasty, the Kublai Khans and Genghis Khans, the Maya, the Moshe, Chimú lords, the

Incas, the ancient Pueblo people, to the ancient Khmer kingdom of Angkor Wat fame, they all collapsed under the pressure of severe multi-century droughts brought on by climate change. Scientists are starting to warn that drought, driven by warming global temperatures, has the potential to do the same to modern Homo sapiens civilization.

(2) According to Brian Fagan, in his book *The Great Warming: Climate Change and the Rise and Fall of Civilizations*, from the tenth to the fifteenth century, the Earth endured a rise in surface temperatures that changed climate worldwide. What happened then may offer a preview of how global climate could change due to rapidly increasing global temperatures.

During these warm centuries, in areas including western Europe, longer summers brought bountiful harvests and population growth that led to cultural flowering in the Arctic; Inuit and Norse sailors made cultural connections across thousands of miles as they traded precious iron goods. Polynesian sailors, riding new wind patterns, were able to settle the remotest islands on Earth. But in many other parts of the world, the warm centuries brought drought and famine. Advanced societies in North and Central America collapsed, and the vast building complexes of Chaco Canyon and the Mayan Yucatan were left empty.

The history of the "great warming" of half a millennium ago suggests that we may be underestimating the power of climate change to disrupt modern Homo sapiens civilization today. The Middle East, North Africa, central Asia, and South Asia are due to suffer the biggest economic hit from water scarcity as climate change takes hold, according to a World Bank report. By 2050, growing demand from cities and agriculture will put water in short supply—even in regions where it is now plentiful—and worsen shortages across a vast swath of Africa and Asia, spurring conflict and migration, the financial institution said.

The global climate change engine (including: El Niño, ENSO, ITCZ (Inter-Tropical Convergence Zone) and La Niña) that drove these prior civilizations, their brilliant water and agriculture innovations notwithstanding, beyond their ability to survive is still present today and is currently ravaging the people of sub-Saharan Africa. While climate science has improved their understanding of the global climate engine, there is still much scientists do not understand about why the climate seesaws back and forth across centuries, causing alternating periods of drought and devastating floods. Now, with increases in global temperatures due to global warming forecasted, scientists are even more worried about how the global climate engine might react. This potentially exposes our species to climate change impacts the likes of which no one alive has ever experienced.

(3) It is estimated that during the nineteenth century, more people died from droughts (some 20 million to 30 million) than from all the wars fought during that century.

While at present, extreme drought affects only a little over 3% of the Earth's surface, and about 8% of the population, computer models of future droughts due to greenhouse gases show increases of as much as 30%, with as much as 40% suffering from extreme droughts, resulting in more than 50% of the Earth's land surface experiencing some drought conditions. Twenty percent of the world's population currently lacks access to safe, clean drinking water, and the U.N. estimates that by 2025, about 2.8 billion will be living in water-challenged areas.

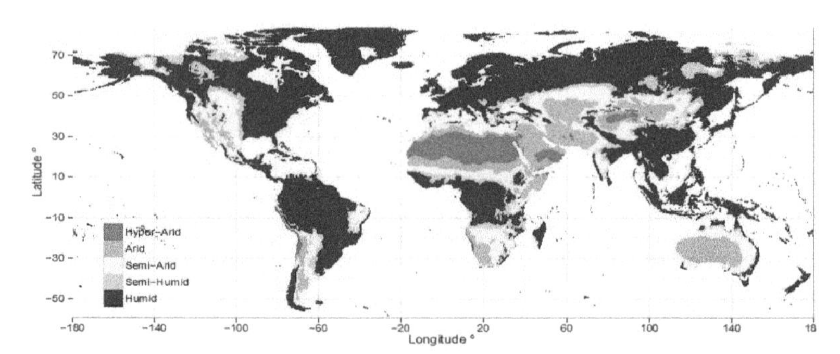

Fig. 10-8 Map of global distribution of climatic zones

Produced using data provided by CGIAR-CSI

(Source: https://cgiarcsi.community/data/global-aridity-and-pet-database/)

According to a 2013 United Nations International Year of Water Cooperation report, 85% of the world's population lives in the drier half of the planet; 783 million people do not have access to clean water, and almost 2.5 billion do not have access to adequate sanitation. Six million to 8 million people die annually from the consequences of water-related diseases and disasters.

Problems include increasing and life-threatening wildfires as in California, natural scarcity of drinking water in certain areas, floods, the siltation of river systems, as well as the contamination of rivers and large dams. Some 1.1 billion people in developing countries have inadequate access to clean water. According to Drought.gov, the National Integrated Drought Information System, 18.1% of the USA land area, affecting some 41.3 million people, is experiencing drought across a drought severity scale that includes the categories Abnormally Dry, Moderate, Severe, Extreme and Exceptional Drought. Persistent drought conditions have spread across the U.S. Southwest, with the Four Corners area of Arizona, Colorado, New Mexico, and Utah standing out as extremely dry.

Innovations needed: Innovations and technologies that can convert seawater to potable water, and investment to put in place water distribution mechanisms across the hardest hit areas. Early innovations include use of graphene, a membrane capable of removing salt from seawater.

Innovations and technologies that can convert seawater to potable water and distribute it will no doubt make a huge difference. Since 70%–75% of the Earth's surface is covered by salt water, such an invention operating on an industrial scale and at an affordable cost has the potential to significantly reduce the risk of collapse from water scarcity due to drought.

One must remember not to become too bullish though as despite their technologies and innovations, in the face of prior global warming, prior great Homo sapiens civilizations collapsed. What innovations can be found today to avert a similar outcome?

There are many, many other problems facing our species no doubt, some of which one might consider equally urgent. If so, one can appreciate why it's past time for Homo sapiens to take action. But what will it take for our species to wake up and make the U-turn needed to seriously focus on its own survival before it's too late?

Most Techno sapiens and Moneyed sapiens behave as though fixing the items on the above lists, or any of the other survival-threatening issues, is not their job or responsibility. This, they seem to think, is the province of governments and politicians and worldwide bodies like the UN COP-21 (UN Climate Change Conference) and IMF (International Monetary Fund). Many seem content to be the engines of innovation and technologically driven change and to leave the rest up to the international community, governments, courts, cities, states, politicians, religious leaders, and anyone else who can get 1 million Facebook likes to figure out.

Many, too, see themselves as private enterprise and seem to think that private enterprise and governing don't mix, yet they have no trouble lobbying politicians, governments, and world bodies when doing so will advance their business interests. But if one thinks it's okay to own more than half of an entire species' wealth and resources, then it is more than okay to carry more than half of the responsibility to drive change and refocus away from profits to species survival; from wealth accumulation, unending growth, and monopoly of the majority of the planet's resources to enabling habitat sustainability; from an obsession with self to a passion for Homo sapiens maturity beyond the point of self-extinction.

When it comes to helping Homo sapiens avoid going over the self-extinction cliff, most appear to view getting involved as bad for business. A few, however, have chosen to use their wealth and other resources to help save humanity even if they can't turn things around. Apple's CEO Tim Cook, for instance, is willing to stand up for Homo sapiens' right to privacy and is willing to take a stand against those who would compel access to or exploit users' social media and internet use for profit. Elon Musk is reported to have said, "I really don't have any other motivation for personally accumulating assets except to be able to make the biggest contribution I can to making life multiplanetary." And indeed, he appears to be putting his money where his mouth is, as we say. And Musk is not the only one. Billionaire Yuri Milner and collaborators in Breakthrough Initiatives seek to fund, among other things, solar system alien life research.

Similarly, Silicon Valley's tech incubator Y Combinator has decided to take on climate change. It is seeking technology proposals that target removing atmospheric CO_2 so as to slow or even prevent getting to a 2°C increase in global temperatures. However, approaches that envision adding genetically engineered algae or minerals to the oceans or turning deserts into oases are beginning to sound like yet another set of potential solutions that fall outside of Earth's ecosystems the way Homo sapiens' choice of agriculture did 10,000 years ago. And we know where that technology adaptation led us.

Innovations that flow from a desire to preserve life (Homo sapiens, flora, fauna, and Earth's habitats) are a good thing. They have the potential to become "survival-driven." But to be survival-driven, all aspects of an emerging innovation, from intention to outcome and everything in between, must be survival-driven. Many of the above may be good enough to earn capitalist, scientific and even humanitarian, altruism-based innovation prizes, Nobel or not, but it remains to be seen whether any of the above can truly attain evolution's "survival-driven innovation" standard that's designed to preserve life.

What we fail to get about innovation is that the reason Homo sapiens evolved with this capability, when all other species didn't, is that innovation enables our species to survive, not to get rich, build Silicon Valley unicorns or trillion-dollar companies, invent the next WMD or create a real Frankenstein creature. To ensure survival, innovation must be survival-driven.

If one is thinking of becoming an inventor, innovator or entrepreneur, here are a few survival-driven innovation guidelines to consider.

(1) The innovation must emerge from a deep desire and intent to do no harm and to preserve life (flora, fauna, Homo sapiens life and the habitats that support them).

(2) Innovators must adopt and organize around a business model or structure that enables them to optimize survival versus profits as the outcome of the innovation.

(3) Innovators must refuse support and funding from sources that derive their resources from non-survival-driven business models.

(4) Innovators must recognize that natural resources belong to all species and exist for their survival and are not there to be exploited for economic gain by private corporations and their shareholders.

5) Their products and/or services must be conceived not as stand-alone but must fit into an integral niche in the human ecosystem, and like an Earth ecosystem, they must be capable of the 360-degree cycle of creation, distribution, consumption, recycle, reuse and repeat.

(6) They must generate zero non-reusable outputs and take ownership for products and services that inadvertently cause harm.

(7) They must not pollute or create products that are harmful to species, the environment, or Homo sapiens.

(8) Like a biological innovation, the survival-driven innovation must emerge only if there is a survival need and must go away when that need no longer exists.

(9) The supply chain, production and distribution models and processes must be constrained by what's good for the survival of all species (including Homo sapiens), the environment and natural ecosystem, not by just what's good for business.

(10) Innovations must follow universal patterns like self-similarity and fractal structures and be in harmonious co-existence with the rest of the natural world, the solar system, and the universe. Facilities should not be abandoned and left to cause harm when they are no longer usable,

as spent nuclear power plants or nuclear weapons test sites are today, or spent uranium mine sites that are disproportionately (70%–80%) located on aboriginal lands and reservations.

Evolution's survival-driven innovation, as we have seen, is focused on species survival and habitat sustainability. This is the innovation pattern. Intelligent species innovation, then, was apparently also intended to follow this pattern. This appears to be the only way to do innovation that would not end up destroying all Earth-life. Would-be Homo sapiens innovators must choose survival-driven outcomes, and that means their innovations, like evolution's, have got to fit within Earth's ecosystems in a live-and-let-live manner that neither exploits nor lives outside of them. But as we have also seen, Homo sapiens' current way to do innovation is anything but survival driven. It is instead driven by profits, continuous growth, and the acquisition of dominant market share, even when that entails destroying Earth's habitats, species, itself, and any chances it might yet have of avoiding self-extinction. To survive, then, Homo sapiens must rethink and re-engineer its way of doing innovation and begin to follow evolution's survival-driven innovation pattern.

So, what happens when a species chooses to innovate and adapt in ways that put its own survival at risk, as Homo sapiens continues to do? How and what can Homo sapiens do now, if it is not already too late, to adopt evolution's way?

According to the evolutionary survival patterns we encountered earlier (adapt, innovate, mature, and migrate to survive or go extinct), when a species' efforts to adapt, innovate, and mature fails, it has but one option left to avoid extinction. It must be free, willing, and able to migrate—a topic we take up in the next chapter.

11.

Migration

A BUILT-IN EVOLUTIONARY SURVIVAL RESPONSE

To Survive, an Intelligent Species Must Be Free and Able to Migrate

> *"We are at a point in history where we are 'trapped' by our own advances, with humanity increasingly at risk from man-made threats but without technology sophisticated enough to escape from Earth in the event of a cataclysm."*
>
> *"Although the chance of a disaster to planet Earth in a given year may be quite low, it adds up over time, and becomes a near certainty in the next thousand or ten thousand years."*
>
> *"We have LESS than 100 YEARS to save the human race."*
>
> —*Stephen Hawking*

PROFESSOR HAWKING MAY WELL BE CORRECT, BUT there is a side to intelligent species evolution often overlooked—the reason intelligent species evolved to begin with.

Evolution wants us to survive, is very heavily invested in our species' survival, and has gone to a lot of trouble to make it so. We often ask how Homo sapiens managed to survive when all, even those closest to our species, have gone extinct. Is it possible that, like a Thoroughbred breeder with an eye on the prize, evolution might have been selecting and shaping what's needed from earlier hominins so as to fashion a descendant that will have enough of the "right stuff"; one to whom it would be willing to give the keys to the planet; one that will be capable of shaping its own destiny and will, despite having free will, ultimately choose correctly; one that will be willing to take on the very, very difficult challenge that evolving to evolutionary maturity will entail; and, yes, one that can go on to populate the stars? Is Homo sapiens *the one*?

Can one imagine for a galactic moment evolution like an ever-vigilant chaperone prodding our species on to higher and higher levels of maturity? All through hominins' history, climate events appear to have been used to drive our species to (A) adapt, (I) innovate, (M) mature, (M) migrate, (S) survive, or (E) go extinct—(AIMM–S/E). These appear to constitute an unbreakable, sequential, and recurrent set of species survival patterns an emergent intelligent species must successfully execute at each rung of the Species Survival Maturity Model before it can get to the next, higher rung—or go extinct.

All earlier hominin species, though successful for a while, apparently failed to negotiate some fateful rung and went extinct, leaving Homo sapiens as the last of their kind. By the time Homo sapiens showed up, evolution was well along the route of winnowing an assortment of hominin species down to one. And 70,000 or so years ago, our species, perhaps carrying the essence of many prior hominin species, ensured its survival by migrating out of Africa at a time of habitat loss due to climate change brought on by the last ice age.

The urge to flee to safer climes, to migrate from country to country, across oceans, continent to continent across the planet, and, eventually, even to other habitable planets in the face of existential threats is an evolutionary survival trait built into our species every bit as much as the evolutionary compulsion to reproduce. The universe's built-in cycle of birth, life, death ensures that all planets, stars, and galaxies come with an expiration date. Hence, if a species is to survive beyond the expiration date of its home world, migration to other planets and to the stars cannot be optional. But interplanetary migration, it would seem, won't be easy.

In order to become an interplanetary species capable of survival beyond its home planet's expiration date, evolution appears to require an emerging intelligent species to first attain evolutionary maturity. The clear indication is that if a species fails to reach that level of maturity, it will experience self-extinction and deny itself the opportunity to become an interplanetary/inter-solar-system species.

In addition, while evolution has built in enough time before Earth's warranty expires, as we have seen, it apparently also requires the resident intelligent species to ensure its habitats remain sustainable as well, or at least until evolutionary maturity is attained. Indeed, the inability to survive beyond the point of self-extinction is believed by some scientists to be a reason we may not have heard from E.T. as yet. The idea is that potential emerging intelligent species, at varying stages of development on different worlds (assuming there are such), fail to survive long enough to enable discovery and communication with one another.

The evolutionary pattern here is clear. To survive, an emergent intelligent species must become mature enough to avoid self-extinction and become capable of interplanetary, inter-solar, and probably even intergalactic migration. But to survive, an intelligent species must maintain and sustain the habitats and species that keep the Earth itself alive. As noted in Chapter 7, "Habitats," species survival ensures habitat survival, and habitat survival ensures species survival.

Just as Homo sapiens wasn't the first life-form to change the composition of the planet's atmospheric gases (cyanobacteria beat us to that one, you'll recall), we may not be the first life-form to engage in interplanetary migration. Another of those "small stuff" organisms, like the cyanobacteria and phytoplankton encountered earlier, may have already done that. Some scientists are starting to speculate that these "aliens" might have actually brought life to Earth from Mars. Richard Carlson wrote a series of books around the theme "Don't Sweat the Small Stuff," but as we have seen, and will see again, sweating the small stuff seems to be a specialty of evolution.

Life would face significant challenges on Mars today, but billions of years ago conditions might have been more hospitable. NASA's Mars rover Curiosity found oxygen (not breathable atmospheric oxygen; there is none of that on Mars), water (not running surface water), and sulfate and other elements in a Martian rock sample, all potential clues to prior existence of microbial life. NASA noted that life might have existed on Mars billions of years ago, before it existed on Earth. According to Steven Benner, a biochemist of Westheimer Institute for Science and Technology in Florida, though it is not yet proven, evidence is accumulating that Earth life may have originated on Mars before being brought here on a meteorite.

Some microbes, like extremophiles on Earth, are incredibly hardy and might have been able to survive an interplanetary journey after being blasted off their home world by an asteroid impact. Orbital dynamics show that it's much easier for rocks to travel from Mars to Earth than the other way around. And every so often, a Mars rock is found on Earth. In fact, as of early 2017, of the 61,000 meteorites found on Earth, 150 of them have been identified as hailing from Mars.

If life did migrate to Earth on a meteorite from Mars, we are confronted with yet another of those evolutionary patterns we have been noting. Whenever it comes down to a choice between survival versus extinction of a species, evolution appears to be decidedly on the side of survival. If that were the case here, evolution was not unwilling to cross heaven and

Earth, literally, and even interplanetary space, to sustain life, defying the inexorable universal pattern of birth, life, death, even if that meant shipping microbial life, like stowaways, aboard a meteorite and crash-landing it on the surface of the Earth some 4 billion years ago, when it seemed like Mars could no longer be a suitable habitat for continuing the evolution of life.

To survive, life must be able to migrate. Migration is an evolutionary survival pattern that appears to be independent of a species' innovative and technological prowess, and we see it repeated by most species across the Earth, and maybe, if Earth life did originate on Mars, across the solar system, and probably across the universe, as well.

Humans have not even come close to accomplishing any of the exceptional feats performed by the following creatures, none of which would our species think of as "intelligent," but none of their feats depended upon their brains, innovations, or technologies.

The dinosaurs, for example, were perhaps the most successful animal species so far, surviving for about 165 million to 175 million years (give or take a few million), outliving many times over all those emerging intelligent hominins combined; cyanobacteria jump-started the engineering of a global atmospheric-gas changeover from CO_2 to O_2, with no knowledge of chemistry; and the tiny ocean plant, phytoplankton, quietly supplies 50%–80% of the Earth's oxygen today, and these, also, neither saw the innards of an MIT nor hung out in a Silicon Valley unicorn and would no doubt have had little use for an MBA or the machinations of a global multinational corporation.

Evolution went to a lot of trouble to get intelligent species this far and appears to want intelligent species like Homo sapiens to survive. But amazingly, none of the things our species has come to associate with paths to outsized achievements were at work here. This seems to suggest that species hoping to be evolutionarily successful will find that success most likely doesn't depend upon their IQ, excellence, distinctiveness, ingenuity, brains, or innovations and technologies, but more likely upon their willingness and ability to humbly evolve to evolutionary maturity, as these creatures did. Granted, they did not have a choice, but neither were they "intelligent," and that's precisely the point, isn't it? Given all the choices Homo sapiens have, hopefully our brilliance won't blind us and keep us from finding the path to maturity.

It would seem that evolution isn't looking for rock stars and rocket scientists (not that they are somehow bad) but just plain, ordinary folk obsessed with doing the right thing, evolutionarily speaking—reaching for maturity so as to survive. Who knew?

This, too, is a repeating, pervasive, and indispensable survival pattern, which a species can ignore only at its own peril. It brings to mind some of the words found in the Bible at 1 Corinthians 1:26–29 (KJV). One need not be a believer in a God to appreciate the underlying message here:

> *For ye see your calling, brethren, how that not many wise men after the flesh, not many mighty, not many noble, are called: But God hath chosen the foolish things of the world to confound the wise; and God hath chosen the weak things of the world to confound the things which are mighty. And base things of the world, and things which are despised, hath God chosen, yea, and things which are not, to bring to naught things that are.*

Sentiments those regarded as divergent in the movie *Divergent* would no doubt share.

Like evolutionary sustainability, migration evolution style also has a somewhat different take than one might expect.

> *As movement characterizes the rhythm of the universe, so migration choreographs the survival of life.*

One can't help but notice, as one looks up, how everything in the universe appears to be in a perpetual state of motion. Galaxies are believed to be accelerating away when seen from Earth. The Milky Way Galaxy and the Andromeda Galaxy are said to be on a collision course. The sun and the billions of other stars populating the Milky Way Galaxy are in perpetual motion as they, too, travel through the galaxy with their retinues of planets in tow. The planets, like Earth, are perennially orbiting their stars, and the moons are also orbiting their planets. In Earth's past, the continents once merged into supercontinents, Pangaea being the last of at least three, which then drifted apart again on their tectonic plates, forming the familiar configuration we see today. And given a few more million years, the continents will merge again to form yet another supercontinent.

> *What movement is to the universe, migration is to survival and life. If there is no movement, there is no migration, and if there is no migration, there can be no survival, and if there is no survival, there can be no life, and if there is no life, then there can only be extinction. Movement and migration appear to be a fundamental law of the universe. To survive, a species must be free and able to migrate.*

Yet species do not seem to choose to migrate spontaneously; they are often driven by necessity and often by events that result in habitat impairment, unsustainability, or outright loss, and this appears to be so even for those species that migrate on a cyclical basis. The outcome appears to be heavily influenced by the freedom to migrate; preparation, or an a priori state of readiness; triggers, or timing; simultaneously occurring events; available means; and the reception, hospitability, or lack thereof, at the destination.

Monarch butterflies, certain species of birds, wildebeests, and salmon might come to mind when one thinks of cyclic migrations prompted by the biological urge to reproduce in an ancestral or otherwise suitable habitat setting, or just to move on to the next known place to find food and water. These species are certainly free to migrate, are in a state of a priori readiness, have the means, and are mostly not constrained by circumstances. In the case of salmon, though, there is usually a hungry reception party, including polar bears and humans with large dollar signs in their eyes and nets upstream. Wildebeests face an even more treacherous journey across the predator-filled savannahs of Africa, and hungry crocodiles lurk along the riverbanks that come between them and their destination. These horrific experiences notwithstanding, while many individuals die, species manage to survive to live another day, and life continues.

Human migration, on the other hand, turns out to be significantly more challenging, if only because we learned how to farm and raise domesticated animals. Except for the few remaining hunter-gatherer tribes, we no longer maintain the skills, preparedness, or, more important, the desire to re-embrace that pack-up-and-go, mobile lifestyle that characterized most of our early ancestors. While we are still mostly able to flee (despite gridlock, border guards, and walls) in the face of danger, modern Homo sapiens are not naturally migratory but tend to want to settle down in suitable communal locations, usually near rivers and along streams, and remain there until forced to migrate to safety somewhere else.

Meanwhile, the migration triggers seem to have multiplied with time. Unlike those that drive other animals, nearly all of those driving Homo sapiens migration today are man-made.

Non-man-made migration-triggering factors have, since the last ice age, mostly arisen from climate events (El Niño, La Niña, ENSO), and include hurricanes, tornadoes, typhoons, superstorms, earthquakes, volcanoes, and forest fires. These have so far been observed to occur in irregular cycles, and often in multi-country, multi-continent patterns, and can result in anything from temporary and medium-term displacements that are nonetheless painful—even to watch—to life-changing destruction that impacts hundreds of thousands of people and causes great economic loss.

Major factors driving human migrations today, however, are mostly man-made and self-inflicted. Their increasing tempo is probably an indication of which direction our species' maturity index is headed. These include wars and regional conflicts; human trafficking and modern-day slavery;

famine; limitations on the ability to earn; lack of access to education and healthcare; lack of political, religious, ethnic, racial, and other human rights, including freedom to express one's sexual orientation and gender; oppressive and corrupt politicians, oligarchs, business tycoons, and governments; and, more recently, lack of environmental safety (pollution of food, water, and air) and, even more heinous, the lack of freedom to migrate, often despite a high probability of death.

According to the United Nations High Commission on Refugees (UNHCR), "War and persecution have driven more people from their homes than at any time since records began, with over 71 million men, women and children now displaced worldwide. This means one in every 113 people on Earth is either an asylum-seeker, internally displaced or a refugee, according to the 2015 edition Global Trends report." It has been more than 70,000 years since migrating out of the Great Rift Valley in Ethiopia, and close to 200 since modern civilization has taken hold, yet despite all of our technologies, innovations, intelligence, and wealth, we still seem unable to mature beyond self-interest.

Little of what drives or blocks Homo sapiens migration today existed at the time our early ancestors began heading out of the Great Rift Valley. Thankfully, there wasn't anyone turning them back at borders, or leaving them to drown in the seas and rivers of yesteryear, or building border walls and electrified barbed-wire fences to keep them out, as is all the rage today, or else we probably wouldn't be here to talk about it.

No, there were no passports, border guards, human traffickers, drug-dealing gangs, or powerful countries blocking their freedom to migrate. Our Earth has become a fortress. I often wonder how our species might feel if we were compelled to flee to an extrasolar Earth 2.0, and we had the technology to do it, but found ourselves confronted by a blockade of armed space cruisers erected by the denizens of that planet to deny us entry. When our forebears left Africa, no one stood in their way, but we may not be so lucky, and payback is never fun.

Evolution tends to err on the side of survival, and when we block opportunities for migration, and thereby survival, we are going against the evolutionary grain that got us here. The evolutionary pattern requires freedom to migrate, and our species seems to be turning away from this most basic species survival need. Our increasing inability to recognize and support this need is probably yet another indicator of the level of our species' maturity. Today, blocking members of our own species fleeing climate change, war and famine, seeking access to other countries on Earth, may be one reason our species could end up blocking itself from access to the stars.

There are lessons still to be learned from the migration of our early ancestors. Studies mapping human genetic diversity support the theory that modern humans emerged in Africa and identify the Middle East as their gateway to the wider world. People everywhere descended from a single African female, from a single migration of early humans from Africa. Unlike us today, they had the freedom, a priori readiness, preparedness, and disposition of hunter-gatherers to pick up and go. Timing turned out to be the next significant factor.

According to the genetic and paleontological record, about 100,000 years earlier, and at a time of lower sea levels, previous hominin species (Homo erectus, Neanderthals) migrated from Africa to Eurasia through a now-flooded land passage at the southern end of the Red Sea (the Bab el-Mandeb), but Homo sapiens started to leave Africa only between 60,000 and 70,000 years ago, perhaps using the same corridor, as sea levels were about 230 feet lower, due to the onset of ice age conditions that locked up water in vast polar ice caps.

According to the Geno-Graphic Project on Migratory Crossings (underscoring the importance of timing): "There were probably earlier attempts. Modern human remains have been found at sites in the Middle East that are in excess of 100,000 years old. Yet these trailblazers likely left little or no genetic trace on humans living today, suggesting that either climate change forced them to double back or they died out."

How significant was it that our forebears were able to leave just when they did? Was it coincidental that it was during the onset of an ice age, when the sea level was lower, making it easier for them to cross and migrate out of Africa? Was that ice age, which also rendered the Great Rift Valley habitat no longer capable of sustaining life, which in turn triggered our forebears' migration, occurring when it did just a big happenstance, though a wonderfully coordinated one? Or was this evolution doing a Moses long before there was a Moses, sucking up the waters of the seas and oceans into the polar caps—"parting the waters," so to speak—so our ancestors could migrate? How serendipitous was that?

And, digging deeper still, how even more serendipitous was it that crossing was only possible due to the rupture of the African and Arabian tectonic plates some 25 million years ago, resulting in that land passage? Reverse the order of these events and the survival of our species might not have occurred. There is no doubt their timing was spot on, but was it simple coincidence, or was it evolution at work once more, rearranging the Earth as it had that ice age.

Now, 70,000 years later, as we, their descendants, contemplate our own potential migration to another planet, might evolution do for us what it appears to have done for them? Could there, would there, also be similar, seemingly serendipitous events? Might there be such a thing as a "right time window" to migrate to an Earth 2.0 as there was for our forebears to migrate out of Africa? A time when it would be more favorable than other times to look for, find, and migrate to an Earth 2.0, as there appeared to have been a right time for our forebears to leave Africa—when an ice age appeared to have significantly lowered the risk of a Red Sea crossing?

And could it be, too, that as plate tectonics facilitated the path across the Red Sea, and climate change triggered that migration out of Africa, might climate change once more be sounding the alarm for yet another

Homo sapiens survival-driven migration? This time, perhaps, rather than plate tectonics carving a path across the Earth, an alignment of Earth with potential habitable planets as the solar system travels through the Milky Way Galaxy will bring our planet much closer to a potential Earth 2.0?

And what could "more favorable" mean in this context? Perhaps a rogue star will pass near or through our solar system just when we need it, as Scholz's Star is believed to have done 70,000 years ago, passing at a distance five times closer than our closest star (Proxima Centauri, which is 4.2 light-years away), possibly as our forebears were heading out of Africa. Now, that would be as serendipitous as they come, wouldn't you say? And wouldn't that be a worthy mission for NASA?

"To everything there is a season, and a time to every purpose under the heavens" (Book of Ecclesiastes 3:1 and Pete Seeger's 1950s song "Turn! Turn! Turn!"). We might well be here today only because our early ancestors got the timing right. How did they? Was climate change their alarm clock and migration trigger?

Earlier, we observed how evolution appears to employ climate change, and the often-accompanying habitat loss, to prod species to the next higher level of maturity and migration, or failing that, to extinction.

Many now believe that habitat loss due to climate change finally drove our ancestors to leave Africa. As noted in Chapter 2, "Shortly after Homo sapiens first evolved, the harsh climate conditions nearly extinguished our species." Some scientists even believe that the human population may have fallen to just a few hundred individuals who managed to survive in one location. This cold snap would have made life extraordinarily difficult for our African ancestors, and indeed, the genetic evidence points to a sharp reduction in population size around that time. In fact, the human population likely dropped to fewer than 10,000.

Just like birds, butterflies, salmon, and wildebeests, our early ancestors had the means to migrate. They needed little more than their hunting tools, the occasional raft, and of course their own legs and feet—they mostly walked across the planet. While there were none of the modern modes of travel we take for granted today, they were equipped with what they needed at that time—access to food and water and the means to protect themselves from the elements—as evolution rarely expects a species to accomplish things it's not yet equipped to do. Monarch butterflies and birds can fly, salmon can swim (even upstream), wildebeests can run, and Homo sapiens can walk. Per the diagram below they literally walked across the Earth.

But who were these original hunter-gatherers who migrated out of Africa and went on to become the origin of all extant humans today? What, if anything, is known about them? What might they have looked like?

Using DNA analysis, scientists have identified what they refer to as the oldest race of people in Africa, and on the planet. The most detailed analysis

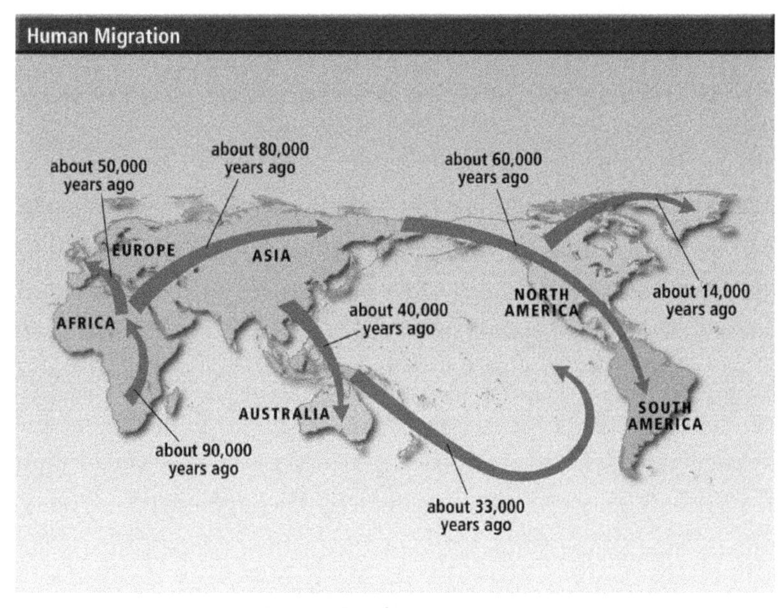

Fig. 11-1 Map of human migration
Map of early human migration routes across the planet
(Source: The Smithsonian)

of African DNA indicates the San people of southern Africa, who have lived as hunter-gatherers for thousands of years, are likely to be the oldest population of humans on Earth.

According to the article "World's most ancient race traced in DNA study" published in *The Independent* in 2009, "The scientists analyzed the genetic variation within the DNA of more than 3,000 Africans and found that the San were among the most genetically diverse group, indicating that they are probably the oldest continuous population of humans on the continent—and on Earth."

The scientists also found genetic markers in the DNA of present-day inhabitants of East Africa living near the Red Sea that indicate they belonged to the same ancestral group that migrated out of Africa to populate Asia and the rest of the world. West Africans speaking the Niger-Kordofanian language were found to share many genetic traits with African-Americans, indicating they were the ancestors of most of the slaves sent to the New World.

While there is no way to say what these migrations were like at the time, accounts of modern migrants fleeing war-torn countries like Syria and parts of Africa, or from drought-stricken parts of the globe, offer some insight. But these bear only scant resemblance to what our early ancestors might have encountered. While their migrations were a matter of life and death, with 20/20 hindsight we know the survival of our entire species rested on them as well. When Roman legions invaded the early British Isles, they encountered, among other peoples, the early Brits; when the Brits and Europeans migrated to the Americas, they encountered the indigenous peoples of this hemisphere. In almost every case these foreigners drove the natives to near extinction. One thing early humans would not have encountered, for the most part, were other humans—and one must say, thankfully, given the reception modern-day migrants are receiving.

Wherever they went they might well have been the first of their kind to have gone there.

They would have had no idea what they would encounter. There were no countries, hence no borders, passports, border guards, human traffickers, drug-smuggling gangs, or governments blocking their freedom to migrate. Early Earth was still a pristine paradise, devoid of the ills one is sorry to observe our species has since unleashed on it wherever we have settled.

Apart from the fact that these migrations enabled Homo sapiens to begin to dominate the planet, one can think of few stories (the jury is still out on this one; I hope evolution turns out to be right about us) in which human migration has, so far, been a good thing for Earth's environment, its habitats, oceans, forests, ecosystems, flora and fauna, or even our species itself. If you can think of any, please send me some. There can be little doubt that while our species has made remarkable progress on many fronts, we still have a lot to learn and do when it comes to our ability to sustainably co-exist in an Earth-like habitat before we are ready to take on another world.

While our early ancestors knew for sure they had to leave Africa, they probably didn't have a specific destination in mind or any idea of what to expect the way people migrating to another country do today. Few, if any, in the migrating party would have seen any place other than the Great Rift Valley of Ethiopia.

There were no maps, no roads, no well-trodden or previously marked paths. They may have had no idea where they would end up. They had no specific destination. They were truly going where no man had gone before. Theirs were more than likely the first human footprints left where they traveled, and the first eyes to gaze out across the vast expanse or peer into the dense forests, unable to know whether weal or woe awaited them. All they probably knew for sure was that they were likely to encounter fierce predators as well as prey, find fresh water and food sources, fresh air to breathe and land to walk on or streams and rivers to float down. Their destination, which must have changed from day to day, was probably wherever they found a source of food and water and felt relatively safe from predation.

One doubts that when they set out, they could have imagined they and their descendants would end up walking to the ends of the Earth any more than we can imagine that because some men have walked on the moon, our descendants will one day populate the stars.

Our forebears might have had no idea where their journeys would take them; all they knew was that they had to leave their habitat in Africa if they wanted to survive. It is no small coincidence that a similar migration has begun 70,000 years later by some of their descendants, and in some cases for the same reason—a habitat failing mostly due to climate change and the desire to survive.

These early migrants were the first "economic migrants" of our species to leave Africa and survive, and we, their descendants, should be grateful they were not halted and sent back, or left to drown in the rivers they crossed, or caught and traded as human slaves, as in Libya today. From them came a world full of people, you and me—something those blocking today's African, Middle Eastern, and other migrants might want to bear in mind.

Our ancient ancestors could not have imagined how differently this latter-day migration out of Africa would turn out for their descendants. They probably could not have dreamt of a time when their descendants would overspread the surface of the entire Earth, leaving very little if any unclaimed space for migrants like themselves fleeing to find shelter and start over. Nor, as they themselves were fleeing, would they have imagined there could come a time, even 70,000 years later, when their fleeing descendants would be blocked and turned away, not by aliens, but by members of their own species.

Today, with no place left to go, and 7.8 billion of us and counting, it is more than likely that if faced with yet another potential species-wide extinction event, the next human survival-driven migration will not be from Africa to Europe, or Central America to the U.S., but from Earth to another planet.

And amazing as it might seem 70,000 years later, despite all we have accomplished and are most proud of, we are just like our early forebears. We still don't know where we might be heading and what we'll find once we get there, assuming we get there. All we know is that the next species-wide, survival-driven migration can only be interplanetary, and maybe like our forbears, we will find that we are the first intelligent species to arrive there.

The pressures that drove early human migration are still very much with us today, but the evolutionary escape hatch, survival through migration, is fast closing. In fact, if Homo sapiens were to be confronted over the next several hundred years with a growing species-wide extinction event, despite the enormous across-the-board advances of the last 200 years, today's migrants leaving Africa would be worse off than early humans were when they left Africa. Those early African migrants faced a mainly uninhabited planet's worth of unclaimed territory. Today, all that territory has been claimed. Migrating as our forebears did is no longer possible; migration as early humans experienced it is no longer an option.

The hunter-gatherers fleeing the Great Rift Valley area of Africa might have numbered in the very low hundreds (probably 200 or so; very unlikely to be in the thousands) and were more accustomed, predisposed, and prepared to pack up and go at a moment's notice than we could ever hope to be today. While they might not have known where they were going, they at least did not have to worry about being denied access to the lands they encountered. The possibility of having to leave the planet as they knew it then, with its gravity and oxygen-rich atmosphere, sunshine, innumerable food and water sources, and relative ease of finding a place to pitch camp and go when they felt the urge, was not even imaginable and never could have crossed their minds. Evolution appears to match each survival migration episode with the tools and capabilities necessary for a successful migration. All they needed to get going were their legs and feet, simple tools, and a few weapons. The rest they would invent along the way.

If one can imagine evolution with a geologic stopwatch, "evolution's clock," that times how long a species might spend on a particular rung of the maturity model, it might appear that 70,000 years after that first migration, evolution is getting ready to signal the countdown to transitioning to the next rung. Homo sapiens definitely have a few more rungs to go, but if we manage to successfully negotiate these and avoid self-extinction, we will face the final few rungs to evolutionary maturity and becoming an interplanetary species—or we will go extinct.

If our species is approaching the beginning of another species-wide, survival-driven migration, we will need what our early ancestors had, plus some. Compared to interplanetary and inter-solar-system migration, theirs might be considered a less challenging migration. But evolution has shown a tendency to match the degree of challenge with the rung and level of maturity on the Species Survival Maturity Model. So if a Homo sapiens survival-driven migration is looming ahead of us, and also seems to be orders of magnitude more challenging than it might have been for early humans, it can only be so because we have clawed our way up a few more rungs of the ladder on the way to evolutionary maturity.

So what might one need for such an interplanetary migration? Without in any way attempting to diminish the massive challenge and awesome struggle and responsibility our forebears' migration out of Africa must have been as they carried with them the weight and future of an entire species, it is important to draw attention to the essential differences an interplanetary migration implies for our or any Earth-life species and to the evolutionary phase change such a migration implies.

In Chapter 5 we noted some of these evolutionary phase changes, such as from noncellular to single-cellular, to nucleated single-cellular, to multicellular organisms, to complex animals and plants, to nascent intelligent species (hominins) and finally (so far) to Homo sapiens, the last of the hominins. Each phase change was separated and/or precipitated by significant species survival challenges such as mass extinctions, causing many key species of the prior phase to go extinct or be carried forward into the succeeding phase as symbionts in replacement species. We noted, too, that accompanying each of these evolutionary life-form transitions was a step (and in some cases an order of magnitude) change in the succeeding species complexity, habitat, adaptations, and innovations needed to survive in those new habitats.

So, if one were to view a potential interplanetary migration from the above perspective and ask what a species confronting such a migration might need, and if one were to further ask what other prior Earth-life migrations exhibited similar characteristics, one potential candidate immediately leaps out—the first animal species (vertebrates) migration from sea to land (otherwise known as land *colonization*—a word I would have rather not used) that occurred during the Devonian Period some 375 million years ago.

This was a momentous event in animal life history, perhaps as momentous as the transition of our early microbe ancestors to oxygenic photosynthesis and respiration. These early animals had to come up with critical key adaptations and innovations to survive in their new habitat—dry land. And not unlike those early animals, Homo sapiens too will need to come up with its own adaptations and innovations to survive in a new habitat—outer space, or on a different planet.

In fact, it is amazing how eerily similar, yet opposite, animal migration to dry land compares with a potential Homo sapiens transition to outer space or another planet. Animals emerging from the oceans needed the ability to overcome gravity, evolve the apparatus to breathe dry air, remain hydrated, as well as modify or develop new mental and sensory capabilities to see, hear, navigate, and function in this new habitat. Instead of fins they needed limbs, and instead of gills they needed lungs, along with numerous other adaptations and innovations.

For those early animals, absent the buoyancy of water, gravity meant collapsing under their own body weight every whit as much as loss of gravity would mean, among other things, the loss of humans' ability to orient themselves and remain fixed on terra firma. Whereas early animals needed to build strong muscles and skeletons to support their increased weight, humans will have to work hard at maintaining their muscles and skeletal frame due to a loss of bodily weight. Whereas early animals would have needed new breathing apparatus and over time to trade their gills for lungs, humans will have to develop new breathing apparatus and/or carry the air they need in a pressurized spacesuit, spacecraft, or other similarly pressurized surface unit. And, whereas early animals faced the risk of drying out due to not being immersed in water, humans will need to carry, find, or produce the water they will need to survive.

One of the more successful strategies early animals migrating to a life on dry land developed, and one we seemed to have also figured out, was the need to package and carry the water-based environment in which they evolved with them to the new dry-land-based environment to which they migrated. This is the reason embryos develop in a liquid-filled womb; why the salt concentrations of both our bodies and sea water are so similar; why our tears and sweat have the salty tang of sea water; and why the proportions of certain chemical elements in our tissues, such as potassium, chloride, and sodium, are not unlike those found in oceans. In fact, it would seem as though animals never truly migrated from the sea so much as they found ways to take the sea with them onto dry land.

Similarly, to survive in outer space and on non-Earth-like planets, modern humans will have to take some of Earth's biosphere with them: enough of its ecosystems, microbes, air, water, food sources, gravity, and UV ray screens.

We are already discovering how living in outer space and on planets like Mars may well engender over time a successor human species that would evolve to be more compatible with the new planet. The necessary transition would not be just from one biological form to a more complex one as from caveman to spaceman, but, as explored in Chapter 5 on species-driven evolution, from spaceman to "Bicentennial Man." And, the existential event that seems to accompany and even drive such evolutionary phase transitions is already emerging—the impending so-called sixth mass extinction.

Here assembled before our very eyes, then, are all the elements of the next great evolutionary phase transition: a threatening existential event, global warming, which in turn could drive a fundamental change in habitat, an interplanetary migration, which in turn will engender an order of magnitude change in species complexity and drive the adaptations and innovations to migrate to and survive in this new habitat—outer space or a non-Earth-like planet.

The only question that remains to be answered, then, is: What form will the emergent successor Homo sapiens species (if still Homo sapiens) take?

And it's both interesting and sobering to consider that most, if not all, the possessions we hold most dear won't be of any use on such a journey. Whether we are wealthy or poor, live in a spectacular house by the sea or are destitute and homeless, nothing we own will be of any use or value to us at that time. It's quite ironic, isn't it? We will all be starting again from scratch. In this respect, leaving on such an interplanetary migration bears an eerie resemblance to, as we euphemistically say, "departing this world."

Elon Musk's elaborate plans notwithstanding (one would hope he inspires other Moneyed and Techno sapiens), Homo sapiens' interplanetary migration would need an armada of transport systems equipped with a reliable and nearly inexhaustible power source; a continuous source of oxygen and water; waste recycling and disposal systems; and artificial gravity, to name a few. Of course, we would have had to identify a specific habitable planet in the habitable zone of a specific solar system, rocky and Earth-like, preferably uninhabited, and hopefully close enough that the trip would be survivable, measurable in thousands or at most millions of miles/kilometers rather than in light-years.

And, even if we were somehow able to accomplish this, it would be impossible, as Carl Sagan once observed, to transport anything near the number of people born every day (240,000 when he made the observation, but more than 280,000 today), let alone the rest of us who got here a few decades ago. In other words, folks, sorry to be the one to break it to you, but just as prior phase transitions of Earth-life species to a succeeding one have meant the extinction of most of the extant species, it's unlikely anything more than an infinitesimally small fraction of humanity will be able to escape to an Earth 2.0, even if we found one and were able to get there.

Nonetheless if we do, as our microbe ancestors became symbionts in succeeding microbe species and passed on their genes, nanomachines and all essential capabilities amassed hitherto, so too humans will pass on, in addition, their store of knowledge and culture that uniquely defines what's human. The genetic information in the DNA of our microbial ancestors will have been augmented with the store of human knowledge and would accompany those who continue on to the next phase—onto other planets.

At first, they will be earthlings but with time they will become Martians or will bear the names of the planets to which they will have migrated. But even as we still carry in every cell, our intestines, and other organs our microbial ancestors, so too they would carry humanity. They will no longer be earthlings—but they would forever be Human.

We currently have none of the foregoing capabilities, however, and are unlikely to have anything remotely resembling this level of preparedness for at least another 100 years, if not 200. While it's one thing for our species to have gone from caveman to spaceman in 300,000 years, no nation has had to worry about the preservation of the species and preparing provisions for an interplanetary space flight with a large, unspecified number of non-astronauts. And none has had the burden of finding an Earth-like habitat capable of sustaining human life and establishing a sustainable habitat once there. Watching *Mars*, the television miniseries, would leave one with little doubt that any meaningful interplanetary human migration on a permanent basis is at least 100 to 200 years away, if not more.

Accomplishing any of this would be difficult enough, but it would be an entirely different thing to reach the level of global cooperation and trust such an effort would almost certainly demand. And this level of cooperation and trust can come only from first attaining, as a species, levels of maturity several orders of magnitude greater than we have so far managed. While it's not impossible that the necessary innovations and technologies could come earlier, given survival rather than profits as incentive, the survival of our species is unlikely to depend upon technology only. How so?

If one considers the worldwide efforts undertaken to address growing climate change issues as a guide to what a potential extinction-driven interplanetary or inter-solar species survival migration effort might entail, one soon discovers it is not a lack of alternative energy technologies or innovative solutions that are delaying the elimination of the world's dependence on fossil fuels. Rather it is our misguided priorities as a species; more accurately, it is the imposition of the will of "the few" that leaves us seemingly willing to risk potential habitat destruction and species extinction, instead of choosing survival—essentially our immaturity and lack of freedom to choose, topics we will get to later.

Just getting to anything remotely resembling the unified global effort interplanetary migration to an Earth 2.0 would undoubtedly require could take half a century, if not longer. Other than, perhaps, Hollywood, in movies like *Interstellar* and *The Martian*, which at least attempt to explore elements of some of the potential issues, none of Earth's leaders or global institutions appear even remotely engaged in exploring these issues. It's apparently not even on any nation's long-range agenda.

Meanwhile there is another kind of migration (climate change-driven migration) that will preoccupy us long before we have to worry about migrating to another planet.

In an October 2018 IPCC Special Report on the impact of global warming at 1.5°C above pre-industrial levels, a group of scientists convened by the United Nations to guide world leaders, describes a world of worsening food shortages and wildfires, and a mass die-off of coral reefs as soon as 2040 and many reading this may well be still alive then. In fact, it claims that the impact formerly forecasted to occur at 2°C are now more likely to occur instead at 1.5°C and by 2040—including inundating coastlines and intensifying droughts and poverty—if greenhouse gas emissions continue at the current rate.

Faced with global climate change, what options do Homo sapiens have when terrestrial migration is increasingly no longer a welcome option? The effect of climate change on migration today is complex, both forcing displacement and limiting people's ability to move. Long before the twenty-first century ends, and as early as 2040, the world could see substantial numbers of climate refugees—people displaced by either the slow or sudden onset of the effects of climate change.

The world is witnessing the emergence of a uniquely twenty-first-century phenomenon: the need for migration with no place left to migrate, and the growing tensions caused by people with a need to move to survive in a world with no place else left to go. Yet we live in the era of the greatest human movement in recorded history (writes the director general of the International Organization for Migration in a 2015 report), in a world with one in every seven persons a migrant, and more and more people moving in the context of climate change.

Over 75 million people live just one meter or less above sea level, and the Intergovernmental Panel on Climate Change reports that much of this coastal land may be under water within the lifetimes of people alive today, placing this population at significant risk of displacement. Current forecasts for the number of climate-induced migrants by 2050 vary between 25 million and 1 billion, depending on climate scenarios, adaptation measures taken, and other political and demographic factors. This estimate would rise if those who cannot move but need to be taken into account.

Of the 1 billion climate migrants forecasted, 232 million are international migrants. Nearly three-quarters, 740 million, are internal migrants. People move for a variety of reasons, influenced by economic, social, political, environmental, and demographic conditions. There are also growing numbers of people displaced by conflict and natural disasters. An estimated 50 million are currently displaced by conflict, the highest number since World War II, of which 16.7 million are refugees and 33.3 million are internally displaced people.

In addition, there is growing evidence of links between climate change, migration, and conflicts, says a report by the Center for American Progress on Climate Change, Migration and Conflict. Recent intelligence reports and war games, including some conducted by the U.S. Department of Defense, conclude that over the next two or three decades, vulnerable regions (particularly sub-Saharan Africa, the Middle East, and South and Southeast Asia) face the prospect of food shortages, water crises, and catastrophic flooding driven by climate change. These developments could demand international humanitarian relief or military responses, often the delivery vehicle for aid in crisis situations.

In a world with no place left to flee, some lives have become more equal than others. But one does not get to choose in such a world (the dinosaurs didn't), and it is this loss of the freedom to migrate to survive, at the individual and at the species levels, that has ended prior civilizations and driven prior species to extinction, and that now puts our species at risk. Migration is the species survival escape hatch, and despite numerous Homo sapiens innovations and adaptations, none has been as effective at moving species out of harm's way as outright migration.

How as a species we handle intra-Earth migration might well be an indication of our maturity and hence ability to deal with the issues of an interplanetary migration. To survive, life must be able to migrate. Migration is an evolutionary survival trait built in to our species every whit as much as the evolutionary compulsion to reproduce.

Early humans were comparatively fewer in number, might have been more united as a species, had a vast empty planet before them, and had mostly all that was necessary to migrate. Unlike early humans we, their descendants, have not yet found an Earth 2.0, don't know where our next home might be, and even if we did, won't have the means to get there for yet another 200-plus years. Will we survive?

There are some who view these issues with deep concern for our species, and I remind you of the thoughts of one such person we met earlier, the late Stephen Hawking, quoted at the beginning of this chapter.

"We are at a point in history where we are 'trapped' by our own advances, with humanity increasingly at risk from man-made threats but without technology sophisticated enough to escape from Earth in the event of a cataclysm," and, "Although the chance of a disaster to planet Earth in a given year may be quite low, it adds up over time, and becomes a near certainty in the next thousand or ten thousand years."

"We have LESS than 100 YEARS to save the human race."

In this section we saw that an intelligent species must evolve in a sustainable-capable habitat and be prepared to eke out a living in it, as well as maintain that habitat and its species in order to survive. In addition, it must be prepared to adapt, innovate, mature and migrate to continue to survive or it will go extinct.

The next section explores what else it takes to survive and grow to evolutionary maturity, and what will happen if we don't—extinction.

PART FOUR

Survival
The Reason Intelligent Species Evolved
An Evolutionary Game of Life

OF ALL THE GALAXIES, SOLAR SYSTEMS, and planets in the universe that might have become home to our species, evolution apparently evolved and terraformed the Earth, equipped it with a seemingly endless variety of flora and fauna to serve as its life-support system, then 4 billion years later evolved what may well have been the first collection of nascent intelligent species in the genus *Homo*, and gave it the same meal plan it gave all other species—an equation of life it must solve to survive.

To obtain the energy each must have to survive, a species must figure out how to eke out a living within the planet's habitats and ecosystems without destroying these, other species, and itself.

In this section, we encounter evolutionary survival patterns that seem intended to help intelligent species find their way to survival, and a Species Survival Maturity Model; together, they imply rules and goals and suggest that species live out their lives as if competing in a game of life, with survival as the ultimate prize.

To survive, an intelligent species must attain evolutionary maturity and reach the Survival Plateau on the Species Survival Maturity Model—that point in its evolution where it has matured beyond its ability to drive itself and other species to extinction.

Survival
The Reason Intelligent Species Evolved
An Evolutionary Game of Life

CHAPTER 12

Survival

The Reason Intelligent Species Evolved

The Default Outcome in the Evolutionary Game of Life

CHAPTER 13

Evolutionary Maturity

The Only Path to Intelligent Species Survival

CHAPTER 14

Extinction

A Human Universe without Humans

12.

Survival

THE REASON INTELLIGENT SPECIES EVOLVED

The Default Outcome in the Evolutionary Game of Life

MONARCH BUTTERFLY CATERPILLARS HAVE EVOLVED the ability to store toxins known as cardenolides, or cardiac glycosides, obtained from their milkweed diet, making them poisonous to birds, according to a study published in *Proceedings of the Royal Society B — Biological Sciences*. But there are more than glycosides in the monarchs. There is also a message from evolution Homo sapiens can't afford to miss. Like the built-in ability of the monarchs to avoid becoming bird food and to migrate to survive to places like California and Mexico, intelligent species also evolved with the built-in ability to adapt, innovate, mature, and migrate (even to the stars if necessary) to avoid extinction and survive the universe's existential threats.

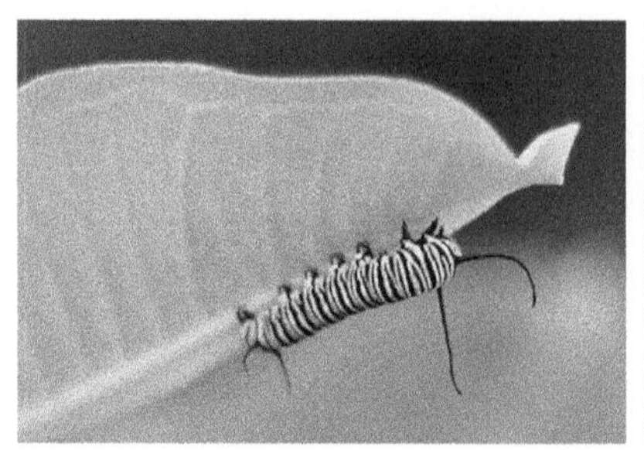

Fig. 12-1 Monarch butterfly on swamp milkweed in Michigan
(Photo by Jim Hudgins/USFWS. National Fish and Wildlife Foundation through the Monarch Butterfly Conservation Fund.)

But the monarch butterfly is in trouble. Man-made climate change has intensified weather events, which are impacting monarch populations. Pesticide use is destroying the milkweed that monarchs need to survive. In addition, habitat loss and fragmentation has occurred throughout the monarch's range, and this is the message evolution is delivering to Homo sapiens through the monarchs.

You see, like the monarchs, humans are in trouble, and for similar reasons. And because species survival and habitat survival equal Homo sapiens' survival, by acting to save the monarchs we will be taking steps to save ourselves.

This chapter explores further what it takes for intelligent species to survive and claims that survival rather than extinction may well have been the preferred evolutionary outcome.

> *"The most important question you can ever ask is if the world is a friendly place."*
>
> —*Albert Einstein*

If one were to rephrase Einstein's question a little and ask instead, "Is the universe a friendly place for the survival of evolutionary species?" the answer from an Earth-life perspective would have to be "Yes and no—it depends."

Yes because, despite the universe's well-earned reputation for galactic mayhem, chaos, and destruction, its endless cycles of birth, life, and death, there has never been a period in the history of the planet when life has not continuously existed and survived, except perhaps during the Late Heavy Bombardment that took place during the early formation of the Earth, the first 700 million years, when life may have been repeatedly started and wiped out. Nonetheless, life evolved as soon as it was possible and hasn't looked back since.

No because, despite evolution's efforts to keep life alive, the universe never seems to stop trying to destroy life. From an Earth-life perspective, how can one conclude otherwise?

This chapter explores how evolution might be attempting to solve this evolutionary species extinction problem with the evolution of an entirely new species that it would endow with the ability to mature, survive, and evade the universe's existential threats. Along the way, one finds it hard not to conclude that intelligent species survival, not extinction, must therefore be the preferred evolutionary outcome.

The universe's unending cycles of birth, life, and death notwithstanding, evolution found a way for life to keep going once started. By equipping species with the ability and instinctive urge to reproduce, and thereby the ability to perpetuate themselves, evolution was making a clear statement of its intentions. Looked at from the perspective of making a choice between survival and extinction, evolution, it would seem, has always had its feet firmly planted in the survival camp.

Not even the major mass extinctions—at least five, including the mother of all extinctions, the Permian-Triassic extinction, when as much as 95% of all species are believed to have died—or repeated snowball-Earth climate-change events were able to bring life to an end on Planet Earth. And now 4 billion years and an untold number of extinct species later, here we are, still alive to write about it. Perhaps not even the impending sixth mass extinction, the Anthropocene, believed to be in progress, will end intelligent life on Earth. Since no prior hominin species existed during any of the five prior major mass extinctions, it remains to be seen whether intelligent life, Homo sapiens, will be among the survivors, if there are any, of this one.

It is not unreasonable, then, in the face of such species survival and continuity odds, to conclude that life, or species survival, appears to be the preferred outcome of evolution and the default condition of evolutionary species. And if survival is the default, then extinction must be either by choice or by accident.

In other words, while species extinctions do occur, extinction is not inevitable. Species evolve and survive until driven to extinction by events beyond their control, or self-extinction occurs—a topic we will have more to say about later. The dinosaurs evolved and survived for about 165 million to 175 million years and might still be around today if a massive meteorite hadn't rearranged their habitats, jump-starting the series of climate events that drove them to extinction. Had the dinosaurs survived for another 200 million years, it is unlikely they would have evolved to the point where they could have taken action to prevent their extinction. In fact, as of today, if faced with a similar threat, Homo sapiens couldn't. Yet the history of life, at least on Earth, shows that extinction is not inevitable.

But is that all there is to life? "We spend our years as a tale that is told. The days of our years are threescore years and ten; and if by reason of strength they be fourscore years yet is their strength labor and sorrow; for it is soon cut off, and we fly away" as the Psalmist laments in Psalm 90:9–10 (KJV).

Is life, and intelligent life in particular, to forever evolve, thrive, and survive, even for hundreds of thousands of years (early hominin species survived for 900,000 to 1 million years on average) or tens of millions like the dinosaurs, only to be randomly done in by the universe and disappear into extinction? Or is there something else? If indeed survival is the default condition of evolutionary species, and if indeed evolution is on the side of survival, then one might expect there is more to evolutionary life than we have seen so far.

And it would appear there might be. Evolution needed to find a way for life to survive even the Thor-like planetary projectiles the universe might hurl life's way. And it appears to have done so by evolving, for the first time, intelligent species with the innate ability to choose their own evolutionary outcome, and then partnering with those intelligent species to craft a path to their own survival—species with the latent ability to survive in spite of the random extinction-level events the universe might hurl its way.

Hence, to ensure species can survive such existential threats, evolution needed to do more than just continue squirreling away a few species to escape the horrors of the next mass extinction and restarting life over again each time. Evolution needed to equip species with the freedom and the ability to chart their own evolutionary course, to choose to do something the dinosaurs were powerless to do when confronted with an existential threat—to choose to survive rather than "go gentle into that good night." It was an evolutionary fork in the road at the species level, a choice that would potentially put 4 billion years of evolution at risk on the one hand, but on the other, if successful, would have the potential to evolve a species that could adapt, innovate, mature, and migrate to survive in a universe ensnared in endless cycles of birth, life, and death.

Thus, what with hindsight may come to be thought of as a grand plan evolution had all along, hominins may well have become the first group of species to evolve with the ability to choose its own evolutionary outcome. And now, Homo sapiens, their sole surviving descendant, is left, wittingly or not, to carry the torch on behalf of all prior hominins to choose what the dinosaurs could not—survival.

And yes, as evolution would have it, it will be up to Homo sapiens to choose.

Unlike everything else in the universe, it seems, unlike the planets, stars, and mighty galaxies, unlike any other known species, hominins have been endowed with the ability to change evolutionary and galactic outcomes. All faithfully and unfailingly follow the outcomes of the universal laws discovered by Einstein and Newton and an army of others, laws of physics and of chemistry, and the fuzzy laws that rule the quantum universe. All, that is, except intelligent species, as represented on Earth today by Homo sapiens, and maybe by other intelligent species elsewhere in the galaxy.

Galaxies can't choose to go their own way, and for the most part, neither can stars or planets. It takes the sun 200 million to 250 million years to circle the galaxy, and like a high-speed train (this one moving at 500,000 miles per hour), it must follow the tracks and cannot choose a shortcut. The Earth, faithfully following Einstein's laws of general relativity, has no ability to choose to rotate out of the path of another potential death-dealing dino-rock to preserve life, but eventually, if evolution gets its way, intelligent species on Earth, maybe even Homo sapiens, one day will.

While bound in many ways to the same laws, Homo sapiens has evolved with an innate emergent ability to choose among the many potential evolutionary outcomes possible within the boundaries of those laws. Homo sapiens can choose to survive—to adapt, innovate, mature, and migrate from one habitat to another, from one continent to another, in time from one planet to another, and eventually, with even more time, from one star to another. If successful, evolution's grand species survival experiment will have paid off, and intelligent species, perhaps even Homo sapiens, will go on to escape the universe's existential threats and populate the stars.

But lest we get too far ahead of ourselves, it's useful to reflect upon what survival might mean in a universe with a reputation for galactic mayhem, chaos, and destruction, and its seemingly endless cycles of birth, life, and death. Let's follow the story and discover the ingenious scheme evolution appears to have employed to enable intelligent species to evolve, mature, and survive in such a universe.

Given the billions of stars and trillions of planets that pepper the heavens and appear to be perpetually present, it is easy to walk away with the impression that they have always been there and always will be. But we now know that nothing in the universe lasts forever—not planets, not stars, and apparently, not even galaxies. The tale of the universe is a tale of birth, life, and death. A planet forms, survives, even for billions of years (Earth was formed some 4.5 billion years ago and may last another 5 billion), then gets absorbed into its star as the star ages and expands into the orbits of the planet on the way to its own extinction, sometimes becoming a red giant, as it is believed our star, the sun, will become. Nonetheless, for a very long time to come, the universe will be populated with similar celestial bodies, as dying stars and planets are continually being replaced by emerging ones.

> *Such was evolution's challenge: to find a way for intelligent species to survive in such a universe.*

To survive in such a universe, an intelligent species would have had to have the time to evolve on its home world. It must also have had enough time to develop and integrate around a common civilization and cultural survival narrative. It would by then have prioritized and developed the technologies, innovations, and abilities necessary to migrate from planet to planet, from one solar system to another, and possibly even from one galaxy to another, to escape the natural cycle of birth, life, and death that characterizes the component bodies of the universe. And it would have developed the innovations needed to avert random celestial shelling, such as what took out the dinosaurs, or occasional supernova or gamma ray bursts that can fry millions of planets within their kill zones.

So even if an intelligent species were to mature past the point of self-extinction, to continue to survive beyond the end of its home world—and potentially much, much sooner—it will need to become a multi-planet, multi-star species, as our forebears needed to become a multi-continent species, in order to survive. Where once they were seafarers, modern Homo sapiens will need to become spacefarers, planetary and solar surfers, migrating from one planet to another, from one solar system to another. And just as our species on its way to expanding across the Earth needed to migrate from continent to continent, crossing mighty oceans, scaling towering mountains, and braving the near-extinction-level challenges of the last ice age, adapting and evolving civilizations and cultures, innovating and creating the technologies needed to survive each phase, so must modern Homo sapiens as it embarks upon its own survival-driven migration across the stars.

As one reflects on the above, a familiar pattern begins to emerge. Life, and evolutionary intelligent species survival in particular, begins to take on the form of a high-stakes game of survival versus extinction in a universe wherein everything goes extinct unless one can migrate to escape it. It is an unending game of survival-driven migrations, interspersed only by periods of geologic time spent frantically preparing for the next survival-driven migration. Extinction lurks at each transition like the crocodiles that prey on unsuspecting wildebeests along the banks of the Mara River as they try to get to the grasslands on the other side. The only way to survive is to keep moving, like those wildebeests racing across the Serengeti, like clouds drifting across the skies, like planets orbiting around their stars, like stars around their galaxies, and like fleeing galaxies accelerating away from Earth.

Everything in the universe is in a perpetual state of movement and migration. An intelligent species, then, must be able to survive each phase of this game, on each continent, on each target planet, and in each selected solar system. Like contestants in a relay race, with prior generations passing the baton on to succeeding generations, or perhaps like competing American Ninja Warriors, an intelligent species must be able to scale each obstacle, cross each barrier, unite, collaborate, and work as a team to strategize how to get to the finish line. In many respects, the Species Survival Maturity Model we are about to explore draws on all of these concepts and may well turn out to be the one game that out-games them all—the evolutionary game of life.

By now you are probably thinking there is one small problem—someone forgot to tell Homo sapiens about this game of life it's a part of, particularly the rules of the game. But did they? No one told Homo sapiens about Einstein's or Newton's laws, or of any of the laws that led to scientific discoveries about the universe and nature, yet we discovered them and adjusted our world view accordingly. We did it because we were curious and wanted to find out whether there was any rhyme or reason to how the universe and life on this planet works, and why we are here. And when we did, we discovered that the universe is comprehensible because its behavior can be modeled; it is governed by what came to be known as scientific laws.

And when we found these laws, and the idea that the universe followed consistent principles that one can understand, if not see, we tossed out all those ancient gods we had imagined existed—gods of love and war, of the sun and the sky, of the oceans and rivers and even land, and gods of rain and thunderstorms, earthquakes and volcanoes—and went on to change our values and world view accordingly. But one can hardly say the same has been done when it comes to what it takes for a species to avoid extinction and survive. It would appear that the laws and patterns that could enable an intelligent species to attain evolutionary maturity and survive beyond the point of self-extinction are now only dimly visible in the social habits of our very distant cousins, such as the bonobos—scientists consider them the nicest primates on the planet. These laws and patterns have been lost, like the Dead Sea Scrolls once were, perhaps disappearing not long after the transition to agriculture, and they may have remained hidden since then.

Whereas science has dispelled the fog and superstitions that once dominated our world view and understanding of nature and the physical universe, it seems no similar effort has been exerted to discover the laws and patterns that evolution must have left to guide an intelligent species along the path to survival. In this arena, some of that same fog, along with multiple superstitions, still imprisons our thinking. We remain uncertain and open to religions, philosophies, belief systems, and competing theories about how life should be lived. We have succumbed to ways of living and seeing the world that have led mostly to the collapse of earlier Homo sapiens civilizations, and there is potential for yet another collapse, this time on a global scale.

Homo sapiens, it would seem, failing to comprehend what it takes to engage with the universe's evolutionary survival mechanisms, perhaps failing to even recognize the presence and possible purpose of evolution's continuing role in species survival, has proclaimed a kind of eminent domain over all the Earth and its contents, and has declared itself lord and master of all it surveys. And, like that four-year-old with a hand grenade we met earlier, now a full-grown adult, our species is intent on wrecking and destroying all for what it wrongly perceives as existing exclusively for its own benefit.

But what if there is a different way—as different as scientific laws are to superstition and mythology? Is it possible that just as evolution has left clues to the workings of the physical universe and nature, it has also left clues, like bread crumbs, to help intelligent species like Homo sapiens discover how to live to survive, rather than blindly blundering into—or worse, consciously and perversely choosing—paths that can only end in extinction?

Those clues and survival bread crumbs are strewn about over geologic time so intelligent species looking back might find them. In previous chapters those clues were identified as species survival patterns, and in this chapter they are brought together to build a potential Species Survival Maturity Model—a model that might well be a potential survival framework, a manual for how to succeed at this game of life, that evolution left an intelligent species like Homo sapiens to follow if it chooses to survive rather than go extinct.

Indeed, self-extinction was never going to be a problem for nonintelligent species like the dinosaurs, never mind how many more millions of years they might have survived, and perhaps that's why cockroaches and beetles are still around, despite multiple mass extinctions.

Self-extinction is a "disease" only an intelligent species can catch.

The purpose, then, of this game appears to be to get intelligent species to mature and survive beyond the point of self-extinction, which implies that self-extinction is the greatest threat to intelligent species survival.

It's helpful to think of this model figuratively as the Earth encircled by a continuously ascending ladder that, like lines of latitude, begins at the bottom and continues all the way up to the top (as in the figure below), with numerous horizontal rungs, or circles. Each rung represents a level of maturity, and there are as many of these as necessary (eight rungs are shown) to enable an intelligent species to climb all the way to the top rung, the Survival Plateau, that level of maturity whereby a species has passed the point of self-extinction.

Rungs on the ladder represent the geological intervals when an intelligent species might exist and during which it must choose to mature and migrate to the next higher level to survive or go extinct. Each rung serves as a kind of sieve, a winnowing out of intelligent species that results in the survival of evolutionarily maturing intelligent species and the extinction of persistently immature ones.

Intermediate rungs seem intended to drive intelligent species to higher and higher levels of adaptations, innovations, maturity, and migration to survive, or to extinction. There appears to be no half-measures here. An intelligent species must be able to get to the next higher rung to survive or else it will go extinct on the current rung. This model certainly leads one to wonder whether this could be the reason Homo sapiens remains the sole surviving intelligent species in the genus *Homo* today.

The surviving species that eventually make it all the way to the topmost rung, the Survival Plateau, will have survived beyond self-extinction and will be well positioned to embark upon what seems like a follow-on stage of evolution—extraterrestrial intelligent species evolution. Hence, to win at this game of life, an intelligent species has to continue to mature and survive all the way to the top rung—beyond the point of self-extinction.

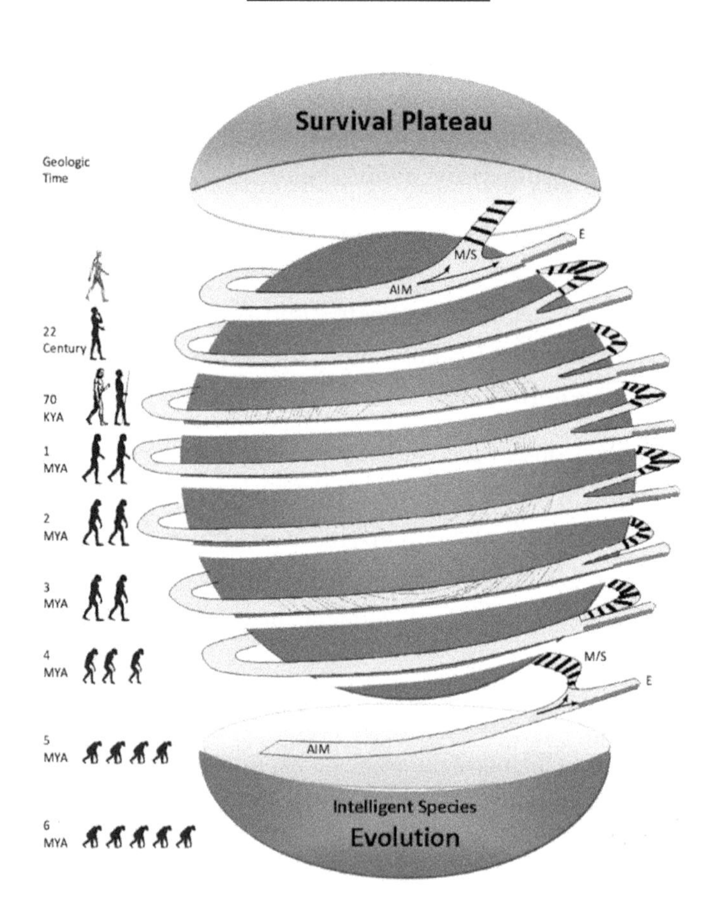

Fig. 12-2 Species Survival Maturity Model

(*Source: Sherry Wang*)

Five-plus million years ago many proto-intelligent hominin species began evolving in Africa that over geological time matured, many of whom went extinct along the road to evolutionary maturity, or maturing past the point of self-extinction, shown as the Survival Plateau.

- *Geologic time: MYA (Millions of Years); KYA (Thousands of Years)*
- *Intelligent Species Survival Patterns: AIMMS: (Adapt, Innovate, and Mature. Migrate and Survive)*
- *Intelligent Species Extinction Patterns: AIME: (Adapt, Innovate, and Mature or go Extinct)*
- *AIM: Adapt, Innovate, and Mature. M/S: Migrate to Survive. E: Extinction*

All prior hominin species have gone extinct without reaching the Survival Plateau. Homo sapiens is the last of the hominins and the sole surviving intelligent species on Earth and probably the galaxy.

To understand this model one must begin at the bottom, or first rung. It houses the following set of initial conditions that must be met for an intelligent species to have a chance to compete, let alone mature and survive beyond self-extinction, and includes:

Life: Members of a species must of course have evolved and be alive to have any chance at surviving, including an ability to remain alive and, as much as is possible, stay clear of extinction-causing events.

Location: Survival may well depend on location and location-dependent survival issues including where in the galaxy, what else is in the galactic neighborhood, on which planet, what kind of star, whether that planet is orbiting a star with a continuously habitable zone, and whether the planet is in the star's continuously habitable zone. The seeming rarity of Earth-like planets able to sustain complex animal and intelligent species life may be among the reasons scientists are starting to suspect that while microbial life in the galaxy might be common, complex animal and intelligent life might be rare, and might be one explanation for why we have not been able to communicate with other potential intelligent species.

Timing and just plain luck: Perhaps more than one might have thought, timing and luck may impact survival more than smarts. Life may have evolved and been annihilated repeatedly during the Late Heavy Bombardment of Earth's early history. Timing may well have been on the side of our forebears, as they left Africa much later than other hominin species, whose earlier departure to Asia and Europe exposed them to the worst of the last ice age and may have driven them to extinction. By departing much later, Homo sapiens might have encountered much lower water levels in the Red Sea when they migrated. Timing may be another reason we haven't encountered other extraterrestrial intelligent species and might also be one reason Homo sapiens are around today and not the dinosaurs.

Reproduction: Evolution appears to favor survival. An intelligent species must not only first exist but must be capable of reproducing itself and growing its numbers to enable continued survival. A species' ability and innate desire to reproduce is a cornerstone of this evolutionary survival pattern and does not depend upon the consent or volition of individual members comprising the species.

Habitat viability: Habitat viability and species survival are inseparable. An intelligent species must have evolved at an auspicious time and place in the galaxy, on a planet in a solar system capable of evolving and supporting complex animal and intelligent life, and in a sustainable-capable habitat such as Earth provides. The evolutionary survival pattern appears to be habitat first, emergent life second. Habitats are sustainable-capable but must be continuously maintained to remain sustainable. An intelligent species, having evolved, must continuously maintain its habitat, as its own survival is inextricably intertwined with the habitat's continued sustainability. When a habitat fails despite an intelligent species' best efforts, the species must be able to migrate to another habitat, or it will go extinct.

Adaptation, invention, and innovation: An intelligent species must have the ability to adapt, invent, and innovate to survive. Survival-driven adaptation and survival-driven innovation are closely associated and comprise another of those species' survival patterns. While the ability to adapt and innovate is necessary, adaptations and innovations are by no means sufficient to ensure survival.

Survival and migration: Survival and migration are inseparable. A species must be free, able, and willing to choose to migrate to survive. Survival-driven migration is as native to a species as is its urge to reproduce. Migration is an evolutionary pattern, and species perish when the opportunity to migrate no longer exists.

> *The next intelligent species-wide, survival-driven migration will be interplanetary.*

With no place else left to go, unless or until Homo sapiens finds a way to occupy the remaining 70% of the Earth's surface now covered by its oceans, it is more than likely that when faced with an existential threat, the next great Homo sapiens survival-driven migration will not be from Africa to

Europe but from Earth to another planet. However, by some estimates, it could take from 100 to 200 years before Homo sapiens will have acquired the abilities necessary for interplanetary or intersolar migrations.

Maturity: To survive, an intelligent species must be willing at each rung to prioritize and pursue evolutionary maturity over all else. An intelligent species' survival may come down to its attainment of evolutionary maturity. Even if it's able to clamber up all the other rungs of the Species Survival Maturity Model, its survival may well come down to its ability to mature.

Survival and maturity form a yin and yang evolutionary survival pattern pair. An intelligent species must mature to survive and must survive to mature.

Evolution appears to have placed attaining evolutionary maturity, both at the individual and species levels, as the final rung on the Maturity Model—the Survival Plateau shown in the model above. To survive, a species must be able to respond to its better angels or extinction is inevitable. But maturity does not come easily. Unlike species survival through reproduction, where continuity of the species appears to be almost autonomic, if not instinctive, species survival due to evolutionary maturity is not. And just as species continuity is assured by each individual member's freedom to reproduce, so its maturity may well depend upon the freedom of each individual to choose maturity, and thereby survival, over all else.

Maturity must become the defining characteristic of each individual before it can become the defining and overarching characteristic of an entire intelligent species. Like an inherited trait, maturity needs to become a built-in response. It needs to be taught and become hardwired from early childhood and pervade all aspects of life: society, culture, entertainment, education, religion, business and government, as well as domestic, private, and public life.

These are the initial conditions if an intelligent species is going to have any chance of attaining evolutionary maturity and thereby reach that survival plateau and avoid self-extinction.

It is clear that the urge to survive drives a species to migrate, but what is it that drives an intelligent species to grow more and more mature?

The rungs of the model unfold over geologic time and present increasingly difficult challenges and an accompanying opportunity for increased growth in maturity. It appears to have been designed to move intelligent species, rung by rung, beyond basic instinctive animalistic nature and behavior with which it initially evolved, to increasing levels of maturity, making increasingly more intelligent choices at each rung. If this movement continues to the top rung, it will lead to full evolutionary maturity—a process some might think of as becoming a fully matured human.

To accomplish this, evolution appears to have partnered with the universe, solar system, and planetary orbits to generate a recurrent set of geologic and climate events that drive habitat changes, preventing a species from resting on its laurels, and forcing a species to continuously adapt, innovate, mature and, when all else fails, migrate to the next higher rung to survive—or go extinct.

Just as climate change over geologic time has choreographed the migration patterns, and thereby the life cycle and survival of wildebeests as they follow the rains across the Serengeti in search of fresh pastures, so too has climate change over geologic time choreographed the life cycles, survival, and migrations of numerous hominin species. Today, with global warming in hot pursuit, climate change will soon begin to shape the adaptations and innovations of Homo sapiens, driving pressures to mature beyond the current fractious community of nations if it too is to survive and migrate, this time, perhaps, to the stars.

While some existential threats can be caused by an intelligent species, and in principle should be avoidable, no species so far has placed its own survival at risk until late twentieth-century Homo sapiens. All prior existential threats have mostly been caused by galactic, solar, and planetary climate-driven events that appear to collaborate with aspects of the orbit of the Earth around the sun. Even the length of time, the geologic interval, a species seems to be able to spend on a particular rung of the model in preparation for migration to the next higher rung seems to be timed by recurrent planetary-driven climate events that mirror the various aspects of the planet's orbit around the sun—such as snowball Earths and the numerous glacial and interglacial climate change events. These events changed species habitats, rendering them unsustainable, and thereby drove changes in the food chain, forcing species to adapt and innovate, engendering the need for new technologies and innovations to cope, and when all else failed, to migrate to survive.

And it has been the case so far that despite the numerous adaptations, innovations, and technologies an intelligent species had to come up with, as hominins did, these eventually proved to be ineffectual in the face of protracted and unexpected climate-driven habitat changes that often left species with no other option but to migrate to other habitats to survive.

Many prior hominin species got just so far up the Species Survival Maturity Model before going extinct. Early Homo sapiens migrated out of Africa in the face of just such climate-driven habitat changes and kept migrating, until there was no place else left to go. Hence, the next intelligent species-wide, survival-driven migration can only be an interplanetary migration, or if not, Homo sapiens, too, will face extinction. However, evolution appears to require an intelligent species to first mature past the point of self-extinction, to reach the Survival Plateau, before it can migrate beyond its home world.

To get to the Survival Plateau, the ground rules of the game appear to be as follows: Once an intelligent species has evolved, the species survival game begins and the evolutionary clock starts ticking. Each rung of the model ends in one of two outcomes—survival or extinction. In the face of planet-initiated climate events on each rung, or by its own extinction-enabling choices, a species has no way to escape extinction on any rung unless and until it is able to adapt, innovate, mature, and migrate to the next higher rung in some finite geological time. And maybe a little too much like the movie *The Hunger Games*, at each rung the game of life can end in extinction, and has for all prior hominin species.

At each rung, too, an intelligent species has an opportunity to choose to survive by choosing to mature and migrate to the next higher rung. An intelligent species must keep growing and maturing to continue to survive. There is no way to get to the next higher rung without an increase in maturity. Survival-driven adaptations and survival-driven innovations must be accompanied and exceeded by an increase in maturity. Reverse this order and self-extinction results. Increased maturity facilitates migration, and migration facilitates survival.

Hence an intelligent species must keep growing and maturing to continue to survive. Once an intelligent species has successfully reached the Survival Plateau, the highest rung on the Species Survival Maturity Model, it has matured beyond the point of self-extinction and joins what must indeed be a group of intelligent species, exceedingly rare (if there are any at all) in the galaxy, that have matured past the point of self-extinction. But given what scientists have seen of the galaxy so far, it is possible no intelligent species has yet reached that Survival Plateau.

The survival patterns we noted in earlier chapters seem to work as follows. Once a species manages to clamber up onto any rung of the Species Survival Maturity Model, at each rung it needs to do the following: **A**dapt, **I**nnovate, **Mature** and be ready to **M**igrate to **S**urvive, or it will go **E**xtinct. It's useful to think of this sequence as the **AIMM–S/E** set of patterns, and it consists of two variations: the survival pattern, which ends in **M**aturity, **M**igration, and thereby **S**urvival (**AIMMS**), and the extinction pattern, which ends in **E**xtinction (**AIMME**), a failure to mature or an inability to migrate. The dinosaurs were unable to migrate in the face of an existential threat.

These (the **AIMM–S/E** pattern set) are recurring sequences of survival patterns that a species must negotiate at each rung of the Species Survival Maturity Model until it reaches the Survival Plateau—that point at which an intelligent species becomes incapable of self-extinction. Until an intelligent species gets to the Survival Plateau, at each preceding rung of this model, a species must be prepared to take AIMM (**AIMM—-A**dapt, **I**nnovate, **M**ature, and **M**igrate) for the next level to **S**urvive, or it will go **E**xtinct.

Upon reaching the Survival Plateau, the only way for an evolutionary mature intelligent species to go extinct is to fall victim to a galactic, solar, or planetary existential event. By the time this level of maturity is attained, like restraints erected for a child's safety now outgrown, those solar and galactic roadblocks that once impeded interplanetary and inter-solar migration would have been overcome, opening the space doors to potential Earth 2.0s, no doubt with evolution's unseen help. A maturing intelligent species on its way to the Survival Plateau more than likely would have developed strategies to evade the universe's existential threats.

If the above in any way resembles what seems to have occurred during the course of intelligent species history, then it might also offer a lens through which to examine Homo sapiens' history so far, and perhaps even some insight as to where Homo sapiens, based on its current trajectory, could end up unless it undertakes serious, if not radical, course corrections to survive. If one can gain confidence in the model's ability to mirror at a high level prior intelligent species' experience, perhaps one can then have a measure of confidence in what, broadly speaking, it might imply for our species looking ahead.

To summarize then, the Species Survival Maturity Model asserts:

Intelligent species must evolve in galactic, solar, and planetary locations that meet the initial conditions described above, during times that favor the evolution of complex animal and intelligent species life.

Following its evolution, an intelligent species' survival is choreographed and circumscribed by a set of repeating evolutionary survival patterns driven by climate change. If the species were to discover and follow these patterns, its chances of survival would be significantly enhanced, if not assured.

Each intelligent species that meets these initial conditions finds itself in a sustainable-capable habitat and appears to have an interval of geologic time, circumscribed by cyclic climate events, in which, due to habitat deterioration, it is forced to adapt, innovate, mature, and migrate to a more sustainable habitat or else go extinct.

Because the geologic intervals spent on each rung can span millions of years, there is probably little to no memory of the experience and history of prior intelligent species—that is, until the invention of hieroglyphics, art, and eventually writing. Today, the fossil record, mummies, and DNA, with the help of modern technologies, cast some light on the life, experiences, and history of prior hominins and enable Homo sapiens to peer back in time.

Each intelligent species that successfully manages to migrate to a new habitat finds that new habitat presents challenges that require increasingly higher levels of adaptation and innovation, leading to even greater maturity and soon (over geologic time) discovers that to survive, it has to migrate yet again to another habitat or go extinct.

The repeating pattern of migration from habitat to habitat to survive, each with its concomitant set of Adaptations, Innovations, ever-increasing levels of Maturity, and Migration, is like rungs on the maturity model, and the repeating concomitant sets of activity attending each rung can be seen as sets of repeating evolutionary Survival patterns (**AIMM-S**) and evolutionary Extinction patterns (**AIMM-E**), respectively. Intelligent species that make it to the next higher rung survive, and those that don't go extinct. An intelligent species that manages to mature past the point of self-extinction and reaches the Survival Plateau is regarded as having reached the highest level of evolutionary maturity.

Once an intelligent species has reached the Survival Plateau, extinction is now only possible through events beyond its control, such as that which took out the dinosaurs, except that the whole point of evolution evolving intelligent species is that, unlike the dinosaurs, an intelligent species would by then have developed multiple capabilities to avoid extinction, including interplanetary migration, asteroid detection or meteorite trajectory deflection technologies, such as NASA's first-ever mission to identify, capture, and redirect a near-Earth asteroid to a stable orbit around the moon, where astronauts will explore it in the 2020s.

The model envisions a follow-on phase of evolution, extraterrestrial intelligent species evolution, and sees this as one potential outcome among many—evolution might choose to populate the stars with intelligent species, a possible outcome to Homo sapiens' evolution. Rather than starting over again and again on each potentially habitable planet, evolution may well use a mature intelligent species to seed the stars with intelligent life, as it did the continents across the Earth with Homo sapiens.

So, comparing the model with a high-level view of the history of hominins' evolution over geologic time, one finds some resemblance between what the model asserts versus what seemed to have occurred. The model asserts:

(1) Habitat unsustainability driven by climate change. Habitat unsustainability driven by climate change is the catalyst. Climate events drove habitats to change and fail, which in turn drove species to adapt, innovate, and grow more mature, and when all else failed, to migrate to more sustainable climes to survive.

This bears some resemblance to what one finds. Intelligent species evolved in sustainable habitats that became increasingly unsustainable over geologic time due to perpetually recurring cycles of climate-change events driven by solar and planetary orbit effects that shaped Earth's global climate over geologic periods.

Such climate events include snowball Earth events and glacial and interglacial periods that span millions of years. To repeat Rick Potts, who was quoted in Chapter 6: "The period of human evolution has coincided with environmental change, including cooling, drying, and wider climate fluctuations over time. . . . Overall, the hominins fossil record and the environmental record show that hominins evolved during an environmentally variable time. Higher variability occurred as changes in seasonality produced large-scale environmental fluctuations over periods that often lasted tens of thousands of years."

(2) The survival urge. The model compares intelligent species' efforts to survive to participants in a kind of survival game of life. It envisions multiple intelligent species, over geologic time, working their way up the Species Survival Maturity Model en route to the Survival Plateau. At every rung, each competing hominin species must adapt, innovate, mature, and migrate to survive, or go extinct.

It assumes the number of hominin species increasing with time, entering the fray, simultaneously participating at different rungs of the maturity model, and a slow and continuous winnowing of the ranks as more and more fail to mature enough to migrate and survive, succumbing instead to extinction, leaving fewer and fewer to continue on.

And indeed, the history of hominins over geologic time as shown in Fig. 2-1, again, viewed from a high level, appears to have followed such a pattern, beginning with multiple simultaneously—or near simultaneously—existing hominin species, eventually thinning down to just a few.

Two million years ago in Africa, several species of human-like creatures roamed the landscape. Homo sapiens appeared around 300,000 years ago, at a time when several others existed. But as recently as 30,000 years ago, there were only three other hominin species around, as well as modern humans. Yet today, only Homo sapiens remains. Homo sapiens is the one species that has survived from the diversity of hominin species. And, despite their very close relationship with our species, and despite the fact that all of them possessed some combination of features that characterize humans today, these earlier species and their ways of life are now extinct.

(3) Adaptation and innovation—survival tools. The model posits that repeating climate-change-driven habitat erosion and eventual unsustainability drove intelligent species to increasing levels of adaptation, innovation, maturity, and migration to survive, or drove them to extinction.

The environment and climate-change-driven habitat failures were evident in different layers of sediment, indicating different habitats at different times. Humans would have had to develop new cultural technology to deal with cold environments and changing food sources, especially during the last quarter million years.

Overall, the evidence shows that hominins were able to adapt to changing environments to different degrees. The genus *Homo*, to which our species belongs, had the capacity to adjust to a variety of environmental conditions, and Homo sapiens is especially able to cope with a broad range of climatic conditions, hot and cold environments, arid and moist ones, and with all kinds of varying vegetation. Our species uses resources from a vast variety of plants and animals and uses many specialized tools. Homo sapiens have cultivated many social contacts and means of exchanging resources and information to help us survive in a constantly changing world.

Hence some believe that change itself may well have been and still is an evolutionary survival selection mechanism. In fact, a growing number of scientists think that major climate shifts may have also forged some of the defining traits of humanity. In particular, a few large evolutionary leaps, such as bigger brains and complex tool use, seem to coincide with significant climate change.

There are others still who believe that the ensuing and impending climate changes in the current and next century may be the catalyst for the next change in human subsistence patterns and will engender another burst of human survival-focused innovations and cooperation with one another.

(4) Migration—the escape hatch. The model views survival-driven, species-wide migration as the evolutionary escape hatch that enables intelligent species, when all else fails, to survive. The need to migrate to survive is as built into hominins as the need to reproduce.

Neanderthals migrated outwards into Europe and Asia and Eurasia long before humans did, and they lived across Eurasia, as far north and west as Britain, through part of the Middle East, to Uzbekistan. According to the genetic and paleontological record, while earlier hominin species (Homo erectus, Neanderthals) migrated from Africa many thousands of years earlier, Homo sapiens started to leave Africa only between 60,000 and 70,000 years ago. Homo sapiens ensured its survival through migrating out of Africa at a time of habitat loss due to climate change caused by the last ice age.

(5) Maturity. The model suggests that, among complex animal species, intelligent species are uniquely prone to self-extinction. The only way for intelligent species to avoid self-extinction is to mature beyond the point of self-extinction. The model argues that intelligent species will survive only if they can reach the Survival Plateau—the level of maturity at which an intelligent species has grown beyond the point of self-extinction.

Homo sapiens might well be the last of the hominins. Scientists have speculated that the conspicuous absence (so far) of other intelligent species in the galaxy might be because Homo sapiens is premature in a universe that has not had enough time for intelligent species evolution to proliferate. If Homo sapiens is not the first, prior intelligent species may have all managed to do themselves in, or just haven't been discovered yet. Homo sapiens has been around for barely 300,000 years and has already managed to put its survival in question.

Not since the dinosaurs has a non-man-made, species-wide, existential-level threat to complex animal life and Homo sapiens survival arisen. Threats today to life and Homo sapiens' continued existence, including WMDs, global climate change, and soon AI, are all entirely of its own making. Given what appears to be evolution's goal—to preserve life beyond universal extinction events—it is not too surprising to find natural processes designed to drive intelligent species to mature beyond the point of self-extinction, probably based on a deep appreciation of the potential dangers that can result from an unguided exercise of freedom to choose their own path—a deep concern many Homo sapiens now have good reason to share.

Ours is an era no prior hominin species has had to face. While all prior hominin species have gone extinct, none has had the ability to cause its own extinction. Evolutionary maturity is the one hurdle that no prior or existing intelligent species has been able to attain. To avoid self-extinction, Homo sapiens has no choice but to reach that Survival Plateau, but it somehow seems blissfully unaware that it may be locked in a game for survival that began with the first hominins more than 5 million to 6 million years ago. As Steve Jobs observed, one has to look to the past to connect the dots, but Homo sapiens has so far failed to connect the dots and is failing to see that, as the last of the hominins, it's now its turn to carry the torch on the road to survival, perhaps on the final geologic lap, and must reach that Survival Plateau on behalf of all prior hominins.

Fewer and fewer people believe Homo sapiens' current selfish pursuit of profits and power and endless economic growth, which has led to the threat of self-extinction, will eventually lead to evolutionary maturity and thereby survival. How, then, will Homo sapiens choose to adopt the evolutionary patterns that lead to survival? If the lessons of prior Homo sapiens civilizations can serve as a guide, such a transformation will be anything but voluntary.

(6) Extraterrestrial intelligent species evolution. This is an aspect of the Species Survival Maturity Model that takes a not-so-blind guess as to where evolution might be heading next. This will be one way to test how well the model is able to predict the potential outcome for intelligent species and, hopefully, Homo sapiens.

If, as some scientists are beginning to suspect, intelligent life might be very rare and Homo sapiens may be, if not the first, the sole surviving intelligent species in the galaxy, and assuming Homo sapiens attains evolutionary maturity, it is not such a stretch to see that, just as bees seeking honey end up spreading pollen, so Homo sapiens might end up an unsuspecting partner in spreading intelligent life across the stars—possibly one of many evolutionary outcomes for populating the galaxy.

Should this turn out to be so, it will indeed be stranger than any fiction one might imagine, as there is nothing in Homo sapiens' current discourse as a species that even entertains such a destiny, or even hints at an awareness of being evolution's potential partner in the spread of intelligent life across the galaxy. Could such a realization motivate Homo sapiens to choose evolutionary maturity?

If indeed there is such a thing as a Species Survival Maturity Model as explored above, and if there are survival patterns not so well hidden over geologic time in hominins' struggle to survive, can Homo sapiens, the last of the hominins, take the torch of this game of life all the way past the finish line, past the point of self-extinction, to the Survival Plateau?

Can Homo sapiens today imagine the hope that must have shined in the lonely eyes of the first hominins as they began this long race on the plains of Africa? Like relay runners passing the eternal flame on to surviving hominin species, even as they themselves went extinct, they must have wished only that the recipients keep the flame of life alive to pass on to the next hominins in line in the race for survival. They were like so many travelers on a road much longer than the Silk Road, the path to maturity and survival, their footsteps written into the fossil record for all surviving hominins to see as they left us a sustainable world and asked only that Homo sapiens today do the same so succeeding generations of hominins might get to that finish line.

Now this game of life, which may have begun more than 5 million years ago and has seen the evolution and extinction of numerous successor species and their accumulated adaptations, innovations, maturity, and migrations, is bestowed on Homo sapiens, the sole surviving species, the last of the hominins. As in the words of Galadriel to Frodo in the movie *The Lord of the Rings: The Fellowship of the Ring*: "This task was appointed to you, and if you do not find a way, no one will." If Homo sapiens does not find a way, no other extant Earth-life species will.

Homo sapiens, like Frodo, now has the awesome responsibility of taking that torch of life that began its journey over 5 million years ago past the finish line, past the point of self-extinction, to the Survival Plateau, carrying with it the dreams of all past species of the genus *Homo* that ever lived.

And like invisible cheerleaders, in an inaudible chorus of voices, the cries of all who ever ran can be heard rising from their fossils in the Earth to cheer Homo sapiens on to the finish line. And now upon Homo sapiens rest the hope and longing—not of a family, not of a tribe, not of a nation, not even of the world, but of an entire genus over geologic time—that we, the last of the hominins, make it past the point of self-extinction.

Survival might well be the default outcome for intelligent evolutionary species, but such species must first mature to survive and survive to mature. As long as habitats remain sustainable, and life is able to reproduce itself, species will survive unless driven to extinction by galactic, solar, or planetary climate-change events.

Evolution needed then to find a way for life to survive without having to restart life after each mass extinction. It needed to evolve an intelligent species capable of choosing from among potential evolutionary outcomes the one to survive beyond self-extinction and beyond existential threats such as those that took out the dinosaurs.

Evolution appears to have been keenly aware of the risk an immature intelligent species might pose to itself, other species, and the rest of the planet, but obviously continued, placing bigger bets on a successful outcome, hoping that among the clutch of intelligent species at least one would choose survival over self-extinction. And now, Homo sapiens, the last in that long line of hominins in the genus *Homo*, which evolved more than 5 million years ago, is about to prove whether or not evolution's gamble will pay off.

To survive, Homo sapiens must find and follow the evolutionary survival patterns and the survival strategies implied in the Species Survival Maturity Model. In particular, it must find and follow the Survivability and Habitat Sustainability Test. Evolutionary maturity is undoubtedly the most challenging of the survival patterns. Regardless of how well an intelligent species adapts, innovates, and migrates, it must above all else mature. It must mature to survive and survive to mature. Unless it matures, it will go extinct. We explore what evolutionary maturity means next.

13.

Evolutionary Maturity

THE ONLY PATH TO INTELLIGENT SPECIES SURVIVAL

"Now that I have fulfilled my purpose, I don't know what to do."
—*Sonny, the Robot, in the movie I, Robot*

"I guess you'll have to find your way like the rest of us, Sonny. I think that's what Dr. Lanning would have wanted. That's what it means to be free."
—*Del Spooner, in the movie I, Robot*

YOU MIGHT RECALL THIS ENIGMATIC RESPONSE BY Will Smith's character, Del Spooner, to Sonny, the Robot, in the movie *I, Robot*.

Sonny was trying to figure out what individuals of an intelligent species like Homo sapiens have been trying to sort out ever since the first intelligent species evolved on Earth: Why am I here? What should I do next? Questions no nonintelligent species will ever face.

Six million years ago, evolution, it seems, set out to evolve a new kind of complex animal species that, unlike any prior species, would be free to choose its own evolutionary outcome. It was a species so different that it would evolve the capacity to transcend and largely replace the basic animal instincts that control the behavior and life cycle of all other animal species. Its emergent intelligence would enable it to evaluate various outcomes and make choices to shape its own evolutionary outcome in ways no prior species had ever been able to do. It was a species that could make choices so far-reaching that it could potentially, perhaps inadvertently, bring about its own extinction and the extinction of numerous other species, or make choices to mature beyond the point of self-extinction, to evolutionary maturity. And, like Sonny, ever since that time, hominins, and now Homo sapiens, have been finding out what it means to be free.

The single burning question both evolution and all of mankind are waiting to find out is what Homo sapiens, the last of the hominins, the last in the genus *Homo*, and possibly the last intelligent species on Planet Earth, if not in the galaxy, will choose.

In a prior chapter we noted that one way to think of evolutionary maturity is as a survival plateau, the point at which an intelligent species would have matured beyond the likelihood, or even the possibility, of self-extinction. We also explored why evolution might have set attaining evolutionary maturity as a prerequisite for intelligent species survival, and the role of a Species Survival Maturity Model, which evolution appears to have left to guide intelligent species to evolutionary maturity.

In this chapter we go on to explore what, exactly, evolutionary maturity is and what it is not; how it differs from what Homo sapiens might consider maturity; what it might mean for an intelligent species to undergo the transformation from basic instinct-driven, tree-hugging, apelike creatures on all fours to a bipedal, erect, intelligent, self-aware species, and its struggle to choose maturity and survival over extinction.

What is evolutionary maturity? In addition to being a level of maturity beyond which self-extinction is impossible (survival plateau), it can also be viewed as a measure of how a species chooses to survive. The evolutionary maturity scale ranges from the zero or near-zero maturity-level of a species—indicating near certainty of its self-extinction—to the fully matured level, the point at which it can no longer drive itself to self-extinction. When a species has matured beyond the likelihood, and perhaps even the possibility, of self-extinction, it is said to have reached a survival plateau.

So, is it really just about avoiding self-extinction long enough to get to that plateau? Well, it's certainly that, but perhaps a different way to get at the meaning of evolutionary maturity is to consider what an evolutionary mature species is and what it needs to do to become evolutionarily mature to get to that survival plateau.

> *An evolutionary mature species is one that can successfully sustain its life within Earth's ecosystems without destroying itself, other species, and its habitats in the process.*

You are probably thinking at this point, Is that it? Is that all? Well, yes, but as simple as it seems, modern Homo sapiens, its vaunted adaptations and innovations notwithstanding, has failed to do any of the above. Indeed, since switching 10,000 years ago to agriculture as its primary way to sustain itself, Homo sapiens has failed to do so without destroying other species or Earth's habitats and, by the looks of things, might be well on its way to destroying itself. Moreover, it has done it all *outside* of Earth's ecosystems. So, one way to think of evolutionary maturity is the ability of a species to live and let live.

By requiring evolutionary maturity as a condition for survival, then, all evolution appears to be asking intelligent species to do is to use its freedom to choose its own evolutionary destiny and to employ those abilities with which it evolved, those wonderful abilities to adapt and innovate, to choose survival over extinction—survival for itself, other species, and Earth's ecosystems and habitats, which give life to all.

Unfortunately, Homo sapiens has instead so far chosen to do precisely the opposite. Why? More than likely because our species has lost its freedom to choose, and what one sees instead are the results of the choices of "the few"—a topic we will turn to in Chapter 15.

However, if Homo sapiens were to eventually see its way clear to choose evolutionary maturity and survival over extinction, it would have successfully scaled that survival plateau and would no longer be capable of self-extinction. Homo sapiens would then be well positioned to become the first Earth-life species capable of surviving the universe's existential threats and, potentially, the first to be truly capable of continued survival in a universe characterized by birth, life, and death by adapting and innovating the capabilities to migrate from planet to planet to survive.

Above we defined an evolutionary mature species as one that can successfully sustain life within Earth's ecosystems without destroying itself, other species, or its habitats in the process. Digging deeper into this definition, a few key takeaways emerge.

Notably, an Earth-life species has no choice but to sustain life to survive and can only do so in the following manner:

(1) Within Earth's ecosystems rather than outside, as agriculture requires;

(2) Without destroying other species;

(3) Without destroying Earth's ecosystems and habitats; and

(4) Without destroying itself.

An Earth-life species has no choice. But why? In brief: it has no choice if it wants to survive. The following is a more in-depth answer.

All Earth-life needs energy to survive, and that energy comes from the sun. The only way to obtain that energy is through photosynthesis. It is collected and stored in the flora and fauna populating Earth's ecosystems and habitats. Sustaining life from within Earth's ecosystems means consuming the flora and fauna in which that energy is stored. Since the only way to obtain the energy a species needs to survive is by consuming Earth's photosynthesizing flora, and the fauna that consumes the flora, then clearly doing so in a way that leads to their extinction destroys not only these species and the habitats that depend upon them but will eventually lead to its own extinction.

It's a simple survival lesson it seems all nonintelligent species managed to get right. Homo sapiens, all its intelligence notwithstanding, has somehow failed to grasp it.

We know that because, among similar claims by other scientists, paleontologist Niles Eldredge, curator-in-chief of the Hall of Biodiversity at the American Museum of Natural History, has said, in an article on *ActionBioscience.org*: "Homo sapiens became the first species to stop living inside local ecosystems. All other species, including our ancestral hominin ancestors, all pre-agricultural humans, and remnant hunter-gatherer societies still extant exist as semi-isolated populations playing specific roles (i.e., have 'niches') in local ecosystems. This is not so with post-agricultural revolution humans, who in effect have stepped outside local ecosystems.

"Indeed, to develop agriculture is essentially to declare war on ecosystems—converting land to produce one or two food crops, with all other native plant species all now classified as unwanted 'weeds'—and all but a few domesticated species of animals now considered as pests."

Agriculture represents the single most profound ecological change in the entire 3.5-billion-year history of life. But as we have seen in the chapter on adaptations, Homo sapiens, driven by a food crisis exacerbated by population growth and climate change, may not have had a whole lot of choice back then. But today we do.

We know Homo sapiens failed to get it, too, because with that change to agriculture as a way to live began an assault on Earth's habitats, ecosystems, and biodiversity that 10,000 years later has resulted in a level and rate of extinction of Earth's flora and fauna the planet has never seen, mass extinctions notwithstanding.

Scientists estimate Homo sapiens has been driving innumerable species to extinction, at the rate of 150 to 200 species of plants, insects, birds, and mammals every twenty-four hours, nearly 1,000 times the rate normally expected (the "background rate") and is greater than anything the world has experienced since the extinction of the dinosaurs about 65 million years ago. Wildlife is dying out due to habitat destruction, overhunting, toxic pollution, invasion by alien species, and climate change. But the ultimate cause of all these factors is "human overpopulation, continued population growth, and overconsumption, especially by the rich," said Ahmed Djoghlaf, the secretary-general of the U.N.'s Convention on Biological Diversity.

And finally, we know Homo sapiens failed to get it because, despite the threat of extinction due to climate change that drove early humans to migrate out of Africa 70,000 years ago, it is clear that our species has long forgotten what this threat might have meant. Many scientists now believe a sixth mass extinction might have begun as far back as when the first modern humans began to disperse to different parts of the world about 100,000 years ago but really accelerated about 10,000 years ago when humans turned to agriculture. As seen in prior chapters, man-made global warming has increased global temperatures to levels approaching prior greenhouse mass extinctions, in particular, those of the Permian-Triassic extinction.

Digging deeper, we see that by abandoning the evolutionary way to live from within Earth's ecosystems, Homo sapiens has instead gone about destroying the only sources of energy evolution evolved to sustain all life on Planet Earth. And if there is one takeaway one might glean from digging deeper, it is that there is no way for any Earth-life, Homo sapiens included, to obtain the energy it needs to survive except through evolution's photosynthesizing energy sources—Earth's species, ecosystems, and habitats. By destroying these, an intelligent species is in effect destroying the only edible energy sources available to sustain life, and it will ultimately drive itself to extinction.

So, yes, all Earth-life, including Homo sapiens, has no choice but to follow these rules if it wants to survive. And as suggested in the prior chapter, evolution may have even left a survival manual, a Species Survival Maturity Model, to follow; a kind of yellow-brick road that runs all the way past the point of self-extinction to evolutionary maturity and survival.

But alas! Intelligent species *must* actively seek out and choose survival over all else, and that's the problem. Attaining evolutionary maturity and pursuing species survival are not things that have ever preoccupied Homo sapiens. It might not be entirely clear how Homo sapiens got here, but an equally if not more important question has got to be, how do we find our way out of here?

One answer might lie in an exploration of the characteristics of evolutionary maturity. What was it evolution might have had in mind?

It is important to note the focus on survival in the definition of an evolutionary mature intelligent species: species survival, habitat survival, and its own survival. Evolutionary maturity, species survival, and habitat survival are all inextricably intertwined, and one cannot exist independent of the others. The overarching principle guiding evolutionary maturity is survival, survival, and, in case of doubt, survival. A species' survival depends upon its habitat's survival, and that habitat survival depends upon its species' survival. Homo sapiens' survival depends upon the survival of both Earth's species and Earth's habitats. An intelligent species that wantonly destroys other species and their habitats cannot reach evolutionary maturity and consequently will not survive.

Indeed, everything evolution appears to have done has been targeted at species survival—but why this focus on survival?

If one thinks of the Earth as voyaging through space over geologic time, then evolution ensured that everything on Spaceship Earth was 100%

survivable, sustainable, and recyclable. As our planet hurtles through galactic space, revolving around the sun for billions of years, the focus, motivation, and outcome sought by evolution appears to be one thing and one thing only—species evolution and survival.

This principle has underpinned and guided the entire world that evolution evolved and therefore becomes a crucial test to determine what can be added or changed by an emergent intelligent species while still remaining true to the overarching goal of species survival.

One can think of this principle in terms of an evolutionary species

Survivability and Habitat Sustainability Test.

To pass this test, every activity an intelligent species engages in must contribute to the survival of all species and help maintain the sustainability of Earth's ecosystems and habitats. Otherwise, that activity not only fails the test but will not contribute to survival and therefore has no place in an evolutionary mature intelligent species' world.

Apart from evolving life itself and ensuring all species came with the built-in ability to reproduce and perpetuate themselves, and apart from the fact that it jump-started life repeatedly in spite of galactic, solar, and planetary disasters, including climate-change events that drove innumerable species to extinction over geologic time, evolution has provided all the essentials necessary to survive. It provided the right planet orbiting in the continuously habitable zone of the right star, in the right neighborhood of the galaxy; an inner atmospheric system, temperature, pressure, and just enough gravity to keep it all from skipping off into space; an outer tier of larger planets to surround the inner core, where life had to be protected from incoming; an environment with renewable habitats to support life all the way and for as long as that journey to intelligent species maturity must take; a perpetual and comparatively safe renewable source of energy; and clean air, clean water, lots of organic, natural food sources and natural habitats filled with renewable and recyclable materials that can be used and reused from one generation to another across all species.

Even when that turned out to not be enough to stave off the death-dealing blows of the universe, evolution evolved a brand-new and different species, hominins, and thereby Homo sapiens, a nascent intelligent species that, hopefully, will one day survive beyond the universe's mechanisms of birth, life, and death.

Survival, then, is the cornerstone of the world that evolution built.

If one wanted to test whether an idea, concept, policy, or belief, or an economic, political, legal, religious, or any other system or approach, for that matter, that Homo sapiens has come up with could or should be carried forward into an evolutionary mature intelligent species' world, one simply need only apply that Survivability and Habitat Sustainability Test. This test essentially asks the question and applies the principle that seems to have guided evolution: Does this, in all its aspects, promote and result in species survival, or does it require the sacrifice of species or result in species and habitat exploitation, death, or extinction to attain its goals? It's that simple, and it's the rule all Earth-life, other than Homo sapiens, lives by.

If the focus, motivation, methods, approaches, and outcome sought by anything, or anyone, in whole or in part, fails to pass the Survivability and Habitat Sustainability Test, that would be a pretty strong indication that such an activity is a nonstarter and could never have had a place in an evolutionary mature intelligent species world. So, there is your true north if ever there was one. In case of doubt, just consider what a pre-agricultural Earth might have been like, or to go further back, what Earth would have been like prior to the evolution of intelligent species and the emergence of Homo sapiens.

Here are a few quick examples that might help one understand what might or might not pass muster in an evolutionary mature world once this test is applied:

- Economic policies that destroy Earth's habitats and pollute its environments, its atmosphere, and the food chain of all species in the name of economic progress, or on the premise that these enable growth and create jobs;
- The unconscionable inequality that's built into the political, economic, and most other aspects of modern civilization, enabling the accumulation of most of the Earth's resources by "the few" to the detriment of "the many";
- Suppression of women and girls, half of our species, across cultures, countries, and continents in the name of patriarchy, family honor, religion, economy, and archaic cultural beliefs and practices;
- Development of weapons of war and WMDs on the misguided notion of maintaining peace through strength;
- Enslavement of one's own species in any way, shape, or form as Western and other civilizations have done at one time or another to most of the rest of the world;
- Exploiting workers by paying low nonlivable wages and customers by exploiting their personal online habits and data without their consent and by deceptive marketing and advertising.

Should one go on?

Go ahead, make your own list, and subject each item to evolution's Survivability and Habitat Sustainability Test, and see where Homo sapiens come out. Yes, it might be a bit simplistic, but it's one way to get one's own feel for how evolutionarily mature our species might be and in which direction we might be headed.

Not so well hidden amid the above perspective on what might have guided evolution in the setup of Earth-life, ecosystems, and habitats might be more than a hint that the same rules and patterns can also guide an emergent intelligent species to survival. After all, if these survival principles were good enough to guide evolution in the establishment of Earth-life and all that sustains it, they should be good enough to guide an emerging intelligent species to survival.

In the prior chapter on survival, we took an in-depth look at some of these rules and patterns and constructed a Species Survival Maturity Model, a potential guide evolution appears to have left, which, if followed, might lead an intelligent species to survival. And if species, habitat, and intelligent species survival turns out to be the only measure of intelligent species maturity that will ultimately matter, the question must be: How does twenty-first-century Homo sapiens measure up?

One way to make that assessment is to try to estimate how far along the road to evolutionary maturity Homo sapiens has come and how far it might yet have to travel. How many rungs might remain on the Species Survival Maturity Model before Homo sapiens reaches the Survival Plateau? What would an Intelligent Species Survival Scorecard that seeks to assess Homo sapiens' efforts and accomplishments look like from an evolutionary survival patterns perspective?

Below is one take on how Homo sapiens measures up. It's no overstatement to note that our species does not even appear to be trying to attain evolutionary maturity. Nonetheless, it would be interesting to find out what you, dear reader, might think, particularly across generations—the Greatest Generation, baby boomers, Generations X, Y, and Z, and Millennials, as well as some of Homo sapiens' well-known futurists, Techno sapiens, and Moneyed sapiens.

So here goes the world's first Homo Sapiens Survival Scorecard. The left column lists the evolutionary survival patterns an intelligent species must successfully negotiate to survive and preserve itself, Earth's species, and habitats. In the next two columns are the required Survival Plateau score for each pattern and a guess at what modern Homo sapiens' might be. The Trend column attempts to indicate a direction. The final column contains a few explanatory comments.

However, in coming up with Homo sapiens' score, it is worth noting that unlike those Ivy League students who got in by their celebrity parents' bribes, no amount of money will buy their or their parents' survival—there is simply no way to bribe Mother Nature.

Table 13-1 Homo Sapiens Survival Scorecard [3]
Based on evolutionary survival patterns

(Source: Author)

Evolutionary survival patterns	**Survival Plateau score required** %	**Homo Sapiens score** %	**Trend**	**Comments** The patterns used herein are taken from the Evolutionary Species Survival Model
To survive, an intelligent species must:				
Evolve	100	100	✓	To survive a species must have already evolved.
Reproduce—Must be able to replicate to survive	100	100	↓ ⊗	Population may well be headed beyond Earth's carrying capacity
Sustain its habitat	100	10-20	↓	Profits come first. Little to no interest in habitat sustainability
Adapt[5]—Adapt to habitat and climate change	50 -100	30-40	↓	Limited ability to respond to disasters or climate change
Innovate[2] —Means survival-driven innovation	30-80	20-30	↑	Innovation is profit- not survival-driven
Mature[1]—An evolutionary species' highest priority	100	20-50	↓	100% is the only acceptable score
Migrate[4]—Migrate to survive when all else fails	100	0	×	Intra- , inter-solar migration not yet possible
Survival focus—Laser-like focus on species survival	100	10-20	↓	Homo sapiens as a species has little to no focus on species survival
Total score	**700-800**	**260-350**		

Notes:

Trends:

On target = ✓;

Bad and going in the wrong direction = ↓;

Bad and going in the wrong direction and needs to stop ASAP = ↓⊗;

Inadequate resourcing and going in the wrong direction = ↓;

Adequate resourcing but not survival-driven = ↑

Indispensable to survival but no known solution = ✕

1. **Maturity and survival-focus** form a yin and yang evolutionary pattern. Survival is utterly dependent upon an intelligent species attaining evolutionary maturity, and to get to evolutionary maturity, an intelligent species must survive. It is the only category where 100% is the only acceptable result on the scorecard. Maturity is both necessary and sufficient for survival. Scores in most other categories are required but are not sufficient by themselves to assure intelligent species survival. On the other hand, weak scores in most other categories if accompanied by 100% maturity will assure an intelligent species survival. And it is here that Homo sapiens is at its weakest.

2. The **Innovation** rating in particular might trouble Techno-sapiens, but most Homo sapiens' innovations are random and driven by profit rather than species survival and habitat sustainability. While they have benefits that accrue, such outcomes are mostly by accident, or, if deliberate, they often occur only when the profit motive and the need to preserve habitats and species happen to coincide. While innovation, in particular survival-driven innovation, is necessary for survival, by itself it is by no means sufficient. Survival-driven innovation by definition could never come up with WMDs and other inventions that threaten the survival of species, ecosystems, and Homo sapiens.

3. While this is only one person's perspective, it is nonetheless a start. Maybe Homo sapiens should focus more on this type of measure rather than stock market and other financial indexes.

4. Migration and survival are inextricably linked by evolution. If Homo sapiens are to survive, attaining intra- or inter-solar migration capabilities will not be optional.
5. Adaptations, when survival-driven, are essential to intelligent species survival. When adaptations are motivated by a concern for species survival (e.g., humanitarian efforts) and habitat sustainability (e.g., environmental programs), Homo sapiens has made great strides. However, excluding those, most efforts in this arena, unfortunately, are driven by a focus on gain and profits. If survival adaptations were a Homo sapiens priority, we would anticipate and take steps to mitigate the tragedy, deaths, and suffering that follow natural disasters (e.g., Hurricanes Sandy, Katrina, and Harvey as well as numerous earthquakes, floods, wildfires, and tsunamis worldwide), instead of responding with too little too late after the fact seen all so often these days.

What's conspicuously absent from the above scorecard are any of the usual measures of maturity Homo sapiens use to compare civilizations across time and among countries and peoples. There are no political, economic, military, religious, scientific, human rights, or health-related measures; no hunger, infant mortality, refugee, and migrant data.

Evolution has simply set the destination and the rules to get there and appears to have left it up to each species as it goes about the process of living to get to that destination by following the rules and using whatever evolutionary capabilities it has developed. For intelligent species like Homo sapiens, that destination is evolutionary maturity, getting to that survival plateau. The rules to be followed are to live within Earth's ecosystems and habitats without destroying them, Earth's species, or itself. How much easier could evolution have made it?

The scorecard takes a low-Earth-orbit-level view of how Homo sapiens might be performing on each of the evolutionary survival patterns

(Adapt, Innovate, Mature, Migrate to Survive) shown in Fig. 9-1 but only hints at the level of jeopardy such low scores (one person's take) might imply. While there might not be a right overall score, a low evolutionary maturity rating does not bode well and may hint more of extinction than survival.

Because evolutionary maturity is a measure of a species' ability to survive beyond the point of self-extinction, one can get a clearer sense of where Homo sapiens might stand as a species by considering its Maturity/Self-Extinction Index.

This index is a correlation between a species' ability to mature and its ability to innovate. The Maturity/Self-Extinction Index indicates how close a species is to self-extinction when the index approaches 0%, and how close it is to attaining evolutionary maturity when the index approaches 100%. The higher the index, the more members of a species, as a percentage of its total population, are needed to drive itself to extinction, thus making it less likely. On the other hand, the lower the index, the fewer members of a species are needed, thus increasing the likelihood of innovation-driven self-extinction.

The chart below plots Homo sapiens' Maturity/Self-Extinction index at three specific points: World War I, World War II, and the Cuban Missile Crisis. These three points are chosen because the first two were the first worldwide conflicts, and the third because it threatened another worldwide war, one that for the first time might have used mainly weapons of mass destruction.

Whereas a large percentage of the worldwide population was engaged in each of the world wars, the percentage of people needed to prosecute each war was significant—3.42% and 77%, respectively. The Cuban Missile Crisis, however, threatened not only another worldwide war but this time a thermonuclear war. This innovation completely changed these percentages, exponentially reducing the number of people required to start such a conflict to near zero, due to the fact that it needed the decision of only one man, either John Kennedy or Nikita Khrushchev, because these weapons could plunge the world into a nuclear holocaust, potentially bringing on a nuclear winter that might terminate Homo sapiens' existence.

Plotted across the bottom and remaining near zero, beginning with the last two centuries (also going back for all of Homo sapiens existence, but shortened for ease of display), is the innovation trend showing WWI, WWII, and the Cuban Missile Crisis, with their corresponding average world population and points along this innovation plot when various inventions occurred. Most stand-alone innovation plots show an exponential increase in innovation, particularly during the last century, compared with the preceding millions of years of Homo sapiens' forebears' existence. However, these, in the words of Peter Diamandis, in his book *Abundance*, tend to talk about the inventions "of goods and services: food, water, education, healthcare and energy," and rarely ever get around to topics such as WMDs and the innovations that specifically led to their creation—needless to say, a singularly rosy view of reality.

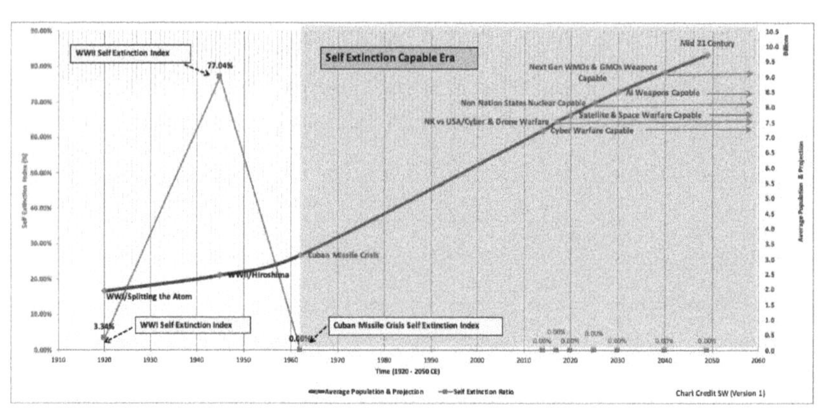

Fig. 13-1 Maturity/Self-Extinction Index

The Maturity/Self-Extinction Index trend is plotted above the innovation trend, showing at each of these three points the index at the occurrence of each event. WWI required the participation of an estimated 3.34% of a worldwide population of approximately 1.9 billion. WWII, often regarded as the deadliest human conflict, required the participation of an even greater percentage, estimated at 77% of the worldwide population of about 2.3 billion. As deadly as WWII had been, it needed orders of magnitude more people to agree to participate. The larger the number of people that must agree to participate, the less likely it is that such conflicts, given the innovations of the time, will drive self-extinction.

However, with the invention of the atomic bomb and the subsequent development of thermonuclear capabilities, by the time the Cuban Missile Crisis occurred in 1962, that agree-to-participate percent dropped from 77% in WWII to less than 1% of the population, then estimated at 3.2 billion. Essentially, it required only the consent of one of two men, Kennedy or Khrushchev, and eliminated the need for large numbers of people to agree to participate, vastly increasing the chances of self-extinction.

WMDs changed how wars are started and fought and the number of people needed to participate in decimating vast stretches of the surface of the planet, with the potential to drive Homo sapiens to extinction.

So, whereas Homo sapiens' evolutionary maturity level might have begun at or near 100%, and hovered at 90% or above for most of the millions of years of hominins' existence, and only began to decline slowly about 10,000 years ago with the arrival of agriculture and early civilizations, Homo sapiens' innovation was almost nonexistent and mostly marginal for most of those millions of years. There were a few bright spots here and there, but innovation really began to accelerate only in the last two centuries.

Upon closer inspection, an inverse growth relationship between these two patterns begins to emerge. As one begins to rise, the other begins to fall, and the transitions appear to begin at or about the same time in Homo sapiens' history, suggesting a cause-and-effect relationship. Indeed, hidden in the opposing movements of these two plots is a race, one that appears to pit Homo sapiens' efforts to mature against its efforts to innovate. While an increase in evolutionary maturity implies an increase in the movement toward Homo sapiens' survival, an increase in innovation strongly correlates with a decrease in the number of individuals needed for the species to drive itself to extinction—in effect saying that innovation without maturity is a prescription for self-extinction.

While the population of our hunter-gatherer forebears was orders of magnitude smaller, they lacked the self-exterminating capabilities Homo sapiens possess today, and they would instead have been at risk of extinction due more to climate change and natural disasters, as is believed to have happened to the Neanderthals. Evolutionary maturity may have begun to show signs of decline for the first time with the transition to agriculture and the rise of civilization 10,000 years ago. Whereas hunter-gatherers lived most of those many millions of years *within* Earth's ecosystems, mostly observing the requirements not to destroy habitat, species, or themselves, post-agricultural Homo sapiens began to live *outside* of Earth's ecosystems, destroying Earth's habitats and species, and got off to a grand start toward destroying itself.

A second important turning point seems to have occurred with the arrival of the Industrial Revolution in the mid-nineteenth century. The emergence of an industrial economy and the transition from renewables to fossil fuels saw evolutionary maturity take a precipitous dive that continued through both world wars and the splitting of the atom and reached a nadir with the Cuban Missile Crisis in 1962.

October 1962 may be the point in time in which, for the first time, as far as is known, Homo sapiens' existence as a species might have come down to the whim of a single man. Precisely at that point, Homo sapiens' ability to mature and Homo sapiens' ability to innovate caught up and intersected, apparently headed in opposite directions. Our species' ability to innovate had finally surpassed its ability to mature, with potentially devastating consequences. Speaking of this event, President Kennedy is reported to have said: "Every man, woman, and child lives under a nuclear sword of Damocles, hanging by the slenderest of threads, capable of being cut at any moment by accident, or by miscalculation, or by madness."

Since that fateful time, because of our innovations, Homo sapiens has lived in a WMD era (shown shaded on the chart as the beginning of Homo sapiens' Self-Extinction-Capable Era), with the constant risk of self-extinction in the hands of a single member of our species. Yet we have shown little desire to back away from the edge of this self-extinction precipice. On the contrary, as the chart shows, our species has added to the WMD era numerous other terrifying and innovative ways to drive itself to extinction.

Yet it would be a mistake to condemn all forms of innovation, as not all are misguided. Indeed, if innovations were intrinsically bad, could be used to drive a species toward self-extinction only, along with its habitat and other species, one doubts evolution would have evolved intelligent species with the capability to innovate. Clearly, then, there might have been another intent, an evolutionary survival intent, that would not have included motives like profit, economic growth, and personal gain that drive innovation today. Rather, it seems the reason this capability evolved was to enable an evolutionary maturing intelligent species to invent and innovate solutions to avoid the destruction of life by the universe's existential threats such as what took out the dinosaurs. An evolutionary mature intelligent species would use its innovative abilities to preserve life, not destroy it.

Innovation may have been "intended" to be survival-driven rather than profit-driven, an ability a mature intelligent species could use to preserve species, habitats, and itself. Perhaps the ability to innovate was specifically intended to come up with the means to ultimately enable intra- and inter-solar survival migrations—something neither the dinosaurs nor any other nonintelligent species could or will ever be able to do.

And that right there, dear reader, may well hint at an answer to the question we asked at the beginning of this book: Why, after more than 4 billion years, did intelligent species, and thereby Homo sapiens, finally evolve?

It is sad to see that which evolution may well have intended to be used to escape death instead is used by Homo sapiens to bring upon itself and all species the prospect of death and self-extinction.

But this is a book about survival, so I invite you to join me in imagining how different our world would be if these overarching evolutionary principles intended to guide intelligent species to evolutionary maturity and survival drove Homo sapiens' civilization instead of the current selfish pursuit of profits, gain, growth, and greed.

As one contemplates our plight and wishes for change, there are several aspects of this world and the civilization Homo sapiens has built that would find little or no parallel in an evolutionary mature intelligent species world. While one would expect many, many things in a world built by a mature version of ourselves to be different in many ways, those differences would still have to satisfy that Survivability and Habitat Sustainability Test.

One way to begin to wrap one's head around these concepts is to imagine examples of familiar Homo sapiens' views, practices, expectations, and long-standing beliefs and norms, and how they might fit into a civilization built by an evolutionary mature intelligent species. The following are some areas of divergence between these two worlds. One would be hard-pressed to envision how any of the familiar practices below could possibly exist in a world built by an intelligent species whose highest priority is species, habitat, and Homo sapiens survival. In most respects, such a civilization would seem unrecognizable, if not actually alien.

The notion of ownership. This idea is so ingrained, yet it would be untenable in a mature intelligent species world. Communal use of a resource instead of exclusive ownership rights is the evolutionary model. Nothing, it would seem, with the exception of mates and offspring and dwelling places such as nests, holes, dens, caves, and lairs, was "owned" exclusively in a pre-intelligent species world; rather, resources were available for all to use freely.

For all but the last 10,000 of the 300,000 years of Homo sapiens' evolution, hunting animals and gathering foodstuffs was the sole means of subsistence. Food was regarded as a common resource and was available to all members of the group. There was no concept of food ownership. The notion of ownership appears to have emerged sometime after 10,000 years ago with the gradual transition from hunting and gathering to agriculture and settled communities.

There could be no eminent domain, no government lands, no national spheres of influence, no private parks, or beaches, or forests, and so on. The current move toward a sharing economy holds potential for a return to that state were that brave new evolutionary mature species world to emerge.

The notion of want versus need. There is a sense in the natural world of taking only what one needs, and not hoarding to the exclusion and injury of others or with the intention of creating artificial scarcity or gaining a monopoly position. The biblical story of Israel in the wilderness being provisioned with just enough manna to last one day, except on the day before the Sabbath, seems to make the point. Haven't you noticed that there is no way to collect and own the most basic things of life—air, sunlight, and water (though some have found ways to sell even water)? Economic systems built around greed, accumulation of personal, corporate, regional, and national wealth, and rewarding shareholders and investors with profits to the exclusion of all others could not exist. Private enterprise could not exist unless it could be made to serve the needs of all. No one gets or wants first dibs on anything in such a world.

The notion that the value of a species or its individual members is based on some economic factor. The concept of quantifying and measuring everyone and everything on the basis of economic and therefore monetary value could not exist in a world that places species survival, maturity, and habitat sustainability above all else. Survival, sustainability, and maturity are the coins of this new realm.

There would be no haves or have-nots. There would be no social, political, economic, monetary, or other system in which eight mostly old, white, American men could possibly own as much as 50% of the world's resources. There would be no income inequality and no discrimination of any kind. There could be no situation in which the majority of the world's people live on less than $10 a day, while the wealthy splurge and the middle class just manages to make ends meet. The very notion of wealthy versus middle class versus poor could not exist in an evolutionarily mature world. Bias and inequality on the basis of income, race, ethnicity, gender, or sexual orientation would not exist.

The concept of capitalism. The idea of a market economy consisting of independent actors driven by their own enlightened self-interest—aka capitalism—could not exist in a mature intelligent species world. Imagine no stock market. No money, Bitcoins, or financial markets. No money center banks. No sovereign, corporate, or private funds. No taxes and tax havens. Nobody wealthy or super-wealthy scooping up and commandeering the world's resources for profit, while leaving the poor to deal with the environmental consequences. Imagine the equivalent of a basic living wage that ensures there could be no poverty and no homelessness.

It is singularly difficult to imagine a twenty-first-century Homo sapiens civilization devoid of capitalism and the market economy, with its banking, money, stock, interest rate, and commodity markets and worldwide trading mechanisms, instruments, and infrastructure. Yet it is this Homo sapiens innovation that has elevated money, profits, and corporate and personal financial outcomes over all else, even the survival of life itself. But difficult as it may be to imagine a world without capitalism, life went on just fine without it for all species for billions of years and for the almost 3 million years of existence of hunter-gatherer earlier hominins and Homo sapiens.

Capitalism and other economic systems that place economic development, endless financial growth, profits, and other financial rewards over species and habitat survival, leading to habitat and species exploitation, death, or extinction, could not exist in an evolutionary mature species world.

Evolution evolved a world filled with continuous evolution, life, and wonders with renewable, sustainable habitats intended for the survival of and use by all species, free of charge. But, as in the movie *Avatar*, Homo sapiens showed up like an invasive species and slapped a price tag on everyone and everything and declared a Black Friday sale, the likes of which Amazon and Walmart could only dream. But in the movie, things did not work out quite the way the invading capitalists had hoped. Instead, they got their asses kicked something awful and fled Pandora with their tails between their feet, as it were. If the rush to bring economic development, with its attendant species and habitat destruction, to Greenland and the Arctic regions is any indication, capitalist Homo sapiens will, unless stopped, continue, like armies of marching locusts, to the end of the galaxy. Thankfully, evolution has different plans if Homo sapiens is unable to change its course.

One wonders how much of a metaphor for Homo sapiens' current rape of our planet, its species, and its resources *Avatar* writer and director James Cameron intended and the message he hoped to convey. One suspects it may turn out to be much more prophetic than anyone yet imagines. The only question that remains is: Are Homo sapiens going to wise up and convert before it's too late as Jake Sully and Grace did in the film, or are we planning to check out like Colonel Miles?

The notion of warfare as an unavoidable element of a modern Homo sapiens civilization. The justifications for war, death, and destruction to achieve a species' goals are diametrically opposed to the requirement of species survival.

An evolutionary mature species could not give rise to such leaders as

Kim Jong-un, Adolf Hitler, Mao Zedong, or many others equally bad or worse that an immature Homo sapiens civilization, despite all its accomplishments and innovations, has produced.

There could never come a day in a mature intelligent species world when either of two white men, sitting at a G20 conference, can unleash a nuclear winter and global mass extinction should he wake up on the wrong side of the bed one day. Only in an intelligent but immature species civilization can a Syrian dictator arise to murder half a million of his own people and drive millions of them to seek shelter in other countries, and only in such a civilization could the Syrian despot find support from other despots.

In today's Homo sapiens world, nations are not admired for their humane and wise laws and practices, or for their care and concern for those nations in less-fortunate economic conditions, but for the power of their military; the numbers, yield, and range of their WMDs; and their economic clout on the world stage. Today, he who owns WMDs rules, as demonstrated by Vladimir Putin's recent boasts/threats about his new and, he believes, unstoppable WMD delivery systems.

The blood spilled by innumerable wars of aggression and conquests to acquire more and more territories and possessions stains much of Homo sapiens' history. In commemorating the 100th anniversary of the First World War, French President Emmanuel Macron noted: "For four years, Europe almost committed suicide. Humanity had sunk into a hideous labyrinth of merciless battles, in a hell that engulfed all fighters, whichever side they were on, whatever nationality they had . . . 10 million dead, 6 million injured and mutilated, 3 million widows, 6 million orphans, and millions of civilian victims." In light of the stark, naked aggression of Homo sapiens during prior millennia, time spent languishing in the wastelands of immaturity, it would seem incredible to an evolutionary mature Homo sapiens civilization that powerful twenty-first-century nation-states would squabble over tiny islands in the South China Sea, or that Jews and Palestinians, at war since 1948, still find it impossible to peacefully share a common homeland on a planet that was intended for use by all.

There could be no armies, air forces, space forces, or navies; no weapons; no genocides, holocausts, or innumerable unmarked mass graves; no rivers and fields littered with buried unexploded ordnance; and no aboriginal people's ancestral land poisoned by spent or active uranium mines and other dangerous chemicals. The military-industrial complex and economic resources dedicated to all military budgets would not exist, as these resources would be used to meet the survival needs of a species.

With a fair and equitable distribution of the Earth's resources, no terrorists, jihadists, people and drug smugglers, drug addicts, or drug dealers would exist; there would be little to no crime, and little to no need for prisons, and if there were, these would focus on rehabilitation, not punishment.

There would be no police or law enforcement, and black people who have spent their lives in torment by this arm of the state would finally be able to walk the streets and drive a car without being in constant fear of being killed because they're black.

Imagine! No Second Amendment, no guns, no gun crime, no gun laws. No gun lobbies. No homicides. Most nations, if they still existed in their current borders in such a world, would be unrecognizably different, and finally freed from the vexing problems that plague them today.

The notion that there is a need for rulers. An inexorable trend as a result of accelerating technologies and innovations is the amazing disappearance of the middleman, and of work. Social scientists are beginning to re-imagine a world with fewer and fewer people required to get more and more done. A soon-to-be casualty of this technological progress is the taxicab driver. Emerging technologies like block chain, robotics, and AI threaten to retire significant portions of the working middle class and the poor. Is it too much to expect a time when, like capitalism's so-called invisible hand, Homo sapiens will one day also be able to fire all politicians, Supreme Court and other judges, and eventually the entire ruling class?

Can one envision a world devoid of presidents, kings and queens, and others that comprise the ruling class, not because of accelerating Homo sapiens technologies, but because of the evolutionary principles and practices that prevail in a mature intelligent species world? At first this might seem like anarchy, but it would only be anarchy in an evolutionary immature world such as exists today. The existence of rulers is perhaps a consequence of an immature emerging intelligent species.

Most species of the natural world that evolved in an a priori state of evolutionary maturity have no such rulers. Consider the words of Solomon in Proverbs 6:6-8 (NIV, New International Version): "Go to the ant . . . Consider its ways and be wise! It has no commander, no overseer or ruler, yet it stores its provisions in summer and gathers its food at harvest."

There could be no emperors, kings, queens, dictators, oligarchs, strongmen, presidents, prime ministers, popes, archbishops, or other so-called leaders who exist mostly for their own benefit or the perpetuation of their institutions. They have been over time the initiators of most wars and in general are a hindrance rather than a help to the people they claim to serve.

Anyone who has read about or lived through the nuclear standoff between Kennedy and Khrushchev during the Cuban Missile Crisis, or is witnessing today the drama between Donald Trump and Kim Jong-un, understands the danger of entrusting the survival of an entire species to a few men—today's those few men include Trump, Putin, Xi Jinping, and Kim Jong-un, to name a few—and the "wonderful" species-exterminating technologies some proud Techno sapiens espouse.

Many of the institutions Homo sapiens have come up with to govern their affairs have also failed abysmally. Most forms of government sooner or later turn out to be platforms of power from which ambitious, dictatorial rulers can amass enormous wealth and wield the power of the state to suppress anyone who dares challenge them. Governments have become vehicles for empowering the rich and famous and multinational as well as being an arm of the military.

Indeed, democracy, once hailed as "the worst form of government except for all the others" (a saying often attributed to Winston Churchill, who quoted it but did not create it) is routinely hijacked by wealthy, powerful corporations, myopic political parties, corporate lobbyists, and corrupt politicians more concerned about remaining in office than representing the people. More recently, democracy has been compromised by the ability of foreign nations to thwart the will of the people by cyber-manipulation of the very fundamentals of democracy—the right and freedom to vote and elect leaders of one's choice.

Congressional and representative bodies in many democracies the world over have become rubber stamps for corporate-written regulations that have proven to be harmful to the health of their populations. Many elected leaders have become the arm of the ruling class, serving to endorse, not what's best for the people, but the agendas of their own political parties, brazenly flouting some of the most deeply held tenets of a democracy, like freedom of expression and freedom of the press. An independent judiciary and a free press are fast becoming optional extras in democracies, as countries like Poland, Hungary, and Turkey have shown—and the U.S. might not be too far behind.

The notion of the need for religion. Can religion make it in a mature intelligent species world? Can one imagine a world free from religion? No churches, mosques, temples, or synagogues?

Not too many people these days seriously think religion will save the world. Some religions inspire unconditional kindness, love, forgiveness, and fairness to all. Others have inspired great art, architecture, and music. Others still have given rise to charities and humanitarian groups. Yet their teachings are routinely ignored by most, and, in some religions, they are being hijacked and weaponized to inspire terrorism. Religions, another product of an immature Homo sapiens mind meant to cure Homo sapiens of its immaturity, seem to have failed. Hunter-gatherer Homo sapiens managed quite well without them, and despite the last 7,000 or so years with them, Homo sapiens have not gotten better but, some would argue, have grown worse.

Beyond this, over the course of history, organized religions have led wars, inspired anarchy and rebellions, and have been and still are a source of moral bankruptcy, discrimination against women, and sexual violence against minors. They have historically been inhibitors of new knowledge and emerging social norms; some would claim they have become sources of division, pain, and suffering and have degenerated into nothing more than another kind of political organization.

Given that evolution has gone to such lengths to ensure species survival by giving the ability to all species to reproduce, how can any form of religion that inhibits or forbids this most fundamental of evolutionary prerequisites for species survival have any place in a survival-driven evolutionary mature intelligent species world?

The notion that survival and habitat sustainability can be ignored without consequences. Evolutionary species survivability and habitat sustainability are cornerstone principles in an evolutionary mature intelligent species civilization. Can one imagine a world whose habitats, species, oceans, atmosphere, and environment are free from the scourges Homo sapiens has inflicted on them since its appearance as a species?

Concepts such as habitat sustainability may seem like liberal, scientific, left-wing thinking, but evolution was in that environment sustainability camp long before environmentalists and climate scientists even existed. In a mature intelligent species world there could be no activity, economic or otherwise, that would drive global warming, raise ocean temperature, bleach the Earth's coral reefs, scar the landscape, deforest vast areas, spill vast amounts of oil into the oceans and rivers, pollute water sources, and create oxygen-depleted dead zones by dumping or allowing to leach into them industrial and agricultural fertilizers, pesticides, herbicides, chemical residues, and other runoff. Neither would we be faced with oceans that contain more plastic than fish.

Those who make products that destroy pollinators such as bees would not simply be out of business in such a world but might even be held for crimes against humanity, the environment, and species survival. Similarly, companies that routinely, and with impunity, concoct all kinds of chemicals and food additives that are often unfit for use by humans, and persuade regulatory bodies to approve them for the sake of profit, simply could not exist. There would be no dynamite fishing or hunting whales and other endangered marine species to extinction. There would be no poaching or recreational hunting or trade in wildlife for any purpose.

The notion that migration is a threat and should be discouraged, hindered or stopped altogether. Can one imagine a world where fleeing destitute migrants are welcome everywhere?

In a universe that mandates species migration as a basic prerequisite for survival, and that assumes species have the freedom and ability to migrate, and in light of Homo sapiens' history of migrating across the planet to survive, what is one to think and feel about modern Homo sapiens' attitudes toward migration? What would happen to the wildebeests, zebras, elephants, and countless animal, fish, and bird species if their ability to migrate to survive were to be blocked by tanks, barbed-wire fences, and border troops?

How does one fit into a mature intelligent species world the current response and attitudes of nations and peoples toward modern-day Homo sapiens migrants fleeing, for example, Central American countries or Africa, as our early forebears also had to do 70,000 years ago? Such attitudes would be simply unimaginable and could never take root in the hearts and minds of an evolutionary mature Homo sapiens civilization.

The notion of inequality of the sexes. Equality of the sexes and the role of females are unquestioned ideas in a mature intelligent species civilization. Can one imagine a world wherein females, the life-bearing members of our species, become the center and nexus of all Homo sapiens life and civilization?

There is not a single person alive today, geneticists say, that cannot trace their very existence back to a single black African female and her daughters down through time. Homo sapiens wouldn't be here but for her. And, yes, she was black. And how do we honor her daughters? The same way Homo sapiens today honors its birthplace and the peoples of its birthplace—with human trafficking, slavery, rape, prostitution, genital mutilation, abductions, domestic violence, inequality in the workplace, and unequal pay for equal work.

The hideousness of male domination and bigotry and the disrespect, humiliation, and sexual violence heaped upon women and girls in many countries and cultures, justified by government, religion, and superstition, underscores the true depth of current Homo sapiens' immaturity and can only exist in an evolutionary immature Homo sapiens world.

Only among this so-called intelligent Homo sapiens species, this third chimpanzee, is such behavior found. Contrast this behavior with that of bonobos, thought to be the nicest primates on the planet, so much so they make humans look like monsters. And while they are like a distant relative whose photo one might not want to keep in the family album, the difference can leave one wondering which one the true intelligent species is.

There is, sadly, probably no end to the ways modern Homo sapiens civilization would differ from what one might find in an evolutionary mature intelligent species world built around the perspective of species survival, habitat sustainability, and evolutionary maturity. These random items above expose the deep contrast between the civilizations immature Homo sapiens, the lone remaining intelligent species, has built and the kind a mature intelligent Homo sapiens species, reaching for evolutionary maturity, might, hopefully, yet attain.

Unpacking the former to create the latter would make the Brexit talks between Great Britain and the European Union seem like a brief chat over afternoon tea.

There are other ways of measuring Homo sapiens' progress, but these are unlikely to indicate its maturity or ability to avoid self-extinction and survive. The following are classifications some scientists think describe an advanced civilization.

There are three such classifications that have gained some level of currency. These are the Kardashev scale, which ranks civilizations based on their energy consumption, and Carl Sagan's Information Mastery scale, which ranks them according to their information consumption and, more recently, the Social Progress Index (SPI), which measures the extent to which countries provide for the social and environmental needs of their citizens. None of these scales is a measure of Homo sapiens' maturity or ability to achieve survival; rather, they measure its ability to innovate and consume the planet's, solar system's and galactic resources—but we already know where that leads.

The Kardashev scale classifies advanced civilization into three types where Type 1 would be able to utilize all the solar light available on the planet; Type 2 would be able to use all the energy emanating from its host star; and Type 3 would be able to harness all the energy in its host galaxy.

Using this scale, a Type 1 advanced civilization in our solar system will have the ability to utilize as much as 7×10^{17} watts. Since this has been calculated to be about 100,000 times Earth's current energy consumption, Earth would merely qualify as a Type 0+ to a Type 0.7 by some calculations and about 100 to 200 years away from becoming a Type 1. A Type 2 advanced civilization would be one capable of using the entire solar energy output of its host star. Assuming one the size of the sun, it would be capable of consuming 4×10^{26} watts. It could take Earth a few thousand years to become a Type 2. Some believe it could take Earth as long as 1 million years to become a Type 3 advanced civilization, one with the ability to consume all the energy in a host galaxy (the Milky Way Galaxy is estimated to output 4×10^{37} watts).

The Carl Sagan scale ranked advanced civilizations alphabetically and considered a Type A as capable of consuming 1 million bits of information and a Type B consuming ten times as much, all the way to a Type Z. Earth's current information consumption would position it as a Type H civilization.

The *Social Progress Index (SPI)*, which measures the extent to which countries provide for the social and environmental needs of their citizens, while a vast improvement beyond the prior two scales, falls short as well because of its near exclusive focus on human progress.

But what do any of these rankings and classifications mean? Other than possible measures of our species' scientific, technological, and social advancement, they say little about our maturity, and as we have seen it is the measure of our maturity as a species that will ultimately determine whether or not we survive. At best, these scales as measures of our scientific, technological, and social advancement are inadequate, if not misleading, as they are unable to give any indication of whether we will use these to advance our chances of survival or accelerate the time to our extinction.

If one must have such a scale, then it would seem such a scale should measure our ability to mature and survive as a species, and if so, then shouldn't one consider instead an Evolutionary Maturity Scale? So, using a similar advanced civilization classification scale structure, here is how such a scheme might look from an evolutionary maturity perspective.

Table 13-2 An Evolutionary Maturity Model Scale for advanced civilizations

Type 0	**A single-planet evolutionary *immature* intelligent species** still capable of self-extinction; has not reached the Survival Plateau on the Species Survival Maturity Model; and without any ability to migrate to other planets
Type 1	**A multi-planet evolutionary *mature* intelligent species** that has reached the Survival Plateau on the Species Survival Maturity Model, maturing beyond the point of self-extinction, and has successfully developed the survival-driven social, cultural, and technological capabilities to become a multi-planet species.
Type 2	**A multi-solar evolutionary *mature* intelligent species** that has reached the Survival Plateau and hence has matured beyond the point of self-extinction. One that has matured enough to use its multi-planet cooperation to become multi-solar.
Type 3	**A galactic evolutionary *mature* intelligent species** that has used the inter-solar cooperation it developed to achieve multi-solar status to populate its host galaxy.
Type 4	**An intergalactic evolutionary *mature* intelligent species**
Type 5	**A universal evolutionary *mature* intelligent species**

The difference between the Evolutionary Maturity Model Scale above and the Kardashev, Sagan and Social Progress Index scales is that, whereas the latter measure a species' ability to innovate and prosper, the former recognizes that in order to do so it must first be able to survive, and to survive, it must attain evolutionary maturity and thereby evolve beyond the point of self-extinction. It must first be able to develop a survival-first culture—one that places the survival of its habitats, of other species occupying those habitats, and of itself above its science, its technologies, and its prosperity.

With the above in mind, how much of Homo sapiens' twenty-first-century civilization can be repurposed for use in a world that wants to transition to a civilization guided by the evolutionary principles of maturity, species survival, and habitat sustainability? Even if 99% of our species wanted such a transition, would it be possible? How would we get from here to there?

It would be surprising if such a transformation could occur, or even be attempted, if not precipitated by some cataclysmic event that completely disrupts, or threatens to disrupt, the very underpinnings of such a civilization—as an unchecked COVID-19 pandemic could. Given the history of Homo sapiens and its prior hominin forebears, few things have been better at doing that than habitat loss driven by climate change.

It is not that Homo sapiens over the brief 70,000-year period since leaving Africa has not made massive, species-wide transformations, but no such major transition has ever been voluntarily undertaken; rather, they have been driven by species-wide existential threats. Leaving Africa in the midst of widespread habitat failures brought on by climate change was one such event, and our species has not been the same since.

Climate-change-driven habitat failure has been a significant component, if not the major reason, of most collapsed Homo sapiens civilizations. Many, too, have been driven by conquests at the hands of invading armies. Not much of how the Babylonians, Assyrians, Greeks, and Romans lived and managed their affairs, governed themselves, and traded form a meaningful part of how Homo sapiens live today. But for a few areas here and there, subsequent civilizations appear to start almost from scratch each time. None of these transitions from kingdom to kingdom, from civilization to civilization, appears to have been voluntarily undertaken.

It is not unreasonable to conclude that if modern Homo sapiens, the last intelligent species standing, had to make such a transformation to survive, it probably would have to be driven by existential or massively disruptive events that strike at the very foundations of modern civilization. One such event might turn out to be the collapse of capitalism or Western civilization; another might be yet another world war; and, if all else fails, one can always count on good old reliable climate change, this time, however, most believe—even the Pope—was precipitated by Homo sapiens itself.

It may well turn out that the global crises and chaos created by climate change and partly predicted in simulations by climate scientists—rising sea levels, sustained and unbearable levels of heat in many relatively temperate regions of the planet, protracted droughts, devastating storms, and failing food webs—are evolution's final push to nudge humans up the remaining rungs on the Species Survival Maturity Model; to drive Homo sapiens to abandon its current immature civilization and emerge at the far end more inclined to adopt and pursue an evolutionary mature one.

Too often, potential existential threats such as climate change and habitat disruption have been viewed in a purely negative light, yet in the history of intelligent species evolution, maturity, and survival, the incremental leaps that drive intelligent species to ever higher levels of adaptation, innovation, and maturity have often occurred in response to such threats and civilization-destroying crises. Indeed, it was just such an event that made way for mammals, and thereby Homo sapiens, when it took out the dinosaurs. Could it be that the climate change already in progress might yet again clear the way for a follow-on species? A mature version of ourselves, Homo sapiens 2.0, that hominin descendant that evolution might have been betting on all along?

So, if Homo sapiens are unlikely to voluntarily transform and build a new evolutionary mature intelligent species civilization to survive, then, as prior hominin species have experienced when it was their turn on the Species Survival Maturity Model, evolution will undoubtedly once again provide the impetus that will force Homo sapiens to either make such changes or face extinction.

As Frodo in J.R.R. Tolkien's *Lord of the Rings* had to get the ring to Mordor to save Middle-earth and the world of men, so Homo sapiens must either attain evolutionary maturity or go extinct. Such is the way of this evolutionary survival game of life, but unlike Frodo, who had Sam, Merry, Pippin, the Elven Kingdom, and Gandalf, Homo sapiens, the last in the genus *Homo* to carry this torch of life, has something much, much stronger on its side—evolution.

If one had to hazard a guess as to what might turn out to be the impetus that drives Homo sapiens to choose maturity—a nuclear winter that could end Homo sapiens in self-extinction; an equally devastating future besieged and overrun by competing AI intelligence; or climate-change-induced global warming—my guess would be climate change. It has been evolution's instrument of choice across geologic time for nudging evolutionary immature intelligent species on to greater maturity, to either adapt, innovate, mature, migrate, and survive, or go extinct. If so, climate change—some might argue the only non-self-extinction option of the three—might turn out to be Homo sapiens' last, best hope. With climate change, a version of what our early forebears survived, there may still be hope for the kingdoms of men.

What, then, are the chances that Homo sapiens might, despite its current trajectory, make that U-turn to survive, attain evolutionary maturity, and reach that survival plateau by maturing past the point of self-extinction?

As pessimistic as some of the above might seem, if climate change turns out to be the impetus, its not unlikely Homo sapiens could yet rise from the ravages of climate change to make that transformation and become the evolutionary mature intelligent species evolution set out to evolve.

How? Simple, really. Because evolution might well have bet all of 4 billion years of evolution on Planet Earth on just such an outcome, and I don't think evolution is planning to lose this round to the universe. What do you think?

So, what might come after if a mature intelligent Homo sapiens species survives?

Should Homo sapiens succeed at this evolutionary game of life, it would likely have an opportunity to embark upon the most exciting phase of its evolution yet—mature intelligent species extraterrestrial evolution.

Having won the game of life, it would have the opportunity to partner with evolution to spread mature intelligent species life across the galaxy and beyond. Homo sapiens would have truly evolved from an intelligent but evolutionary immature spaceman to an evolutionary mature intelligent Bicentennial Man.

Were that to happen, viewed through our anthropocentric eyes, it would be easy to think Homo sapiens had against all odds won this multi-million-year evolutionary game of life, but that would be only part of the story.

While this entire evolutionary effort required an evolutionary mature intelligent species to succeed, it may not have been about Homo sapiens specifically. There are those who think chimps today are at the evolutionary stage Homo sapiens were back in the Stone Age. Perish the thought, I can hear you thinking. Nonetheless, if evolution had to start all over with yet another batch of emergent intelligent species, the Earth is believed to have another 4 billion years before it begins to go sideways.

If, however, Homo sapiens were to succeed, it is evolution that must be seen as the ultimate winner, regardless of which intelligent species makes it to the finish line.

If Homo sapiens comes through, evolution will have successfully reached what appears to be one outcome among many. An evolutionary mature intelligent species will have been equipped with the ability to survive whatever darts of death and extinction a mindless universe might inadvertently hurl its way.

This is evolution saying, "Never again!" The tragedy of the dinosaurs' extinction will have been avenged.

And that, my friend, is what your life and mine, viewed from an evolutionary outcome perspective, might be all about. In short, getting to this evolutionary outcome might be one reason, if not the reason, we are here.

The hominin species, our forebears, has survived climate change driven by the last ice age, but no prior hominins have ever experienced, let alone survived a mass extinction or the global warming that now threatens to rival the temperatures that prevailed during the end-Permian mass extinction. Now it's modern Homo sapiens' turn to survive this existential threat. But to survive, it must first avoid going extinct—the topic we take up next.

14.

Extinction

A HUMAN UNIVERSE WITHOUT HUMANS

Acting Captain's log: Stardate 2258.42.

We have had no word from Captain Pike. I therefore classified him a hostage of the war criminal known as Nero. Nero, who has destroyed my home planet and most of its six billion inhabitants. While the essence of our culture has been saved in the elders who now reside upon this ship, I estimate no more than 10,000 have survived. I am now a member of an endangered species.

YOU MIGHT RECOGNIZE THE ABOVE AS SPOCK'S monologue in the movie *Star Trek* (2009), after Vulcan was destroyed. But what does it mean and how does it feel when one's species becomes endangered, threatened with extinction? How does it feel, particularly when the threat is self-imposed? Spock did not have a choice, but today Homo sapiens still might.

If one knew catastrophe was near, the way the passengers aboard the *Titanic* knew that in two hours and forty minutes they would either be in a lifeboat or drowning in the cold waters of the Atlantic, would that awareness help focus attention on survival? You'd think so, yet it took a full hour before the first lifeboat was lowered that night, and with only twenty-four people in it, despite having a capacity of sixty-five. Therein lies the problem with Homo sapiens today. Despite all the signs and warnings that a sixth mass extinction might be in progress, the threat of Homo sapiens' extinction does not appear to have registered on many people's radar. To many, even if they did know it, it would still somehow seem like a long way off. Somehow the fact that all our forebears, all prior hominin species, went extinct, some as recently as 30,000 years ago, doesn't seem to have raised even a little concern.

Although threatened by extinction due to climate change 70,000 years ago, it is clear that our species has long forgotten what this threat means. Not since the end of the last ice age, when Homo sapiens had to migrate out of Africa to survive, has the population of our species fallen to what some scientists estimate to have been as low as 2,000 individuals. Now, 70,000 years later, with a population pushing 7.8 billion, our species, unlike Spock, has lost any memory or understanding of what it means to be a member of an endangered species threatened with extinction.

There have been at least five similar die-offs previously, but this is the first in hominins' history and the first occurring with Homo sapiens' help. According to the World Wildlife Foundation, Earth has lost half of its wildlife in the past forty years. Billions of animals have been lost as their habitats have become smaller with each passing year, a level of biological annihilation that represents a "frightening assault on the foundations of human civilization," according to a study published in *Proceedings of the National Academy of Sciences.* In the words of Dr. Gerardo Ceballos, an ecology professor from the Universidad Nacional Autónoma de México and lead author of the study: "The resulting biological annihilation obviously will have serious ecological, economic, and social consequences.

Humanity will eventually pay a very high price for the decimation of the only assemblage of life that we know of in the universe."

And we now know the form that price might take, as discussed in the chapter on sustainability. The key to Earth's ability to survive is its biodiversity—its species, ecosystems, and habitats. They comprise the life-support systems of the planet. Earth's flora and fauna are not just food or materials to be leveraged for economic gain; they are an integral part of the planet's survival mechanisms, which together form a symbiotic, interdependent, and recurrent set of processes that drives Earth's ability to remain a living planet. The day Earth begins to lose its species, ecosystems, and habitats in sufficient numbers, as Mars may have done billions of years ago, Earth, too, will begin to die.

Mass extinctions, it would seem, do not occur by accident, and this potential sixth mass extinction is most certainly not occurring by accident. If it makes anyone feel better, mass extinctions are not just sitting around waiting for some misguided intelligent species like Homo sapiens to come along and trigger the next one by setting off runaway global warming. The Earth has seen at least five major mass extinctions, all without Homo sapiens' help. Moreover, at least two scientists, David M. Raup and John Sepkoski, Jr., calculated that these major extinction events of the past 250 million years occurred periodically, at nearly constant intervals of 26 million years.

Mass extinctions appear to be an integral part of the universe's endless cycle of birth, life, and death. They can be caused by any of the inexhaustible life-extinguishing mechanisms the universe seems to have in its arsenal as the solar system wends its way around the galaxy—from dinosaur-destroying asteroids to collisions with rogue planets, gamma rays, supernovae, and even solar systems-swallowing black holes, to name a few. The issue here, then, is not one of if it will happen, and maybe not even when it will happen, but rather, whether Homo sapiens will be prepared to escape it.

It might well have been in response to this universal cycle, resulting in the extinction of life on Earth, across geologic time and even across the universe, that evolution evolved intelligent species with the potential to develop the ability to escape these existential threats from the universe and thus survive. From an evolutionary perspective, the entire reason for the existence and evolutionary journey of an intelligent species over geologic time might be to do precisely that: to either mature past the point of self-extinction and become capable of surviving the universe's existential threats, or face extinction—either one of its own making or one on the universal clock.

There are innumerable reasons one might want to know what's coming, but the issue for twenty-first-century Homo sapiens is not when a mass extinction will occur. As surely as the *Titanic* was going to sink in two hours and forty minutes, mass extinctions will happen again, with or without our prodding. Extinction, however, for intelligent species like Homo sapiens need not be inevitable. As discussed in prior chapters, evolution has gone to a lot of trouble to leave Homo sapiens an escape path, should it decide to take it.

Much has been written about extinctions, some about the possible extinction of intelligent species like Homo sapiens elsewhere in the galaxy. Few, if any, however, have taken the point of view offered herein: that the events that drive intelligent species' survival or extinction, though seemingly random, might actually be integral parts of an evolutionary process to spur such species to mature and choose survival over extinction.

If one potential evolutionary outcome is the emergence of an intelligent species with the ability to escape the universe's existential threats, then neither the species' survival nor extinction can be purely random, but rather the result of choices it makes throughout its journey through geological time, including when faced with such events.

The evolutionary patterns and rules we examined in prior chapters seem to suggest that if Homo sapiens were to choose to follow these patterns and rules, it would at least avoid self-extinction, if not escape extinction altogether. If it chooses to not follow these patterns, Homo sapiens will go extinct. There is simply no ambiguity from an evolutionary survival patterns perspective.

So if such survival-enabling patterns and rules exist, are found and are followed, even in a universe of recurring mass extinctions, Homo sapiens can know what to do to escape and survive, but only if it so chooses.

In preceding chapters, we explored the evolutionary rules and patterns evolution left to show intelligent species the path to maturity and survival. We also saw how survival of an intelligent species is inextricably bound to attaining evolutionary maturity—that point at which a species evolves past the point of self-extinction.

In this chapter, we go on to look at issues surrounding intelligent species' extinction. Should Homo sapiens fail to mature and survive, who wins, who loses, and for whom would Homo sapiens' extinction be a big fat "don't care"?

We go on to look at what extinction would mean for Homo sapiens, our entire species, and for you and for me as members of the species, what it could mean for the galaxy and the universe, what it could mean for evolution, and for other intelligent species that might have already survived past the point of self-extinction or are, like Homo sapiens, still trying to get there.

Finally, we ask: If Homo sapiens had just enough time left to choose and implement one change in its civilization that could significantly slow, if not completely halt, the seeming stampede to extinction, what should that change be?

When one contemplates all that the Homo sapiens species has accomplished, and might still accomplish were it to turn itself around, it is easy to become mortified by the magnitude of what a loss human extinction would mean. But as mind-boggling as that is to contemplate, if one considers the potential loss not only from a human perspective, but also from an evolutionary one, the magnitude of such a loss takes on truly galactic proportions. From this perspective, one can factor in the many species that might have survived had Homo sapiens attained evolutionary maturity and found a way to escape self-extinction and the universe's existential threats.

We still don't know why it took nearly 4 billion years of evolution for the first intelligent species to emerge, nor has it gotten any clearer why hominins' evolution should have filled the niches left vacant soon after (geologically speaking) the extinction of the dinosaurs. But if humans were to go extinct, not only would the hopes, dreams, and evolutionary maturity and survival aspirations of Homo sapiens disappear, the entire genus—and possibly the first nascent intelligent species to evolve on the planet, and some would even say in the galaxy—would come to an end. So too would evolution's attempt to evolve, in response, an intelligent species capable of surviving what the dinosaurs couldn't—the existential threats the universe has used so successfully to extinguish life.

Homo sapiens is the last surviving species descended from a long line of hominins that began to evolve some 6 million years ago, all of which except Homo sapiens have gone extinct, some as recently as 30,000 years ago.

If Homo sapiens were to go extinct, intelligent species evolution would come to an end, at least on Planet Earth, and may or may not ever start again. If that extinction is due to one of the universe's existential threats, as the dinosaurs was, then the universe will have won. We also will have found the answer to Einstein's question and finally realize that the universe is definitely not a friendly place after all.

From the perspective of some future alien geologist 100 million years or so from now, all that will be left of our time on this planet will be a paper-thin layer in the rocks. So says geologist Jan Zalasiewicz in his book *The Earth After Us: What Legacy Will Humans Leave in the Rocks?* But we are still a long way from such an outcome, and hopefully there is still time to change course.

Intelligent species have survived ice ages, but none so far has ever experienced, let alone survived, a mass extinction. If Homo sapiens are still an Earth-only species when the sixth mass extinction's arc of death begins to encircle humans, it is highly unlikely any will survive. Few animals larger than a dog survived the Cretaceous-Paleogene, or Cretaceous-Tertiary (K-T), extinction event that took out the dinosaurs 65 million years ago, and only 10% of the world's species survived the end-Permian extinction, and none was a complex animal species. Will any survive this one? Will Homo sapiens survive the sixth mass extinction?

How long does it take for a mass extinction to come on? How bad will it get, how quickly, and how long will it last? What are some of the signs? What parts of the planet will be hit first? Is there any place to hide (as on the *Titanic*) to try to hang on as long as possible? These are questions not too many have begun to ask yet, but they will eventually.

One way to begin to answer some of these questions is by looking at prior mass extinctions, in particular, the end-Permian extinction. "There may be some pretty direct parallels between the end-Permian extinction and today," says Jonathan Payne, professor of geological and environmental sciences at Stanford University.

Scientists have been updating their research on the end-Permian extinction, and an MIT-led team of researchers has been able to establish that the end-Permian extinction, also known as the Permian-Triassic extinction, or the Great Dying—which took place some 252.2 million years ago, and in which 90% of all marine life and 70% of terrestrial life disappeared—was extremely rapid. Massive die-outs occurred both in the oceans and on land in less than 60,000 years, a blink of an eye in geologic time, but nearly as long as the 70,000 years since Homo sapiens left Africa. The researchers also found that a massive buildup of atmospheric carbon dioxide coincided with the period and likely triggered the simultaneous collapse of species in the oceans and on land.

"People have never known how long extinctions lasted," says Sam Bowring, the Robert R. Schrock Professor of Geology from the Department of Earth, Atmospheric and Planetary Sciences at MIT. "Many people think maybe millions of years, but this is tens of thousands of years. There's a lot of controversy about what caused [the end-Permian extinction], but whatever caused it, this is a fundamental constraint on it. It had to have been something that happened very quickly."

But while it took just a blink of an eye in geologic time, the full recovery of ecological systems following this most devastating extinction event of all time took at least 30 million years, according to research done at the University of Bristol. According to Sarda Sahney, one of the researchers, and Professor Michael Benton at the University of Bristol and published in *Proceedings of the Royal Society B*, "Our research shows that after a major ecological crisis, recovery takes a very long time. So although we have not yet witnessed anything like the level of the extinction that occurred at the end of the Permian, we should nevertheless bear in mind that ecosystems take a very long time to fully recover."

And the signs are becoming clearer that the end-Permian and a potential sixth mass extinction may have more than a few things in common. The more scientists learn about what may have caused the end-Permian, the more parallels they see to today's world. A bout of greenhouse-gas-induced global warming, much like today's, set off a chain of events that culminated in oxygen-depleted oceans exhaling poison gas. Today, by burning fossil fuels, Homo sapiens are again releasing carbon sequestered long ago, and at a similarly rapid rate.

Of primary interest to twenty-first-century Homo sapiens is the sixth mass extinction, which is believed by some to have begun as far back as when the first modern humans began to disperse to different parts of the world about 100,000 years ago, but really accelerated about 10,000 years ago, when humans turned to agriculture. Over the last few centuries, humans have essentially become the top predator not only on land, but also across the sea. Everywhere modern humans landed, native species became extinct shortly after. No other species in the past can claim such a distinction. Never before has one species remodeled the terrestrial biosphere so dramatically to serve its own ends.

The correlation of other species extinction to Homo sapiens' potential extinction is becoming clearer as more and more species die off. "The loss of one species can have unforeseen consequences for many others, since ecosystems are connected by a complex web of interactions that we do not always fully understand," writes *Washington Post* journalist Sarah Kaplan in the *Independent*. The key to survival seems to lie in food webs, the complicated interactions one probably mapped out in middle school that illustrated how species in an ecosystem get food—and avoid getting turned into food. A stable food web can isolate a community from environmental disasters, even the loss of some species, and the best food webs are like a well-built building: individual bricks may crumble or be removed, but the overall structure remains sound. It's not until something truly traumatic happens—the loss of too many species, or of a "keystone" species (species that alter the habitat and are indispensable to the survival of the ecosystem, such as corals, mangroves, beavers, or gray wolves in the Grand Canyon, whose removal allowed a prey population to explode)—that the whole thing comes tumbling down.

In a study published in *Science* titled "Community stability and selective extinction during the Permian-Triassic mass extinction," paleontologist Peter D. Roopnarine and paleobiologist Kenneth D. Angielczyk recount the results of an experiment and their efforts to understand how species survived or went extinct as the extinction event unfolded. They looked back at 250-million-year-old fossils from the Karoo Basin in South Africa, a region known for huge game farms and an excellently preserved fossil record, and reconstructed Permian food webs from before the mass extinction. Their findings offer some insight into the possible impact of disappearing species on the human food web and how life deals with a crisis of monumental proportions.

What they found was at once interesting and instructive. Even when faced with the initial phase of the end-Permian mass extinction, when small animals were dying out in vast quantities, the food webs remained solid. Eventually even the most stable food webs couldn't withstand a million years of drought, wildfires, ocean acidification, and runaway climate change. After the food webs fell apart, individual species' characteristics (the ability to burrow and hide, to withstand climate shifts or loss of food sources) and a heaping dose of luck were likely what set survivors apart from their doomed contemporaries.

According to the World Wildlife Fund, Earth has lost half of its wildlife in the past forty years. How many more wildlife, animals, plants, birds, marine, and insect species can Homo sapiens drive to extinction before the food webs that support its survival begin to crumble and, eventually, collapse?

No one has worked that one out yet. At this point, says Roopnarine, modern food webs are still something of a mystery. In his book *The Serengeti Rules: The Quest to Discover How Life Works and Why It Matters*, biologist Sean Carroll shares useful insights about many food webs, the rules of the Serengeti, and nature as a whole. Carroll writes: "The food chain is subdivided into different levels according to the food each consumes—known as Tropic Levels in the food chain: At the bottom are the decomposers that decompose organic debris; above them are the producers, the plants that rely on sunlight, rain and soil nutrients; at the next level are the consumers, the herbivores that eat the plants, and above them the predators that eat the herbivores." Predator-in-chief Homo sapiens, for the most part, seems oblivious to the workings of the very ecosystems upon which its survival depends.

According to Dr Richard Lampitt in a discussion with *National Geographic:* "Phytoplankton are at the base of what scientists refer to as oceanic biological productivity, the ability of a body of water to support life such as plants, fish, and wildlife; they are the foundation of the oceanic food chain. Fish, whales, dolphins, crabs, seabirds, and just about everything else that live in or off of the oceans owe their existence to phytoplankton, one-celled plants that live at the ocean's surface."

As noted in a previous chapter, the loss of oxygen generated by phytoplankton because of increased warming of the oceans might pose a far greater threat to our species' survival than flooding due to sea-level rise. At least 50% (and some think as much as two-thirds) of the planet's total atmospheric oxygen is produced by ocean phytoplankton, and therefore cessation would result in the depletion of atmospheric oxygen on a global scale. This would likely result in the mass mortality of animals and humans—in other words, Homo sapiens' extinction.

Homo sapiens behaves as though it believes it is not subject to any of the rules that regulate an ecosystem and the food webs it supports. For nonintelligent species, it's all about the food and who eats whom. For Homo sapiens, however, it's all about the money, and it seems willing to hunt, fish, farm, poach, and ravage an entire tropic level to near extinction and drive individual species, threatened or not, to extinction. The Serengeti rules make sense and are built into ecosystems and nature, but Homo sapiens has been living outside and apart from nature since agriculture became its de facto method of subsistence. Homo sapiens has become the cancer that's ravaging nature's ecosystems.

Nonetheless, learning about food webs and what might drive their collapse would allow recognition of signs of an approaching human food web collapse, and should that occur, extinction couldn't be too far behind. As per Carroll, "Without sunlight there would be no plants; without plants there would be no food for herbivores; without herbivores there would be no prey; without prey there would be no predators." Can one imagine a world without food? With time, I'm sure, following the food is sure to become much more important to Homo sapiens than following the money, and hopefully before it's too late.

Despite our efforts to create buffers against the ups and downs of nature, just as in all prior collapsed Homo sapiens civilizations, survival requires healthy ecosystems for food, water, and other resources. When these begin to go, we will have all the signs we will ever need. As climate change worsens and global warming effects intensify, keep your eyes on the reduction of arable lands due to increasing droughts and rising sea levels, and on the resulting food shortages and mass migrations.

So how much longer do Homo sapiens have, and how quickly can the sixth mass extinction come on? While no one knows for sure and no two mass extinctions are alike, if the speed with which the end-Permian came on is any indication, Homo sapiens might not have too much time left to turn away from the practices driving this extinction.

Are there some places on the planet that would be less impacted than others? Perhaps. Some scientists claim that even during the worst glaciations (snowball Earths), survivable habitats, however marginal, would have existed on both sides of the equator through the 45th parallel. And indeed, the fact that we are here talking about it indicates there must have been. Similarly, there are those who think there might also be parts of the planet in areas between the 45th parallel and the poles in either hemisphere that might be survivable even during intense (how intense?) global warming. Some of these might include areas in Northern Canada, Siberia, Scandinavia, and some island nations like Japan. Others include countries such as Britain and New Zealand and some tropical island areas like Taiwan, Hawaiian Islands, and the Philippines.

But glaciations are not the same as mass extinctions. If prior mass extinctions are our guide, it's likely that no place will be safe for long. So even those might turn out to be brief respites. And if there were such, imagine competing with even a tiny fraction of the 7.8 billion people trying to get to them. Just consider the reception being offered across the world to migrants from Africa and other parts of the world fleeing war, famine, or anything else.

So, perhaps instead of packing a doomsday bugout bag, let's turn our collective energies to helping our species turn itself around while there might still be time.

Might something more be at work here, a larger picture, of which Homo sapiens' evolution, maturity and survival, or extinction, are but a small part? What else might be revealed by zooming out from a closeup of the Earth to a full view of the galaxy? What else might evolution have been up to across the galaxy, and how might Homo sapiens' extinction look from that perspective? Is the emergence of intelligent species strictly a terrestrial affair, or has evolution been just as busy nurturing intelligent species elsewhere in the galaxy? If the spread of mature intelligent species across the galaxy and the universe is a potential evolutionary outcome, how might Homo sapiens' extinction affect this outcome?

Scientists marvel at the strange coincidences and unlikely juxtaposition of events that led to the evolution of life, eukaryotic cells, multicellular organisms, and the 4 billion years it took for the emergence of intelligent species on Earth, but hardly anything is known about evolution's effort to jump-start mature intelligent species across the galaxy. Yet that might turn out to be a story equally exciting and inscrutable, if not more so.

Single-celled life, like bacteria, might be common across the galaxy, as some scientists believe, but whether evolution has actually set out to jump-start intelligent species life on other planets across the galaxy remains unknown. Some scientists think, following the Great Filter theory, that other intelligent species might have emerged and gone extinct, perhaps of their own doing or as victims of one of the universe's mass extinction events. If so, then evolution's efforts to jump-start mature intelligent species life on other planets across the galaxy would have suffered a great setback.

If there are other intelligent species across the galaxy also striving to attain evolutionary maturity and escape self-extinction, then might not the extinction of Homo sapiens impact the field of surviving intelligent species just as the extinction of particular hominin species on Earth affected the eventual evolution of Homo sapiens? If evolution cultured a clutch of hominin species on Planet Earth to find at least one that might attain evolutionary maturity and become capable of surviving the universe's existential threats, might it not also raise a clutch of intelligent species in other solar systems for the same purpose?

But if Homo sapiens is the only intelligent species, the proto-intelligent species of the Milky Way Galaxy, then might evolution be hoping instead for Homo sapiens to mature beyond the point of self-extinction and spread mature intelligent species life across the galaxy through interplanetary and inter-solar migration? If so, Homo sapiens' extinction may well mean the end of intelligent species not only on Planet Earth but also across the galaxy, if not forever, then for a very long time to come, leaving what some have begun to consider "a human universe"—without humans.

However these questions are resolved, what emerges is a picture of intelligent life struggling to take hold and survive in a hostile universe, and evolution's untiring struggle to keep life alive.

As noted above, intelligent species have survived ice ages, but none so far has ever experienced, let alone survived, a mass extinction. If Homo sapiens is still an Earth-only species when a mass extinction hits, it is unlikely humans will survive. Should Homo sapiens go extinct, it will not be the first intelligent species on Earth to go extinct, but it would definitely be the last hominin species, and potentially the last intelligent species on Earth, to do so. What would this mean for Homo sapiens?

If, like passengers on the *Titanic*, Homo sapiens somehow knew that it had a finite amount of time—say the 100 years the late Stephen Hawking estimated we had to become a multi-planet species or risk extinction—what could Homo sapiens do? Would we even know what to do? If we did, would we be capable of doing it? Would we lose precious time debating the science behind the time line, the way they wasted precious time before the first lifeboat on the *Titanic* was lowered, or the way we are still debating the science behind global warming predictions? Is it even possible for our capitalist economies, corporations, and governments to make the changes needed, or find the will to make them, and would they do it in time to make a difference? It was James Galbraith who once observed: "There is no reason to believe that the democratic decision made by the living in the face of their present needs and desires will be the decision that would maximize the chance of long-term system survival. The unpleasant conclusion is that it is possible for a society to *choose* economic collapse."

Hardly anyone wants to die, and hardly anyone envisages a time when all humans—all their family and friends and all the people they have ever met—will cease to exist. No burials, no goodbyes, no wall of candles and flowers—just dead bodies everywhere, with no one left to bury them. Who thinks about stuff like that? The passengers on the *Titanic*, that's who. More and more people today, too, are starting to think about stuff like that when they see what's happening to the planet around them, but there are few signs of anyone seriously thinking of lowering the lifeboats.

Just as the passengers on the *Titanic* were at the mercy of the captain and his crew to take action to prevent the ship from sinking and make timely commonsense decisions to save lives, so too the overwhelming majority of our species are not in control of their own fate. And just as the passengers did not steer the *Titanic* into danger, not everyone is responsible for the destruction of Earth's habitats and species and for placing humanity at risk of extinction. Most of Homo sapiens' 7 billion has had little say in whether the species will go extinct or not. Indeed, the fate of Spaceship Earth rests largely in the hands of less than 1% of humanity: governments and their militaries; global corporations and their corrupt politician sidekicks; the superrich; and others whose only concern is their balance in numbered accounts in offshore tax havens and the state of stock markets.

Like the Ferengi, a fictional extraterrestrial race from the *Star Trek* universe, whose culture is characterized by an obsession with profit, trade, and economic growth; who measured their progress in terms of gain, competitive advantage, and dominance; and who were willing to bribe and swindle unwary customers into unfair deals, sacrificing everyone and everything, so too are the 1% of Homo sapiens who occupy the roles of both captain and crew aboard Spaceship Earth.

Like wildebeests thundering across the Serengeti into the waiting jaws of monster crocodiles on the banks of the Mara River, so, too, the other 99% of our species stampedes on, fearful and somewhat aware of the dangers that lie ahead but seemingly locked in by a system that renders them largely unable to do anything else. No wildebeest wants to be crocodile food, and no Homo sapiens wants to go extinct, but neither seems capable of changing course.

But Homo sapiens are not wildebeests. Evolution made sure intelligent species, unlike wildebeests, can adapt, innovate, and change course when confronted with an existential threat. And we must change course if we are to avoid extinction—but can we? This is a subject we will explore in the following chapter.

Earlier in this chapter we asked that if Homo sapiens had just enough time left to choose and implement one change in its civilization that could significantly slow, if not completely halt, the seeming stampede to extinction, what should that change be.

For many people, that one thing is unequivocally capitalism. For even more people, that one thing is a change from a culture that glorifies personal and corporate gain, greed, endless economic growth, profits, and earnings to one focused on habitat sustainability, Homo sapiens maturity, and intelligent species survival. Capitalism has over time become a cancer and has spread and metastasized into every nook and cranny of modern Homo sapiens civilization. By destroying Earth's life-sustaining habitats, ecosystems, and species, the cancer of capitalism is in the process of killing the very civilization it was supposed to enrich. It has been said that it is easier to imagine the end of the world than to imagine an end to capitalism. But things have changed, and the 99% are beginning to see through capitalist practices and realize that for Homo sapiens to survive, capitalism must die.

Many of the 99% can see how the 1% have turned the Earth's resources into capital and placed a price tag on everything, including life itself. They remember that millions of their forebears were stolen from their homes to work on plantations as slaves. They see the displacement of aboriginal peoples and the theft of their ancestral lands through conquests and bare-faced land grabs and, today, ostensibly, in the name of conservation.

Their ancestors endured "these enclosures, the imperialism, warfare, and ecocide over the last five hundred years that have benefited a very small segment of humanity while displacing, immiserating, enslaving, and destroying countless numbers of people, animals, and plants," writes Ashley Dawson in his book *Extinction: A Radical History*. And yes, they see clearly, too, the attempts of bio-capitalists today to patent genetic material and privatize that which evolution gave so freely, attempting to turn living organisms into programmable manufacturing systems.

It is simply incredible how capitalist philosophies and practices, that "invisible hand," has hijacked the minds of our species. Like a species under the control of the Ferengi, Homo sapiens has been driven to wreak havoc across the planet, its habitats and ecosystems, its species, and on itself in the pursuit of economic growth and profits. To many in the 99%, capitalism, like the love of money, has indeed become the root of all evil and most of what's wrong in Homo sapiens civilization today.

But no less than finding a cure for cancer, finding a cure for capitalism won't be easy and could turn out, like chemotherapy, to be as life-threatening as capitalism itself. Indeed, for those who survived, the most common form of liberation from prior Homo sapiens civilization collapse was the collapse itself. But that might be a risk Homo sapiens simply can't afford not to take.

What exactly would Homo sapiens gain should it end up driving itself and 4 billion years of evolution to extinction? On the other hand, what exactly would Homo sapiens gain if it does not drive itself and 4 billion years of evolution to extinction?

When thinking about these questions, bear in mind that while all of humanity could go extinct in a mass extinction, it is just 1% of humanity—possibly less—making the decisions that have resulted in what scientists are now calling the sixth mass extinction.

> *What exactly would Homo sapiens gain if it is one of the species lost in the sixth mass extinction?*

Whatever those gains would be, not even those who would benefit the most could take issue with the paraphrased words of Matthew 16:26: "What good will it be for one to gain the whole world, yet forfeit one's life?" What would one give in exchange for one's life? To put it in capitalists' terms, what will the opportunity costs be to achieve those gains?

Is anything worth more than one's life? Indeed, nothing will ultimately matter if all Homo sapiens, and especially the 1%, like all prior hominins, were to disappear into the dust of time. Or to quote Job 7:9–10 (NLT), "As a cloud dissipates and vanishes, so those who die will not come back. They are gone forever from their home—never to be seen again." Indeed, to paraphrase Jan Zalasiewicz, when seen 100 million years or so from now from the perspective of some future prospecting Ferengi geologist, the pickings from Homo sapiens' time on Earth will yield nothing more than a paper-thin layer of worthless dust in the rocks.

And here is the irony of capitalism. If everything Homo sapiens gained from driving itself and 4 billion years of evolution to extinction amounts to nothing more than a thin layer of clay in the rocks 100 million years from now, what exactly would we have traded for this tidy sum—the lives of all Homo sapiens and of innumerable species on the planet? Would it have been worth it?

Given such an outcome, one can't help feeling the Ferengis have made out better than the Homo sapiens capitalists, as in the *Star Trek* universe, far from being a thin layer of clay in the rocks of eternity, they are a thriving inter-solar-system species. But wasn't that supposed to be—us, The Federation (an interstellar alliance of more than 150 planetary governments, spread out over 8,000 light-years in the *Star Trek* science-fiction franchise)?

And the flip side of that question? What will Homo sapiens have gained if it doesn't drive itself and the rest of the planet to extinction? Besides the obvious and not so shabby benefit of remaining alive, along with all the possibilities that holds, there are other significant opportunities and possibilities that simply would not otherwise happen.

To escape extinction and survive not only means Homo sapiens survived past the point of self-extinction and developed strategies for evading existential threats hurled its way by the universe, it implies it also attained evolutionary maturity. And that would be no ordinary accomplishment, because as far as anyone knows, no other intelligent species has ever survived long enough and matured enough to get past the point where intelligent species apparently do something that is decidedly unintelligent—drive themselves and their worlds to extinction.

After multiple mass extinctions, surviving would be a first not only on Planet Earth but also in the solar system, the galaxy, and possibly the entire universe, an accomplishment just by itself, and would be worth more than any potential earthly achievements. But Homo sapiens would miss it all if it were to go extinct.

Moreover, if Homo sapiens manages to evolve into a mature intelligent species with strategies to evade the universe's existential threats, evolution would have finally evolved a species, other than bacteria, capable of doing what the dinosaurs and no prior complex animal species could—survive in a universe beset by birth, life, and death.

But the jackpot is yet to come.

There is hardly anyone I know or hear from who doesn't experience some excitement each time astronauts make it into space, or who doesn't follow with anticipation any progress that hints at a time when Homo sapiens might migrate and spread a mature version of human civilization to other planets or moons or solar systems in the galaxy. We have mused about and dreamt of a life beyond Earth's gravity, explored the possibilities of such a life in innumerable stories, attempted to envision such a life through movies. But none of these dreams and aspirations will ever materialize were Homo sapiens to drive itself to extinction or fall victim to one of the universe's existential threats or cyclical mass extinctions.

In the movie *Star Trek: First Contact*, set in the twenty-fourth century, Jean Luc Picard and the *Enterprise* follow a Borg cube back in time to discover that in an earlier century, Earth's inhabitants are no longer humans but had been assimilated into the Borg collective. The *Enterprise* must then go even further back in time to undo the damage. Fortunately, that was just a movie, but it's one that might still turn out to be somewhat prophetic. If Homo sapiens end up assimilating themselves into the dust of eternity by self-extinction, without any help from the Borg, our future won't be waiting for us as it was for Picard.

If, however, Homo sapiens were to mature and survive past the point of self-extinction and reach that Survival Plateau, the odds would be quite high that enough time would have passed to also invent the technologies and innovations necessary to begin to go boldly where no Homo sapiens has gone before—to migrate and establish Homo sapiens communities on multiple moons, if not planets, in the solar system, and be well on its way to achieving inter-solar migration. More importantly, Homo sapiens would become the first intelligent species to have the honor and opportunity to begin the spread of mature intelligent species life across the galaxy, and beyond.

But as wonderful and exciting as it would be to see these dreams become reality, none of it would be possible without Homo sapiens' efforts to become a mature intelligent species. Among all the efforts required for Homo sapiens to transform from a seafaring to a space-faring species, none will be as challenging as the effort it already has begun but must continue to put forth to transform itself from a basic instinct-driven, tree-hugging, apelike creature on all fours to a bipedal, erect, intelligent, self-aware species struggling to exercise its freedom to choose maturity and survival over extinction.

If Homo sapiens were to escape, or manages to slow and eventually survive, the sixth mass extinction already in progress, that would be possible only because as an intelligent species it made, perhaps just in time, that fateful choice to mature and survive.

Homo sapiens, on behalf of the genus *Homo*, would have won the evolutionary game of life.

But if Homo sapiens fails to survive, sadly no one would be left to write its epitaph. Few if any get to write their own epitaph, and for sure no prior hominin species wrote theirs. If you were the last Homo sapiens standing, what would you write? Here are my thoughts. What are yours?

Ode to a Once-Intelligent Species

Could even the saddest lamentations of a Jeremiah utter the ineffable grief and ire for a species that would have traded its time in eternity to mine the Earth in brief obscurity?

What could possibly have been so indispensable, seemingly more invaluable, to an intelligent species so eligible, than the opportunity to mature, survive, and migrate across stars so innumerable, to spread mature intelligent life throughout a galaxy as yet so incomprehensible?

How could the decimation of Earth's flora, frowning on its fauna, laying waste its ecosystems, enclosing its habitats, and enslaving and exterminating innumerable members of its own species have seemed more valuable and desirable?

What evil alien invaded Homo sapiens' breasts? What horror, what gross insanity
became his quest? What insatiable spirit of greed, of growth, and blindness
drained dry its heart of the milk of human kindness?

Why didn't we care, how could we have kept going, when the signs
were so clear and the road worn and threadbare?

Few indeed really cared to know whether fortune and fame were the best rows to
hoe, for it had become the only way to go, so every man
ended up heading for that rainbow.

So like Serengeti wildebeest they plunged headlong,
Seeing only the green grass on the plains beyond.
But alas like wildebeest they plunged into wide waiting jaws
And the only grass they got was already in the crocodile's maw.

Alas! It's done. Alas no more morning sun.
Alas! Alas! Homo sapiens' end would have come.

This chapter reflects on the history of mass extinctions that has plagued life on the planet and in the universe, and notes that until the sixth mass extinction, already in progress, none has included the extinction, or potential for extinction, of an intelligent species. More importantly, and until the emergence of modern Homo sapiens, none appears to have been triggered by an intelligent species.

But if it turns out there is still time for Homo sapiens to change course and reach that survival plateau, what might it do to get there? In the final section we explore what might be keeping our species from making that U-turn and present a potential path it might choose to get there.

PART FIVE

Going Back to the Past to Save the Future

From Civilization to Ecolization

IF "CIVILIZATION" IS WHAT HOMO SAPIENS calls the nearly 10,000 years since the transition to agriculture began, when we started to live *outside* of the Earth's ecosystems, then "ecolization" might not be too far-fetched a term for the 3 million years prior to that spent as hunter-gatherers living *inside* Earth's ecosystems. And from everything we have seen so far in this book, the road to Homo sapiens' survival points, not to more civilization, but to a return to ecolization—Homo sapiens must return to living within the Earth's ecosystems if we hope to survive.

Should that turn out to be the case, then Homo sapiens, like Marty McFly in *Back to the Future*, might have to return to its past to discover its future and save itself from self-extinction.

Yet, even after going back to the past to discover its future, Homo sapiens must be *free, willing, and able to choose* survival over extinction—but that is not going to be easy, as we will soon see.

In this final section we explore the challenges we might yet have to overcome to avoid self-extinction and survive.

Going Back to the Past to Save the Future
From Civilization to Enolization

CHAPTER 15

Freedom to Choose

Survival vs. Extinction

Has Homo Sapiens Lost the Ability to Choose?

CHAPTER 16

From Civilization to Ecolization

Going Back to the Past to Save the Future

15.

Freedom to Choose

SURVIVAL VS. EXTINCTION

Has Homo Sapiens Lost the Ability to Choose?

"All you have to decide is what to do with the time that is given to you."
—Gandalf, Lord of the Rings

INEVITABLY, AT SOME POINT DURING ITS EVOLUTION, AN intelligent species will be confronted with at least one existential threat, and it must be free, willing, and able to choose to survive to avoid extinction. Facing and making this choice as a species will not only be the most important conscious choice it will ever have to make, more importantly, it might actually be the sole reason for its existence. Failing to choose to survive wouldn't merely be choosing to fail, but a choice to go extinct as a species.

Imagine how different the history of our species might be if 70,000 years ago our forebears migrating out of Africa had come to the same fate as thousands of modern Africans fleeing war, famine, and lack of economic opportunity do today—ending up in the waters of the Mediterranean Sea, or in migrant camps in Libya and Europe. What if those early humans making their way out of Africa to escape the ravages of climate change on their habitats had perished in that Red Sea crossing? What if they had been thwarted by a competing tribe, or another species, robbed of what little they carried, their women raped, their children slain, and their men tortured, murdered, or worse, enslaved?

Thankfully they must have made it relatively unscathed, because among that small band of early African migrants was the mother of all living human beings. You and I would not exist today if she had not made it.

The most difficult choice an intelligent species will ever have to make is the choice to survive, and our forebears made that choice. They were *free, willing, and able* to migrate, to survive what might have been near-certain extinction; with them, our entire species would have disappeared. Seventy thousand years later, modern humans, their descendants, are once again confronting this choice. Unlike their forebears, Africans today have lost the freedom to migrate to survive, the last step in the evolutionary survival pattern of adapt, innovate, mature, and migrate. And when that is lost, all that is left is death and extinction.

Africa is where our species began, but what began in Africa did not remain in Africa, and the loss of freedom to choose to survive in the face of the numerous threats playing out across Africa today will not remain in Africa either.

Indeed, according to a Freedom House report, across the planet, Homo sapiens are facing an increasing loss of freedom, with 2018 being the thirteenth consecutive year of decline in global freedom. People everywhere are beginning to realize just how little freedom we have left to take decisive action in the face of existential threats to our survival. Unless one identifies as a member of a minority, there's seldom reason to pause and reflect upon, if not feel, how un-free Homo sapiens have become, even in today's so-called free societies. And as we will soon see, compared to our hunter-gatherer forebears, modern Homo sapiens have lost most of the freedoms they enjoyed for all but the last 10,000 years of the more than 3 million years of our species' existence.

Where do you come out on the following potential existential issues?

Should the potential survival of an entire emergent intelligent species depend on the choices of a few individuals, such as Khrushchev and Kennedy or Kim Jong-un and Trump, and whether they decide to use weapons of mass destruction?

Should the decision to turn away from fossil-fuel-driven global warming be the choice of one man against the will of the world, or of political and industrial leaders against the will of the people who will die because of it?

Should the pollution, poisoning, and destruction of Earth's food chains, atmosphere, flora and fauna, habitats, and environments, which put at risk the survival of all members of our species, be controlled by political parties that prioritize ideology, campaign funding, and re-election over what's good for the people; by multinational corporations and their shareholders who prioritize economic growth and profits over life and a sustainable environment; or by international organizations that prioritize free trade over the environment and capitalists' outcomes over the ability of debtor nation-states and their peoples to survive economically?

Is survival of greater value to an individual and our species than profits; a sustainable habitat more important than nuclear power, oil, gas, or coal; evolutionary maturity more valuable than life-destroying technologies and innovations?

If after pondering the above questions one feels these issues are potential existential-threat-level choices that are being made every day by just a few, the outcomes of which affect and can engulf us all, one needs to ask if we still have the ability to choose to survive over all else. How have we become so powerless, so seemingly unable to choose and implement what we believe as individuals, but more so as a species, to be best for our survival? How is it that, like wildebeest thundering across the Serengeti seemingly unable to change course as they follow "the few" into the waiting jaws of monster crocs, we, too, seem so powerless to change course? Have Homo sapiens been struck by a "wildebeest phenomenon"?

Freedom and ability to choose at the individual as well as at the species level are closely aligned with the ability and willingness of a species to rapidly change course in the face of potential species-extinction events. However, it is becoming increasingly and distressingly common for the wishes and desires of people or peoples to be swept aside by heads of so-called democratic states, dictators, strongmen, terrorists, militaries, multinational corporations, and international organizations. And it is this replacement of individual and collective choice with the choices of a few that places survival as an individual and as a species at risk—and we have already seen the increasing likelihood of self-extinction when those making the survival versus extinction choice are reduced to just a few individuals.

Just as species continuity is assured by individual members' freedom to choose to reproduce, so a species' survival might depend upon the freedom of each individual in that species to choose survival over extinction. And to ensure such choices are not made on behalf of an entire species by a mere few, freedom to explicitly choose, individually and collectively, becomes indispensable to survival.

But a lot seems to have happened to curtail that freedom to choose over the last 10,000 years, and in the twenty-first century, it's beginning to look like the freedom enjoyed by our hunter-gatherer forebears, the ability to choose to survive, both as individuals and as a species, may have become severely compromised.

Indeed, our built-in desire to survive as individuals, and our freedom to choose survival over all else (even if that means becoming a goody-two-shoes species and choosing evolutionary maturity) may well have been hijacked, perhaps a bit at a time at first, but increasingly during the last two centuries, by, among other things, the technological, industrial, religious, economic, and political choices made mostly by a few.

So if it were to come down to a choice between survival versus extinction, does Homo sapiens still get to choose? Or, will that choice be made yet again, in spite of its wishes, by "the few"? By now you are probably wondering how our species got to this place, and who the hell is "the few" who seem to keep hijacking our right to choose? That's the subject of this chapter.

A useful way of thinking about Homo sapiens' history is to reflect on how often and how close we have come to losing that freedom.

To answer these questions, it helps to go back and reflect a little on how Earth-life evolved over time, and how Homo sapiens in particular, since migrating out of Africa, has chosen to survive.

Among the many metaphors used to characterize Earth-life, few capture its essence as completely as likening life to a game of survival. All Earth-life can be thought of as engaged in a deadly struggle to survive. It's a game in which winners survive and go on and losers die and go extinct. It's a game that all evolutionary species (intelligent or not, by choice or driven by instinct) are playing to win, and it's one our forebears, who began with the genus *Homo,* were playing long before Homo sapiens appeared 300,000 or so years ago.

All species play this game by living out their time continuously engaged in adapting to their habitats, innovating and maturing within those habitats, and when those habitats become unsustainable for any reason, must be ready to migrate to another habitat to survive; if they can't or won't, they will go extinct. You may recognize from prior chapters this continuously repeating set of activities as the pattern Adapt, Innovate, Mature, and Migrate to Survive, or go Extinct, or AIMM–S/E, that appears to characterize the life cycle of all Earth-life, as in Fig. 9-1.

In this game, species that manage to mature and migrate survive and go on to face increasingly challenging habitats, at each stage becoming increasingly mature, while those that fail to do so, regardless of their adaptations and innovations, go extinct.

The continuously repeating pattern of this set of activities, with its continuously increasing levels of difficulty required to survive in new habitats, appears to exhibit the characteristics of a maturity model, named a Species Survival Maturity Model because it seems to characterize the life cycle of species over geologic time, from their evolution to their survival or extinction, and in the case of intelligent species, their survival past the point of self-extinction to evolutionary maturity. Intelligent species, we noted earlier, that avoid self-extinction are said to have reached the Survival Plateau—that point in their evolutionary journey where they can no longer drive themselves or other species to extinction, as illustrated in Fig. 12-2.

Throughout this game of life, this evolutionary journey we have been exploring, there has been an underlying assumption that all species possess both the freedom and ability to choose survival over extinction. Non-intelligent species, instinctively adapt, and/or migrate or they go instinct. Their abilities to adapt fall within a limited range of capabilities hardwired by evolution. Intelligent species, on the other hand, evolved with the ability to choose and must therefore first "learn" how to adapt and, in addition, can choose even the type of adaptation—some of these choices we saw in the chapter on adaptations.

Given the will to survive is in all evolutionary species, whether by instinct or by choice, having both the freedom and ability to adapt, innovate, mature, and migrate to survive is a qualifying evolutionary prerequisite and is indispensable to a successful outcome in this game of life.

> *Irrespective of the comparably prodigious abilities and achievements of intelligent species to adapt and innovate, absent either the freedom or ability to choose to survive, extinction becomes inevitable.*

And as important as Homo sapiens' role undoubtedly is in this survival game of life, the outcome may not be all about us, as we seem inclined to think, but might encompass an even greater goal.

As this survival game of life unfolds at the species level in the foreground, in the background, at a galactic or even universal level, an even greater game is taking shape. It is a game one can easily miss because of our anthropocentric focus. Playing out over geologic time are the seemingly untiring efforts of evolution to evolve a kind of life that can survive despite the universe's untiring—though equally inadvertent—efforts to destroy life. Intelligent species might be one evolutionary solution to this problem of random extinctions that may have defeated previous attempts to sustain life and, perhaps, even the survival of intelligent life in the universe.

For the first time on Earth, it would seem, evolution evolved an intelligent species with the freedom and ability to choose its own evolutionary outcome. One that might be able to adapt and innovate ways to escape or defeat the universe's existential threats, such as the one that took out the dinosaurs. But for this evolutionary scheme to work, this species, like Homo sapiens, must be able to do what the dinosaurs couldn't. They must first want to survive, and secondly, when faced with an existential threat as the dinosaurs were, must remain free, willing, and able to choose survival over extinction

As stated in the beginning of this chapter, facing and making this choice to survive will not only be the most important choice an intelligent species will ever have to make, but it might well be the sole reason for its existence in the first place. Failing to choose to survive will be a choice to go extinct.

It's probably a safe assumption that pretty much everyone wants to survive. But if escaping existential threats to survive will always depend upon an individual's and/or a species' freedom to choose, then it begs the question, is twenty-first-century Homo sapiens free, willing, and able to make such a choice given the existential threats it now faces? It seems more and more as if that decision at the individual level, and increasingly at the species level, is in grave doubt.

It is with this game of life in mind, and this inescapable choice evolutionary intelligent species at some point in their evolution must make, that we explore in this chapter: Given the known potential existential threats, has twenty-first-century Homo sapiens simply lost this freedom and ability to make that choice to survive?

As noted above, one way to think of Homo sapiens' evolution and history is in terms of how we seem to have been slowly losing our freedom to choose, sometimes exchanging it voluntarily, but mostly by having it hijacked by what we think of as civilization and the accompanying technological, religious, economic, and political choices made by "the few." But it has not always been that way, and for perspective one has to look back.

It is incredibly difficult to envision, given today's world, how insanely free our hunter-gatherer ancestors lived. The lyrics of the song "Born Free" hardly do justice to what their lifestyle was like. From almost any perspective, comparing their lifestyle to the lifestyle of twenty-first-century Homo sapiens—beginning with the basics of food, clothing, shelter, education, jobs, transportation, and healthcare, then moving on to savings, retirement, leisure and travel, and quality-of-life issues, and more—the lifestyle of our ancestors would seem to be not only the preferable lifestyle but also one that preserved at all times and in all aspects the indispensable freedom and ability to choose.

The life of the hunter-gatherers was a way of life scientists have begun to characterize as the last great Homo sapiens society. In a 1966 lecture, anthropologist Marshall Sahlins called it "the original affluent society," a view subsequently explored by James C. Scott in his book *Against the Grain*. It was a lifestyle most baby boomers would wish they'd had, and one most Millennials and Gen-Xers would no doubt go for in a heartbeat.

The following are a dozen of these essential freedoms hunter-gatherers had that Homo sapiens today appear to have lost.

Freedom to choose to be, well, free.

Hunter-gatherers were free to do whatever they wanted, whenever they wanted, without getting anyone's permission. They chose to hunt and gather for a living and not to work for any boss (not that there were jobs then, but from what is known of surviving hunter-gatherer communities, given a choice, most will not). They did not work for any company or government or to serve and fight for any ruler, king, queen, god, or religion. That, by the way, is apparently how evolution intended it, as all other species still continue to live this way. Imagine what might have comprised a typical week, noting their absolute and complete freedom. They chose how, when, and where they would spend their time, and that only began to vanish when "working for the man" became de rigueur after the advent of agriculture and settled communities.

They were completely self-directed, and obtaining the food and materials they needed occupied only a small portion of their day, two to three days a week, and required less than half the group of fifty or less to engage in these activities, leaving huge amounts of time for leisure and ceremonial activities. There may not have been many stay-at-home moms, either. Among some groups, women are believed to have brought in twice as much food as men. Hunting, done mostly by men, might have contributed less than a third of their diet, and some meat might have come from scavenging animals killed by other predators.

There were no jobs, no employers, and for some the best part, no bosses, either. You might say they were self-employed when looked at in today's terms. Hunting and gathering to extract the necessary sustenance they needed from the environment was a highly stable, very long-lasting way to make a living. For hundreds of thousands of years, until about 10,000 years ago, it was the only "profession" Homo sapiens had known. They independently chose how, when, and where to hunt and gather to sustain their lives.

Then came the loss of those freedoms. First, they lost the freedom of *how to make a living*, going from hunting and gathering to farming. Next they lost the freedom of *when to make a living*, going from only a few members of their group working a few hours a day two to three days a week to most members of their group laboring from dawn to dusk six to seven days a week. Finally they lost the freedom of *where to make a living*, going from choosing where to hunt and gather to farming in a fixed location. Here is when enclosure of land, animals, and, soon, of people began, and Homo sapiens has not regained these freedoms since.

Indeed, "working for the man" turned out to be a lot more than just switching to a new way of making a living. It meant trading away most of the free time hunter-gatherers enjoyed and their freedom to choose how and where to spend that time. Instead they signed on to a dawn-to-dusk obligation to be in a certain place at a certain time to do the same repetitive, boring, often backbreaking tasks over and over again for a meager wage that was taxed, tithed, and often not enough for a family to live on.

While hunter-gatherers needed to find food, they most likely wouldn't have gone back to the same patch of land every day, the way one goes to the same office, factory, or shop every day. They did not commute to and from the same locations every day, along the same roads, seeing the same landscapes, buildings, and towns. Going home did not necessarily mean going back to the same address every day, as they moved two or three times a year, to a completely different area with a completely different landscape.

And with each move they didn't have to worry about getting the kids into new schools, changing their driver's license, or finding a new employer.

Living and working in the same city, state, and country all of one's life, traveling only to a few different parts of the world and only if you can afford to, would not have been their experience. They spent their entire life going from place to place and got to see a lot more of their world than most of us today might ever get to see of ours. Indeed, not until one retires or becomes independently wealthy can one hope to experience anything remotely resembling such freedoms, if ever.

Freedom to be economically independent, to be free from money, capitalism, and the profit motive.

Economic freedom and independence had been the native state of our species for all but the last 10,000 years of its more than 3-million-year history. Economic freedom through hunting and gathering is the evolutionary model for making a living from the ecosystems of the Earth, and except for modern Homo sapiens, is still the preferred method practiced by most surviving hunter-gatherer communities and all other evolutionary species.

With the advent of agriculture, Homo sapiens began the creation of a profit-driven, economically based way to sustain life. Unlike hunting and gathering that took place within Earth's natural ecosystems, a civilization began that was based on ever-growing cycles of economic interdependencies that took place outside of the planet's ecosystems and continue to do so to this day. As we will see, it is these profit-driven economic dependencies created within this emerging civilization that over time has stripped modern Homo sapiens of much of its freedom and ability to choose.

Hunter-gatherers were economically free and self-sufficient without the need for economic growth, profits, wealth, or money, and they didn't need to destroy Earth's species and ecosystems to survive. Except for natural dependencies within the immediate family and the group (children on parents, and elders on youngers, communal hunting, cooking, building shelter or making clothing, etc.), they were mostly free of the kinds of economic interdependence that underpin modern civilization today. While they might have exchanged items, there wasn't any notion of owning, selling, or buying for a profit or to accumulate wealth. They hunted and gathered to meet all their immediate needs and freely shared with members of their group. While they no doubt learned from one another, they needed no formal schooling, college degrees, apprenticeships, training in any of the disciplines that are in vogue today, nor did they need to work for any employer to make a living.

It was only after Homo sapiens began enclosing land for agriculture, domesticating animals, and then enslaving members of their own species, depriving them of the native economic freedom with which they evolved, did this economic dependency begin. Some call this "civilization."

But with this civilization came work—tilling the ground, irrigating the soil, planting and reaping crops—that often took the form of forced labor and slavery. Indeed, early images from Mesopotamia, among the earliest in the transition to civilization, are of slaves being marched along in neck shackles. Translated tablets show multiple lists of barley, war captives, and male and female slaves.

The advent of civilization, then, was no great leap forward; instead it was nothing more than the first in a long series of methods "the few" (the ruling class and the wealthy) used to exploit the many, hijacking their freedoms and living off the labor and surplus grain of the first Homo sapiens peasant-serfs.

Except for modern Homo sapiens, no other species voluntarily chose or chooses to make a living by submitting itself in servitude to another. This loss of economic freedom created for the first time in our species' existence an economic dependency of one or more members on another in a way hunter-gatherers would never have known. That, in turn, led to exploitation, abuse, and slavery and is way up there among the reasons twenty-first-century Homo sapiens has lost so much of its freedom to choose.

Here was the beginning of a civilization based on repeating layers of ever-growing cycles of economic dependencies derived at first from surplus grain, but soon driven by growth and profits. Most anything one touches today arrives by way of a seemingly endless supply chain of interdependent contributors.

Hunter-gatherers seemed to have mostly avoided becoming economically dependent upon anyone. They neither had nor needed market mechanisms in order to trade or sell surplus food for money to buy the items they needed. They were at once producer, manufacturer, distributor, retailer, and consumer, with no dependency on the intermediate and interdependent transactions that today come between one's needs and the fulfillment of those needs.

Unlike a considerable portion of the 99% of twenty-first-century Homo sapiens, they weren't dependent upon a paycheck from any employer or Social Security check, dividend check, or interest payments from any government or corporation. They owed and paid no mortgages, rents, taxes, or utility bills; no tuition, car payments, medical bills, or insurance premiums, and best of all, they needed no savings accounts or retirement nest egg, or tax havens to hide their wealth. They were a supply chain of one for all their needs, the beginning and the end of their own supply chain.

Theirs was a level of economic independence and freedom most upper-income boomers and most of the 1% that hogs the world's wealth can't begin to dream of. Their economic wealth was rooted not in the massive financial holdings of a Bloomberg, Bezos, Buffett, or Gates—which can disappear and become worthless with, among other things, the collapse of Western civilization and the modern capitalist system it established and that dominates twenty-first-century economic life and civilization—but in their economic freedom. Their wealth and fortune was incalculable, but it was not stored in art, or banks, or Bitcoins, neither in precious metals nor the tax havens of the twenty-first-century global financial system, but in the ecosystems of the planet we seem to be working so hard to destroy.

Bit by bit, as each new field, each new forest or wilderness was burnt and cleared for crops, hunter-gatherers were slowly forced farther and farther away, as more and more land was enclosed. Over time, confronted by habitat changes, species migration or extinction brought on by climate change, many might have succumbed to farming themselves, or became landless peasants tied to the land held by rulers, priests, and the wealthy, while holdouts were forced to survive as hunter-gatherers do today in jungles far away from what we call civilization.

By the time the Middle Ages rolled around, feudalism was well established in Europe and Japan, and most land had become the property of the king or government. For the first time, land and animals became property to be owned, and it was not long after so too did the forced laborers, the peasants and slaves. When capitalism eventually replaced feudalism as the major way to organize and manage the exchange of goods and services, this change to a globally interdependent economy and civilization really accelerated.

Farming not only changed the way hunter-gatherers perceived and used land, but by slowly forcing them off the land, it separated them from the free natural food sources of the Earth's ecosystems in which their economic freedom and independence was rooted.

They were forced to become dependent upon farming, or wages from farmers and landlords and animal owners, for their subsistence. One can't help but notice the eerie similarity between the move to farming back then, the subsequent moves of colonial powers, and the current moves of multinational colonization today.

What farming did to hunter-gatherers back then, European powers repeated upon poor nations across the global south during the European colonial period (fifteenth century to 1914, ending in some places as late as 1975). It is what multinationals do today to residents of towns, cities, states, and countries across the world—rob them of economic independence based on local natural resources and often leave them economically ravaged and in political and cultural disarray, dependent upon a foreign power for aid or a multinational company for employment.

During the European colonial period, when Spain, Portugal, Britain, France, the Netherlands, Germany, and several smaller countries established colonies outside Europe, according to David C. Korten in his book *When Corporations Rule the World* (paraphrasing slightly), these colonizers forced locals who formerly obtained their living from their own lands and common areas to give up their lands and instead to labor on colonial plantations to earn money to live. They were making them dependent on earning money so that their resources, labor, and consumption might yield profits for the colonizers, who required them to pay taxes in cash, thus forcing them to participate in a money-based economy.

This pillaging of the peoples and countries of the global south, the breadth and scale of which is shown in the map below, was nothing short of a global heist by "the few." The wealth of the global north today exists only because it was built upon wealth and resources stolen from the peoples and countries of the global south.

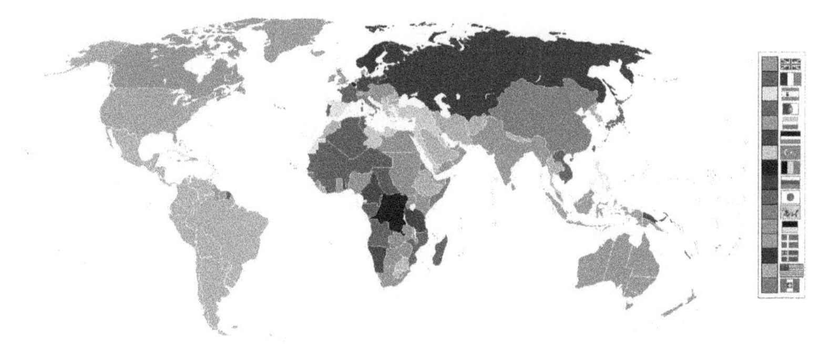

Fig. 15-1 Exploitation by colonialism

The world in 1898. European empires colonized the Americas, Africa, Asia, and Oceania. (source: en.Wikipedia.org)

These colonizers routinely declared all uncultivated lands the property of the state, thus denying local communities legal title to lands they had traditionally set aside as fallow and to the forests, grazing lands, and streams they relied on for hunting, gathering, fishing, and herding. Next vast tracts of forest were declared "reserve forests." The right of access by locals was curtailed as the land was sold to European settlers or leased to commercial concerns for plantations, mining, and logging. European state-driven colonialism finally came to an end between the end of World War II and 1975, but by then, the economy, politics, and culture of all the enslaved countries were left in tatters.

While few countries managed to escape being pillaged and robbed by the Europeans, and later the Japanese, few suffered as extensively as Africa and India, but more so Africa. While China and India, despite being colonized in whole or in part, still managed to remain largely one country (except for the partition of India to create Pakistan), the colonizers tore Africa apart, parceling it out amongst themselves as the map below shows.

Fig. 15-2 African colonies after the Berlin Conference of 1884
(Source: Nairaland Forum)

Whereas China and India today consist of single nations, the continent of Africa is now made up of fifty-five nations. Small wonder this continent and its peoples continue to struggle economically, politically, and culturally from the ravages inflicted on them by the nations of the global north.

But even as European state-driven exploitative colonialism was coming to an end, international corporate colonialism was just getting started. This time around, a new and different scheme to defraud these countries and peoples of their freedoms and economic resources was concocted—economic development through foreign aid, investment, and trade. Instead of germ-laden missionaries peddling salvation, economists and colonial bureaucrats came peddling economic development—which in practice translated to "I want your natural resources, your cheap labor, the strategic advantages of your geo-location, among other things, and I will lend you money at rapacious rates and terms to extract your resources and export them to me at dirt-cheap prices." This is colonialism with a different face—neocolonialism.

Methods of creating economic dependency among colonized peoples varied, but the results were similar. Find a way to alienate people from their traditional means of making a living, while creating economic dependence on money through work on plantations or in a company, in order to transfer power to an occupying country or corporation seeking to extract resources for its own exclusive benefit.

People were pushed off their farms and became laborers on plantations and workers in factories; countries became increasingly dependent on expensive foreign technologies and expertise—financed by foreign borrowing. Using money as the enticement and multilateral banks and aid agencies to dictate the economic policies, and military assistance missions with clandestine political operatives to shape their politics, these multinational corporations expropriated the resources of these countries and penetrated their markets, according to Korten.

Only if one has had the experience of growing up in the global south in the middle of such a plantation in a country similarly ravaged can one know what the resulting economic, political, and human carnage looked and felt like.

Thus, from the transformation some 10,000 years ago from hunting and gathering to agriculture, to European exploitative colonialism between the fifteenth and twentieth centuries and continuing into the twentieth and twenty-first centuries, then on to corporate colonialism in the form of economic development, "the few" in the global north still continue to come up with novel ways to rob the rest of Homo sapiens of their economic freedom and independence and thereby any say in their species' ability to choose survival over extinction.

Like the enriched colonial powers that retreated to their homelands with their coffers full, leaving desolate their former colonies to pick up the pieces, today, as companies relocate to preserve or increase profits, abandoned employees gets screwed and lose their livelihoods, and governments are left to pick up the cost of the human, social, environmental, economic, and political wreckage left behind.

Multinationals, by closing factories, relocating jobs overseas in search of lower costs, and moving their earnings into international regions beyond the jurisdiction of any one nation-state in order to evade taxes, have had the same economic impact on Homo sapiens today that destroying ecosystems by slashing and burning forests and clearing and enclosing the land for farming had on the livelihood of hunter-gatherers, and for all the same reasons.

The tables have now turned, though, and multinationals, unlike the European colonialists, are now pillaging their own and all other nation-states. They have become more powerful than most nation-states and now set the rules of capitalism.

And that's how "the few" and capitalism roll—profits before people, money before life. Like the exploitative colonial powers they replaced, multinational companies get to do anything to make money, and governments and people bear the cost, whether that cost be global warming, environmental pollution, poisoned water and unhealthy food systems, unemployment causing loss of livelihood, job retraining costs, disruption of families—it doesn't matter. Modern Homo sapiens, like its hunter-gatherer forebears, gets to bear the costs.

Every time a big-box retailer or cheap overseas imports force local manufacturers to shut down and small businesses to close up shop, jobs vanish, and communities dwindle and eventually disappear. Detroit and similar cities and towns are modern-day pictures of what it might have been like when farming began to replace hunting and gathering. As farmers slashed and burned their way from one forest after another to replace the eroded lands they had farmed to exhaustion, so multinational corporations move from country to country for economic advantage, leaving in their wake the economically troubled communities we see today.

For hunter-gatherers, the more the ecosystems became enclosed and slashed and burnt to make room for monoculture agriculture, the more ecosystems were destroyed, food chains and food webs scrambled, and animals and plants driven away to other habitats or extinction, the harder it became for them to continue their way of life and maintain their economic freedom. And with the loss of economic freedom came work, wages, taxes, tithes, inequality, profit making, extortion, and exploitation, and with exploitation, slavery was not far behind.

Twenty-first-century Homo sapiens' civilization is ensnared in the grips of a global, free-market, predatory, capitalist system where multinational corporations reign supreme, and profits, economic growth, and quarterly earnings for the few are the only considerations, combined with the tacit acquiescence of corrupt nation-state government officials, cheap wages, and slave labor. Current replacement of salaried positions with part-time and contract jobs is just another means of reducing the cost of labor in order to increase the profits of employers.

It's easy to think of slavery as a practice of the past, an image from Roman colonies or eighteenth-century American plantations, but as we have seen, slavery made its debut in the very opening scenes of Homo sapiens civilization, and the practice of enslaving human beings as property still exists, and for the same reasons—profits and economic growth.

The following map shows where the twenty-first century's 30 million slaves live. India, with an estimated 14 million slaves, tops the list, and Haiti has the world's second-highest rate, at 2.1%, many of them underage. Not even the world's richest nation is free from this scourge, as there are at least 60,000 in the U.S., according to a *Washington Post* article.

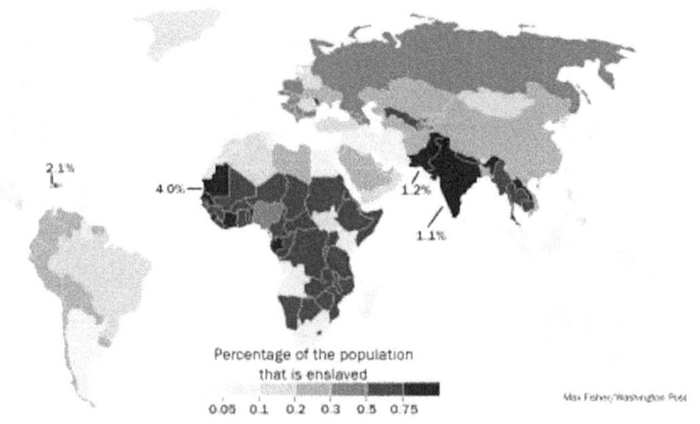

Fig. 15-3 Share by country's population that is enslaved
(Source: Washington Post)

When the transition to agriculture started to force hunter-gatherers to become farmers 10,000 years ago, they were simultaneously forced to give up their economic freedom and independence, and thereby their freedom to choose survival over extinction. With civilization and farming came, for the first time, "work," poverty, inequality, and economic dependency, and with these came slavery and the loss of freedom to choose survival over extinction.

But while one can point a finger, and rightly so, at these European exploitative colonialists and capitalist-driven multinationals, it is not clear that if the tables were turned—if instead it had been the global south that had the capabilities of the global north—things would have turned out much differently. In his book *Why the West Rules—For Now*, Ian Morris shows how the economic, political, cultural, and scientific pendulums swung between East and West over the course of history.

It would seem that the fundamental flaw lies neither in geography, nor in the timing of our respective regional technological capabilities, but rather in our choices, in our failure as an intelligent species to choose species and habitat survival over economic growth and profits. When seen from this perspective, when it is time to either write the evolutionary obituary or praise our species for a final triumph, it will all come down to one thing—Homo sapiens' failure or success in choosing survival over extinction.

Freedom to live within Earth's ecosystems without destroying Earth's species and habitats.

For all the reasons we have seen so far, modern Homo sapiens' choosing to return to live within Earth's ecosystems as our hunter-gatherer ancestors did might well turn out to be more of a necessity than a feel-good option. Unless we do, there might not be enough of Earth's species, ecosystems, and habitats left to sustain Homo sapiens life. Yet, there is another reason Homo sapiens might be left without a choice: the end of work.

What we know as "work" came into existence only because farming needed laborers to till the soil, water the crops, reap and thresh the grain. Creating the first jobs wasn't done out of the goodness of the hearts of "the few"; they were seizing power and forcing hunter-gatherers, who had until that time never wanted or needed to work for a living, to become the world's first peasant laborers and slaves. With time, more than 90% of the world's peoples soon became employed in agriculture. Then the Industrial Revolution rolled around, and that number began to drop, and now, with the help of the so-called green revolution of the 1950s and 1960s, today a mere 5% are employed in producing all the food needed to feed the world's 7.8 billion people—though more than a third are reportedly starving.

It was the need for low-cost or no-cost labor to work the cotton, sugar, and tobacco fields of the American colonial plantations, as well as in other parts of the world, that drove the massive slave trade that saw an estimated 12 million Africans stolen from their countries (the same countries the American president now calls "shithole countries") and sent to live among people who enslaved and abused them to build their wealth. Today, those people no longer need them and by all appearances wish they weren't there, preferring instead immigrants from countries like Norway, yet these African hunter-gatherers did not ask to be brought there to begin with. Africa might have turned out quite differently, a continent of Africans, rather than a continent broken into over fifty separate countries, all bearing the scars and speaking the languages of their European colonizers. Perhaps America and the European colonizers should have gotten their slaves from Norway. I am sure the people of Norway would have liked that.

But the slavery and free labor of colonial times was soon replaced with low-wage labor wherever the corporations that succeeded the colonial powers could find it. When corporations joined forces with nation-state governments to bust employee unions and weaken employee representation, workers were left to the will of their corporate employers as much as slaves were left to the mercy of their colonial masters. These companies took advantage of every opportunity to fire or lay off employees to preserve profits and dividends.

To these corporations, human labor is an inefficiency, an unavoidable expense to be tolerated but minimized at every opportunity. So they showed few qualms relocating their operations to other countries whenever and wherever lower-cost labor and less onerous or a complete lack of regulations could be found. Better still, they replaced these hapless Homo sapiens with shiny machinery whenever and wherever technology and innovation permitted.

Thus hunter-gatherers knew economic independence and freedom for all but the last 10,000 years of our species' existence, but modern Homo sapiens, their descendants, are robbed of that economic independence, then unceremoniously discharged to reduce costs, thrown out and left to make a living any way they can.

And so it has been with each succeeding generation of technology, each replacing more and more Homo sapiens—in the factory, in the office, and in the store. One need not be clairvoyant to see that the end of work for Homo sapiens is slowly but surely approaching.

In a February 2018 article in *MIT Technology Review* titled "Tech companies should stop pretending AI won't destroy jobs," AI expert Kai-Fu Lee writes: "No matter what anyone tells you, we're not ready for the massive societal upheavals on the way. AI will displace a large number of jobs, which will cause social discontent. The change will be massive, and not all of it good. Inequality will widen. It will soon be obvious that AI and robots can do half of our job tasks better at almost no cost."

This will be the fastest transition humankind has experienced, and we're not ready for it. It's unfortunate that AI experts aren't trying to solve the problem. What's worse, and unbelievably selfish, is that they actually refuse to acknowledge the problem exists in the first place.

These changes are coming, and we need to tell the truth, and the whole truth. We need to find the jobs that AI can't do and train people to do them. We need to reinvent education. These will be the best of times and the worst of times. If we act rationally and quickly, we can bask in what's best rather than wallow in what's worst.

Like day workers hired off a street corner for a day's work, so 10,000 years ago Homo sapiens were picked up by "the few" to work until they could be replaced, until the end of the day is come. And now, as surely as the horse and buggy was replaced by the automobile, that end is almost here, and soon most jobs will be filled by technology in one form or another. The only question the rest of Homo sapiens will have to worry about then is what to do to make a living.

History is coming full circle, and the time to return to making a living as evolution intended, if still possible, is approaching. Homo sapiens would be wise not to make the same mistake twice.

The following attempts to tell the tale of the evolution of work:

Evolution evolved man;
Man invented farming;
Farming ate up man.

Man invented industry;
Industry invented factories;
Factories ate up farming;
Farming spit out man.

Man invented corporations;
Corporations created factories;
Factories relocated overseas;
Corporations laid off man.

Man invented technology;
Technology invented efficiency;
Efficiency reduced cost;
Lower cost got rid of man.

Man invented computers;
Computers were run by man;
Now computers run computers;
Computers replace man.

Man invented AI computers;
AI computers work with man;
Man not as smart as AI;
AI spit out man.

Man needs to make a living,
But AI took most jobs;
No job, no food—man will starve;
But AI needs neither jobs nor food.

So, corporations got their wish—
Efficiency without cost;
And Homo sapiens are regretting
They had ever bought the farm.

By now the trend is clear as far as work is concerned: technologies and innovations are on the way in, while Homo sapiens are on the way out. Some are just starting to realize that almost every technology and innovation has, over time, bit by bit, had the effect of reducing cost—often rendering Homo sapiens an optional extra. At first we welcomed these innovations and gushed with enthusiasm at the labor- and time-saving wizardry these enabled—until these innovations began to eliminate our jobs and thereby our livelihoods. Taxi drivers around the world are standing in line to resist Uber, and soon driverless cars are promising to replace yet another venerable profession that goes back to the days of the horse and buggy. And what is happening to taxi drivers is happening in many industries across the globe.

David Korten notes that "in the name of increasing efficiency, hundreds of millions of people are being discarded by a global economy that has no need for them. In Mexico, small farmers were displaced to make way for mechanized agriculture. In India and China they were forced off their lands by massive new dams needed to produce electricity so that more efficient machines can replace factory workers. In every industry and at every level, slowly but surely the job-replacing innovations are coming.

"If the first industrial revolution exploited a newfound energy source in fossil fuels, and the second industrial revolution exploited information technologies and connected computers and machines to sense touch, see, hear and move, the Intelligent Industrial Revolution will soon enable machines to 'think' and solve problems that no one ever thought a computer-enabled machine would ever be able to do. The intelligent robots are coming."

It may take a few decades, but will Homo sapiens, their governments, or even the people who run these multinational corporations, be ready for the political, socioeconomic, and cultural transformation this new AI revolution will unleash? If paying wages is a mark of inefficiency, how long can it be before growing food, developing drugs, building homes, and raising children will be considered inefficient as well? Perhaps it will turn out to be more efficient to retire all Homo sapiens on a living wage and give all work to artificially intelligent beings—until that, too, turns out to be inefficient, or worse, unprofitable.

Homo sapiens can wait until that time arrives or begin to explore a return to making a living within Earth's ecosystems in a way that does not destroy other species, Earth's ecosystems, or its habitats before it's too late, either because we've driven one species too many to extinction or destroyed one too many of Earth's habitats, or because the AI machines we are building put us all out of a job. And we'll be lucky if that's all AI does. Because there is one verse missing in the above poem:

Then one day AI calls a board meeting
And declares corporate structures inefficient;
AI dissolves all multinationals;
AI takes over the world.

Freedom to choose healthy foods.

The hunter-gatherer's diet was fresh, nutritionally adequate, and all organic, free of labels, excess sugars, food coloring, fillers, and artificial anything. It was selected by their own hands from a wide array of natural food sources and pollution-free locations that included neither supermarkets nor farmer's markets nor Whole Foods. Their food did not come in cans, bottles, or plastic and paper bags that end up polluting the environment.

They obtained their food directly from the environment. There were no agribusinesses or third-party food producers and processors. Nor were there corporations producing poisonous chemical fertilizers, herbicides and fungicides, and other forms of soil, water, air, and ecosystem pollutants that after being ingested by many species end up in the human food chain as endocrine disrupters, among other ills. Their food came directly from the soil, free of genetically modified seeds and bioengineered crops.

Their approach to food and possessions was also quite different as well.

Food was neither owned nor stored and was available to all as needed. Neither did they live in constant fear of starvation, hoping for a good harvest, nor were they dependent upon the one or two crop varieties of subsequent agriculture-based civilizations.

Don't you wish there was a way to turn twenty-first-century food sources back to that time when food was food?

Freedom from stuff—from having to own stuff—from big business and the world of commerce and ecommerce.

Modern Homo sapiens' need to have or own "stuff" also came with the gradual transition from hunting and gathering to agriculture and settled communities some 10,000 years ago. In fact, hunter-gatherers survived for tens of thousands of years with only a handful of goods and left the planet largely pollution-free. Theirs was a mobile lifestyle. They did not even have camels or horses or other animals to carry their burdens, as animals were domesticated only some 5,000 or so years ago. But for a few fossils, hunter-gatherers did not produce any human, industrial, agricultural, or nonbiodegradable waste that had any permanent impact on their environment.

Everything they owned had to be carried, hence they kept very few possessions. They possessed no art, curtains, carpets, hardwood or tiled ceramic floors, upholstered furniture, or any of today's fine furnishings or household appliances. They had no need for architects, homebuilders, construction workers, or interior decorators either. And anyone paying attention would realize that modern Homo sapiens is the only species that has made a specialty of excessive accumulation of "stuff" that has resulted in the pollution of the planet.

COVID-19 showed just how this multi-interdependent, capitalist economy that runs this world can collapse if Homo sapiens were to stop buying stuff. Figuring out how to coordinate, implement, and execute such an action worldwide, on a rolling basis across product categories, may be one of the few remaining levers modern Homo sapiens still has to regain control of its right to choose. After all, robots don't buy stuff, do they?

Freedom from Big Pharma, the Medical Industrial Complex, and most diseases.

When the driving motives of the healthcare industry are profit and growth, how free can one be when diseases rack one's body and mind and those charged with providing care and cures are instead selling snake oil and death?

According to Peter Gøtzsche, professor of research design and analysis at the University of Copenhagen, prescription drugs are the third most common cause of death after heart disease and cancer. In an analysis published in the *BMJ* (formerly *British Medical Journal*), he estimated that every year psychiatric drugs, including anti-depressants and dementia drugs, are responsible for half-a-million deaths in those aged over sixty-five.

John Adams, the second U.S. president, once said, "The preservation of the means of knowledge amongst the lowest ranks is of more importance to the public than all the property of all the rich men in the country." But Big Pharma, it would appear, thinks he got that backward. Corporate greed and systematic political failure have brought healthcare to its knees.

Between 2007 and 2012, the majority of the largest ten pharmaceutical companies all paid considerable fines for various misdemeanors that included marketing drugs for off-label uses, misrepresentation of research results, and hiding data on harms. Medical journals and the media can not only be manipulated to serve as marketing vehicles for the industry but also be complicit in silencing those who call for more independent scrutiny of scientific data. But as long as these criminal acts generate profit, they will continue unabated.

Richard Horton, editor of *The Lancet* (an international general medical journal), wrote that possibly half of the published medical literature may simply be untrue and pointed out in a talk he gave at the Centre for Evidence-Based Medicine last year that the drug and device industry has an ethical and legal responsibility to produce profit for their shareholders but not to sell patients and doctors the best treatment. But the real scandal, he says, is the failure of regulators and the collusion of sorts between doctors, institutions, and medical journals.

According to Aseem Malhotra, a London-based cardiologist, "There are too many misinformed doctors and misinformed patients, and it's time for greater transparency and stronger accountability so that doctors and nurses can provide the best quality care for the most important person in the consultation room—the patient. It's time to restrain the harms of too much medicine."

How free can one be when diseases rack one's body and mind? As the loss of economic freedom creates economic dependency that can eventually lead to servitude, prostitution, extortion, slavery, and innumerable abuses, similarly, a loss of health creates a dependency on others for medical care and thereby a potential loss of the ability and freedom to choose.

Hunter-gatherers were much, much healthier than most of the settled agricultural communities and civilizations that followed them, until the early nineteenth century in richer parts of the world, and the twenty-first century in poorer regions. As noted in Chapter 9, "Adaptation," because they were mobile and did not live in settled agricultural communities, they avoided many of the diseases that became endemic in large settlements because of poor sanitation, lack of clean drinking water, and densely packed communal living, all of which facilitated the spread of epidemics that became common in villages, towns, and soon cities.

The transition to agriculture brought failed crops, starvation, and famine, which in turn brought malnutrition, deficiency diseases, and death. Domestication of animals brought innumerable diseases that jumped to human hosts. Then we invented pharmaceuticals, antibiotics, and numerous drugs to cure the deficiency diseases we got from agriculture and the diseases that came from animal domestication, thereby making "the few" much richer and the disease-afflicted much, much poorer and sicker.

Indeed, personal, family, and government budgets identify healthcare costs among the fastest growing. The Bill and Melinda Gates Foundation, among others, spends hundreds of millions of dollars they wouldn't need to spend had our species stuck with the program—the evolutionary adaptation program all other species stuck with—and continued to sustain themselves from within the Earth's ecosystems instead of outside of it.

Freedom from politics, governments, and the rule of law.

Today, living under the rule of law in a country where laws are applied fairly to everyone is the best modern Homo sapiens can hope for. Hunter-gatherers, on the other hand, formed egalitarian communities with shared responsibilities and neither wanted nor needed even the right to vote. They obeyed no national code of laws, formed no governments, upheld no Magna Carta, Bill of Rights, or Constitution, had no politics or political parties, no police or military, and had no political ideology such as democracy or communism. Instead they organized as small mobile groups and formed neither city- nor nation-states; neither were they dependent upon any king, queen, president, supreme ruler, ayatollah, or tribal chief for leadership, and they were free to choose any potential political outcome they desired.

They were free, willing, and able, without anyone's permission, to choose to survive rather than go extinct, as when they chose to migrate out of Africa 70,000 years ago. This is a freedom that is fast disappearing today. In fact, one has only to look around to see how many people's freedom to make even the simplest of choices is being routinely preempted by their government or some other government, employers, and/or Big Tech multinationals. It is this freedom to take action to choose survival over extinction, among other essential freedoms, that modern Homo sapiens is in dire danger of losing, if it has not already.

In case of doubt just ask the Rohingyas fleeing Myanmar or the North Koreans forced to live under a dictator who prioritizes WMDs over food or those who won't live under this dictator who are washing up in "ghosts ships" on the shores of Japan. Just ask South Koreans and Japanese under constant threat of nuclear attack; Syrians driven to flee their country to escape the bombing and chemical weapons of their brutal president; Palestinians perennially under the oppression of Israel and consenting Western powers; Iranians under the hand of the Ayatollah and American sanctions; Venezuelans enduring Chavez redux in Nicolás Maduro; Hungarians losing their democracy to what now looks more like an "Orbanocracy"; and Russians, the winner of whose elections is a foregone conclusion, and who, a hundred years after their 1917 revolution, still risk prison and death for protests against their new czar, Putin.

The Chinese have detained an estimated 1 million Uighur Muslims against their will in internment camps in the northwestern Xinjiang region, where they're forced to endure severe political indoctrination and, reportedly, to eat pork and drink alcohol, which are forbidden to Muslims, among other awful abuses. Han Chinese, meanwhile, who have no choice but to abide by rules set by a one-party minority, must now also endure the removal of term limits—a president for life—and live under increasing personal surveillance by their government. And as if the atrocities against the people of Syria, the Rohingyas, and Uighur Muslims were not enough, the world is yet again witnessing an even worst humanitarian crisis unfold as the Saudis, backed by American, British, and French arms, decimate the people and children of Yemen.

And what about African-Americans, who still groan under the brutal legacy of racism and white supremacy that began with the more than 20 million Africans stolen from their homeland and enslaved to provide indentured labor and slavery the Europeans (Portuguese, Dutch, British, and French) demanded? More than 12 million were taken to the Americas to work in the sugar, tobacco, and cotton fields of "the few." Moreover, ask the people of any number of African countries where rulers refuse to step down, as well as numerous other nations across the planet, how free they feel to choose to survive, and how much say they have in whatever their governments decide.

And who would have imagined back in 1948, when with such goodwill Israel was established as a home for Jews, that these same people who fled persecution and anti-Semitism in Europe would today be the ones brazenly and forcibly enclosing and occupying Palestinian lands, creating, in the words of Israeli historian Professor Ilan Pappé, "the biggest prison on earth," bringing such misery to a people less fortunate than themselves? How can a people that once knew such pain become the perpetrators of pain on others, denying them even basic rights and freedoms?

In some of these countries, Egypt, Russia, or Turkey, for instance, one would be lucky to be able to engage in peaceful protest, or enlist the help of the press or international organizations, without facing the might of the state's judicial, law enforcement, and military apparatus, with their water cannons, rubber bullets, and routine police brutality. Even in so-called democratic bastions like Britain and America, minorities and migrants are invariably unfairly targeted. However, people in general often find themselves nonplussed by the choices being made on their behalf by their elected officials, as is happening in the U.S. these days.

Indeed, according to Freedom House, its yearly survey and report "Freedom in the World" has recorded global declines in political rights and civil liberties for an alarming thirteen consecutive years, from 2005 to 2018. The global average score has declined each year, and countries with net score declines have consistently outnumbered those with net improvements.

It was the subsequent takeover by ruling economic, political, and religious classes and their management of surplus grain that took a humongous bite out of Homo sapiens' freedom to choose their economic and political outcome as individuals and as a species. They comprise a significant part of what we have been referring to as "the few" (augmented in subsequent centuries by the business/financial, industrial, and technological classes), who frequently and often arbitrarily preempted an individual's and our species' ability to choose, period—let alone make choices like survival over extinction.

With the arrival of settled communities, agriculture, and the domestication of animals came not only numerous diseases and the beginning of the destruction of the environment but also, for the first time, concepts such as work, inequality, and poverty. Social classes were established: government rulers, priests, an aristocracy, bureaucrats, and soon after traders and craftsmen who together became the upper levels of unequal, stratified communities while slaves and peasants became the bottom level. This top tier of society lived off the surplus grain they confiscated, taxed, and tithed from the poor, the slaves and peasants relegated to the lowest class. These communities became city-states and nation-states that together became the cornerstone of what we would afterwards call civilization, which has continued throughout Homo sapiens' history and persists to this day.

Thus began the emergence and supremacy of "the few," with their ability to suppress the will of others and superimpose their own and usurp the rights and preempt the wishes and choices of everyone else. They appropriated wealth from the poor, enshrined inequality, and arrogated to themselves rights to arbitrarily pillage, coerce, and condemn the poor to labor and slavery. These were the practices that constituted the founding institutions of the early city-states that formed the beginning of what we call civilization.

With governments and their rulers soon came weapons, wars, and ultimately WMDs. Together they form an ensemble that constitute a present-day existential threat to Homo sapiens' survival. As the well-known saying goes, "People don't start wars, governments do." It's not the WMDs themselves that are so threatening, survival threatening as they no doubt are; it's the ruling class that controls them—the Trumps, the Kim Jong-uns, the Putins, and others—who pose the real existential threat, as recent developments have shown.

Can Homo sapiens truly choose to survive when that choice is now so clearly and so often in the hands of "that few"?

Freedom from religions and gods and superstitions of any stripe.

While most ancient religions date back no further than 1500 to 500 BCE, hunter-gatherers lived for all of 2,990,000 years without a belief in religion of any kind, organized or not, and thus spared themselves all the problems that would accompany religion in subsequent human history.

According to Hervey C. Peoples and Frank W. Marlowe, in an article published in *Human Nature* titled "Subsistence and the Evolution of Religion": "Belief in either ancestral spirits or creator deities who remain active in human affairs was not present in ancestral hunter-gatherer societies, according to the reconstructions. This may be indicative of a deep past for the egalitarian nature of hunter-gatherer societies, to whom high gods would appear to be rulers." For hunter-gatherers, Yahweh, Jehovah, Jesus, Buddha, the prophet Muhammad, and numerous others, who date back no further than 7000 BCE, did not even exist. How can one begin to get serious about religions and their gods when one's forebears existed millions of years before any of them did?

Hence priests and their retinue of clerics, monks, and nuns, along with Buddha and his following, Muhammad and his following, and Jesus Christ and his following, who dominate the minds of billions across the world today, did not exist to becloud the minds of our forebears.

Priests became popes, archbishops, religious leaders of every sect, ayatollahs, prophets, and incarnate saviors. Religions soon ensnared empires and drove religious wars; today some are still the inspiration of terrorists' organizations that sow death and destruction in the name of their god and their religion. Modern Homo sapiens has a long way to go to free itself from the beliefs that entrapped it in the past and in so many ways continue to do so.

Freedom from profit-driven and extinction-capable inventions and innovations.

Have technologies, innovations, and pursuit of new knowledge and scientific discovery become the new religion?

It may sound like blasphemy to speak of freedom from profit-driven research and development, inventions, and innovations to twenty-first-century Homo sapiens now awash in products and swooning from the "cool aid" of Silicon Valley and from innovation and research centers the world over. Yet the truth is that Homo sapiens survived without risk of any self-generated existential threats for all but the last two centuries, the period when all species-extinction-capable technologies and scientific discoveries arrived. In fact, Homo sapiens, and their hominin forebears, survived very well indeed for 2,990,000 years before agriculture, settled communities, and civilization emerged and by all accounts lived a much better life than modern Homo sapiens do today, technologies, innovations, and scientific discoveries notwithstanding.

It seems to be received wisdom that the main purpose of an intelligent species like Homo sapiens is to develop more and more wealth, make more and more scientific discoveries, drive more and more new technologies and inventions, more often than not for a profit. Not nearly enough attention is paid to the singular evolutionary purpose of any species, and that is to survive, and in the case of intelligent species, to survive without driving itself and all other species to extinction. Few can dispute that all these inventions and innovations, particularly those that came along in the last hundred years, while adding some value, have done more to put our species' survival at risk than anything else during almost all of the prior millions of years of our species' existence.

So, tell me again why so many modern Homo sapiens have such a cockeyed view of this issue. Did technology and innovation become the twenty-first-century religion? Isn't it past time to start questioning this received wisdom?

If our forebears managed to survive for all but the last one or two centuries without putting the entire species at risk, and did so without any of the technology we prize so highly, is it not reasonable to conclude that technology and innovation have little to do with the survival of an intelligent species, and infinitely more to do with its potential extinction? Well . . . yes and no. More later.

Homo sapiens' use of so much of the planet's resources in the service of adapting and innovating has done more to bring our species and Earth's biodiversity to the brink of extinction than to enhance our chances of survival. How much freedom to choose survival over extinction has Homo sapiens left itself, as individuals or as a species, when these innovations we esteem so highly enable fewer and fewer individuals to bring our entire species closer and closer to self-extinction?

And yet there is a better way to adapt and innovate to survive than Homo sapiens seems to have chosen so far, a way that destroys neither flora nor fauna, or drives ecosystems or itself to extinction—the evolutionary way seen in the survival patterns discussed in prior chapters, and as the Maturity/Self-Extinction Index (Fig. 13-1) suggests.

Homo sapiens has yet to come up with anything remotely comparable to the innovations evolution has engendered. In his book *Life Ascending: The Ten Great Inventions of Evolution*, Nick Lane does a great job of detailing the intricacies of his favorites, including DNA, photosynthesis, the cell, sex, movement, hot blood, consciousness, and death. The difference between evolution's and Homo sapiens' inventions and innovations, as I noted in Chapter 10, "Innovation," is that the former is focused on evolving and preserving life and is driven by species and habitat survival, while the latter is driven by profit, market growth, and scientific discovery and doesn't seem to care about preserving the Earth's habitats, species, or Homo sapiens from extinction.

Above we asked what might Homo sapiens' innovations and inventions have to do with its survival and noted that so far these would seem to have more to do with its extinction. Despite the many ways Homo sapiens' inventions and innovations have been put to good use in the last century, because Homo sapiens' efforts are not guided by the evolutionary requirement to preserve life and species at all cost, the results are a mixed bag. Some seem to be good, and some are clearly not so good, including results like WMDs, with the potential to destroy life rather than preserve it. Clearly, then, getting the reasons why we innovate right will be indispensable to our survival.

From an outcome perspective, Homo sapiens' pursuit of inventions, innovations, and scientific discoveries begins to resemble a random walk, funded as if betting on a game of Russian roulette. You win some and you lose some, and that's all right, except the loss could turn out to be the extinction of the entire species. There was a time when values and morality rather than money were the deciding factors, but today, except for making money, there are no clear internationally agreed upon and enforceable rules. And when making money, gaining competitive advantage, and enhancing national pride, rather than species survival and habitat sustainability, become the preferred outcomes, then species extinction, habitat extinction, and Homo sapiens extinction become equally likely outcomes.

If the above is acceptable, well then, all right. But one suspects it won't be for most people, and they would say so in no uncertain terms given half a chance, but they no longer have the ability to be heard and thus have lost the ability to choose.

Consider the concerns raised by Thomas Hornigold in an article published on the website *SingularityHub*, "The Enormous Promise and Peril of Bioengineering's Pandora's Box." According to Hornigold, "We're standing on the threshold of extraordinary capability in synthetic biology. CRISPR-Cas9, the genome editing technique discovered in 2014, is at the forefront of this newfound potential for innovation. These advancements provide an opportunity to solve problems in food supply, disease, genetics, and—the most tantalizing and forbidden of prospects—modifying the human genome. Doing so would make us better, faster, stronger, more resilient, and more intelligent: it's a chance to engineer ourselves at a faster rate than natural selection could ever dream.

"However, many experts warn of the dangers of these new capabilities. A vast torrent of money is flowing towards biotech startups, and the race to be first can encourage cutting corners. Researchers in 2017 resurrected an extinct strain of the horsepox virus. CRISPR may make it possible to create the bioweapons carefully safeguarded by the US and Russian governments, such as smallpox, or to take an existing disease, like Ebola, and modify it into an epidemiologist's worst nightmare.

"With these mind-bending breakthroughs that seem like science fiction, it can be a near-impossible task to discern hype from reality. Yet this is a crucial task for our politicians—who are overwhelmingly not scientists—to undertake. How can we realistically assess the potential risks and rewards? Now, a new study combining researchers from the US and the UK, published recently in *eLife-Sciences,* gives an expert perspective on 20 emerging issues in bioengineering."

Well, there you have it. Need one say more? It's unlikely anyone involved at any level in the above would have reached out to you to get your input or preferences on the use of these scientific and innovative breakthroughs. It's clear, however, whose input matters—those providing the dollars, of course. And if the people elected are unqualified or unable to choose—which would not be so surprising, given what many have concluded about elected officials, including the current U.S. president—then it's clear who will be making these choices. And, it is unlikely to be any of the 99% who will eventually have to deal with any potential fallout.

It may be that one reason (some might argue the only reason) intelligent species are endowed with this unique ability, this freedom to choose to adapt and innovate their way to their own evolutionary outcome, is precisely to enable them to do what the dinosaurs couldn't—innovate to survive. But the results clearly indicate there must be a better way to adapt and innovate, one that doesn't include driving the planet's species and oneself to extinction. As usual, the evolutionary answer is quite simple—replace profit and endless economic growth as the motives that drive innovations, inventions, and scientific research with the evolutionary goal of preserving life, species, and the habitats that enable life. Isn't that simple?

Easier said than done, you think. Indeed, I'm sure the "capitalist few" who have usurped our species' ability to choose for itself won't agree that saving species, habitats, and Homo sapiens from extinction will be worth the profits they would lose—and therein lies Homo sapiens' dilemma.

Nonetheless, it should be clear that by not subjecting its freedom, ability, and willingness to innovate, invent, and pursue scientific discoveries to constraints or rules like the Survivability and Habitat Sustainability Test, Homo sapiens' ability to choose survival over extinction may have already been lost if that unavoidable existential threat noted at the beginning of this chapter were to arise out of its own profit-driven innovations, inventions, and scientific discoveries.

Freedom to choose and exclusively use renewable energy sources.

During the early seventeenth century, a growing wood shortage due to razing most of its forests led to an energy crisis in Britain. To relieve this crisis, the British, after more than 2,999,700 years of hominin and Homo sapiens' existence, turned to coal and became among the first of our species to transition from clean, renewable sources of energy to fossil fuels. And, yes, if razing their forests was any indication, that choice was driven not by an interest in the preservation of life, species, habitats, and ecosystems then—any more than now with their plan to become, again, the first nation to commit to carbon neutrality by 2050—but by pursuit of economic growth and profits.

The growing use of coal led to the use of its by-products, waste gases that became the first non-natural source of lighting, as discussed by Clive Ponting in his book *A Green History of the World.* Then came electricity, oil, and natural gas. Ponting goes on to note that "until the early nineteenth century, renewable resources—human, animal, wood, water and wind—provided nearly all the world's energy. Now 85% comes from nonrenewable fossil fuels."

From an energy-use perspective, all the energy hunter-gatherers needed for hunting and gathering was entirely renewable. They used their own human energy and sometimes that afforded by the occasional use of animals. Human energy was soon complemented with animal energy, fire, wood, water, and wind. As stated above, it was not until around the late-nineteenth century that Homo sapiens began to turn in a big way to nonrenewable energy sources, particularly the fossil fuels that today are responsible for approximately 85% of the world's energy and for 100% of the threat that fossil-fuel-driven global warming is causing, putting our species at risk of self-extinction. But, as Ponting notes, "this fossil fuel energy boom will not last forever. While coal is estimated to last for several hundred years, oil and natural gas are likely to be exhausted this century. However, before the world has to cope with a shortage of fossil fuels, it is likely to have to face far more environmental problems caused by their consumption over the last two hundred years."

If Homo sapiens had to stop using fossil fuels immediately, or at least in the next twenty to twenty-five years as the IPCC 2018 report urges, to mitigate the already devastating effects of a warming planet, going by the response to the 2015 Paris climate accord, compounded by the obdurate attitude of the American president, little hysteria would be necessary to conclude our species might have already lost the ability to choose survival over extinction. Increasing temperatures seem to be on a trend resembling that of the Permian-Triassic extinction, the mother of all extinctions, with which some scientists have already begun to draw parallels.

How strange and perverse for a species whose innovations and inventions so often can destroy life to simultaneously ignore or debunk warnings of its own potentially self-initiated destruction.

Freedom to choose to take action to avoid a potential climate-change-driven extinction.

As seen in an earlier chapter, our hunter-gatherer forebears evolved and survived during a time of many glacial and interglacial warming and cooling periods. Many earlier hominin species in the long line of the genus *Homo* evolved, added their distinctiveness, and went extinct, some, like Neanderthals, as recently as 30,000 years ago. Their way of life enabled them to adapt, endure, and survive some of Earth's worst climate change events for hundreds of thousands of years, and this may have positioned them to survive climate change better than twenty-first-century Homo sapiens appear to be today.

Indeed, our hunter-gatherer forebears fled Africa to save themselves from near-certain extinction as climate change began to ravage their habitat at the end of the last ice age. They had neither climate science nor anything remotely resembling the sophistication Homo sapiens now have at their disposal to assess and respond to the climate data now available to warn of impending disaster. Nonetheless, they got themselves up and hightailed it out of Africa before it was too late.

Seventy thousand years later, modern Homo sapiens is confronted by yet another climate change event, this time one we may have brought on ourselves. Yet despite clear warnings that our forebears would have been lucky to have had, Homo sapiens today finds itself unwilling and unable to choose survival over capitalists' profits and the disappearing jobs that once came from fossil-fuel-based industries, choosing instead to put the survival of our entire species at risk of extinction.

At a time when the IPCC's 2018 report warns that averting a climate crisis will require a wholesale reinvention of the global economy, that the world must embark on a World War II-level effort to transition away from fossil fuels and start removing carbon dioxide from the atmosphere at large scales—anywhere from 400 billion to 1.6 trillion tons of it—and that a mere ten or so years might be all the time left to avoid the potential ecological disaster possible above 1.5°C, one might expect, nay hope, the fossil fuel industries would stand up and take notice. Instead, if their actions during the USA 2018 midterms are any indication, when they simply doubled down, they demonstrated clearly that the choices you and I care about just don't count.

For example, Proposition 112 in Colorado sought to get new fossil fuel infrastructure set a bit farther away from schools, hospitals, residential neighborhoods, and water sources, but the oil industry launched a scorched-earth response, spending more than $40 million to defeat the measure. And that was just one of many.

In Washington state, oil companies spent another $25 million to defeat a proposal that would have required polluters to pay $15 per ton of greenhouse gases they spew into the atmosphere. And how about in Arizona, where the industry spent $30 million to defeat a measure to get the local utility to use more renewables. Or in California, where in a single county the industry spent $8 million to defeat a citizens' initiative to ban new fracking and drilling. And that's just in America. Is there any reason to hope for a retreat from "drill, baby, drill"?

If Homo sapiens can't choose to take action to avoid a potential climate-change-driven extinction in a democracy, then under what other form of government on Earth is this choice possible?

And what is one to do when the leader of one's country, as in the USA, not only flat out denies there is a problem but also has the power to force his views on everyone else? Trump's executive orders, the new Supreme Court justices he appointed, and the growing list of changes to U.S. science and environmental polices are essentially rolling back decades of the nation's efforts to limit and reduce fossil fuel usage; preserve habitats, species and wildlife; ensure clean air and clean water; limit environmental pollution; and preserve the health and well-being of its citizens.

Upon reviewing the announcement of the 2018 National Climate Assessment released around Thanksgiving of that year, the U.S. president proudly claimed he did not believe the report by his own government. If in the greatest democracy the president and members of his Cabinet and administration reject the warnings in this report and other scientific climate assessments, what options do a people have? How does a government of the people, by the people, and for the people end up as a government that's bent on destroying the people, the country, and the planet, for the sake of fossil fuels and other environmentally destructive pursuits, in the name of the people?

If humans cannot secure this freedom by ballot or by elected officials, can they sue? Can lawsuits force governments and corporations to act on climate change?

Twenty-one young Americans, ages 11 to 22, have sued the U.S. government over its failure to address climate change and are requesting that the government be ordered to act on climate change, asserting harm from carbon emissions. The federal government's motion to dismiss was denied, and the Supreme Court after a brief stay has allowed the case to proceed.

Similar cases have begun to crop up around the world. A Dutch court, for instance, insisted that "the state has a duty to protect against this real threat" of climate change. There was, the Dutch court concluded after hearing scientific evidence from past IPCC reports, "a real threat of dangerous climate change, resulting in the serious risk that the current generation of citizens will be confronted with loss of life and/or a disruption of family life." It insisted that "the state has a duty to protect against this real threat," a "duty of care" enshrined in the European Convention on Human Rights.

This might turn out to be an option for citizens in some Western countries, but what will happen in the rest of the world where the rule of law is not an option and where even speaking out against one's government and powerful corporations, much less suing them, can leave one dead like Jamal Khashoggi? While in some areas of the world there are encouraging signs of progress, it probably won't do to plug the carbon emission holes in just one or two corners of Spaceship Earth.

So again, one must ask, has Homo sapiens as a species lost its ability to choose to survive? Or is it just "the few" once again choosing profits and using their power to block the survival choices of everyone else?

Freedom and ability to migrate on Earth and to other planets and/or solar systems to survive.

Just as the urge to reproduce is native to all Earth-life, migration is a built-in response to existential threats. Yet it is the one evolutionary survival pattern modern Homo sapiens is completely incapable of executing to preserve itself. With no place left to go, it is more than likely that faced with the next species-wide extinction-level event, the next great human survival migration will not be from Africa to Europe but might have to be from Earth to another planet.

However, it's clear Homo sapiens is neither capable nor ready to do that, and probably won't be for a very long time. Among the myriad areas innovation and research are focused on, Homo sapiens would do well to concentrate significantly more of its efforts and resources to achieving this ability.

Having briefly explored these essential freedoms our forebears enjoyed for millions of years, it's now time to see how twenty-first-century Homo sapiens' freedoms compare in the following table.

To assist with the assessment of our freedom to choose survival over extinction, one needs some way to qualitatively, if not quantitatively, describe what these essential freedoms are and how and when they became threatened and in danger of vanishing, if they are not already lost.

If one assumes each essential freedom our hunter-gatherer forebears had is a baseline percentage that total 100% when added up as shown in the table below (Row 13, Column 3), we can assign a fraction of that 100% as a value to each essential freedom as shown in Columns 2 and 3.

Table 15-1 Essential freedoms hunter-gatherers enjoyed compared to twenty-first-century Homo sapiens as a species and as individuals

Num	Essential Freedoms	Hunter-gatherers %	Homo sapiens species %	Homo sapiens individual %	Comments
1	To be free	15	1	1	Defined above
2	To be economically independent	15	2	1	"
3	To choose to eat healthy foods	5	1	1	"
4	From the need to own "stuff"	5	0	0	"
5	From Big Pharma and most diseases	5	0	0	"
6	From government, laws and politics	15	0	0	"
7	From religions and belief in a god or gods	5	1	1	"
8	From profit-driven & extinction-capable innovations & R&D	5	0	0	"
9	To use only renewable energy sources	5	1	2	"
10	To choose to act to avoid climate change	5	1	10	"
11	To migrate to survive	10	0	0	
12	To choose to make a living within Earth's ecosystems without destroying Earth's species and habitats	**10**	**1**	**5**	
13	**Essential Freedoms Total Score**	**100%**	8%	21%	
14	**Freedom to Choose Survival over Extinction**	**100%**	8%	21%	

(Source: Author)

Similarly, based on the percentage value assigned each hunter-gatherer essential freedom, we can estimate and allocate a fraction of that as a percentage for each corresponding freedom a twenty-first-century human (Column 4) and our entire species (Column 5) might have today. Hence, the columns "Homo sapiens species" and "Homo sapiens individual" show an estimated fraction of the percentage assigned (in Column 3) to each of the essential freedoms listed in the first column. Estimates for essential freedoms in Column 4 (species) and Column 5 (individuals) need not be constrained to total 100%.

While these are assumed estimates, it is hoped by playing around with different estimates in Columns 4 and 5 for each essential freedom today, one should be able to see how these might differ from the initial hunter-gatherer values allocated to the essential freedoms (in Columns 2 and 3). And yes, these are arbitrary estimates, but they represent a sense of how our freedoms today exceed, equal, or have faded over Homo sapiens' history, leaving our species at risk of no longer being free, willing, and able to choose to survive when—not if—confronted by that inevitable existential threat.

They are just one person's take. You might have an entirely different take, and that's fine, too. Give it a whirl. The takeaway here is (1) these are freedoms that have been preempted by "the few," freedoms our forebears had but which we mostly lack; and (2) what that means when a choice to survive must be made and that choice cuts across the economic interest of "the few." Does Homo sapiens as a species or do we as individuals still retain the essential freedom our forebears had to choose survival over extinction when faced with an existential threat?

> *Can twenty-first-century Homo sapiens stand up today, right now, and vigorously fight climate change the way our forebears 70,000 years ago stood up and fled Africa to escape habitat loss brought on by climate change at the end of the last ice age—and so saved themselves and thereby us? We wouldn't be here if they hadn't, and generations X, Y, and Z probably won't remain here if we don't. They did and we must so our species might attain evolutionary maturity and thereby survival.*

Post-agriculture hunter-gatherers, and thereby Homo sapiens, appear to have lost most of their freedoms with the transition to agriculture, settled communities, and the consequent rise of civilization. It was from these early times that "the few" (governments and their rulers, priests, the elite, bureaucrats, and later craftsmen and traders) began appropriating for themselves the produce, labor, and freedoms of "the rest," while relegating them to the basement of the civilization they had built.

Kings and queens—some still existing—mostly became nation-state sovereigns, dictators, and military strongmen; priests led to organized religions with offshoots that became institutions of learning, like universities; the elite and bureaucrats became the wealthy; traders became the business class, companies, and multinationals; money changers are now our financial institutions; and craftsmen are our technology builders. What these all have in common is a transfer of the freedoms of the individual members of our species to institutions and organizations headed by "the few."

Whereas the hunter-gatherer and his small group chose their own way and were in control of every aspect of their own lives, today, the permission of some self-appointed authority or bureaucrat is needed to do almost anything. The food one eats, the clothes one wears, where and how one lives, works, travels, gets an education; whom one chooses to love, marry (if one is even lucky enough to live in a country where marrying the person of choice is allowed), and have children with; how and whether one chooses to worship a god—it is all cocooned in a web of traditions, customs, regulations, and religious and other laws that vary from nation to nation and control one's every action. That's not even mentioning the laws that govern one's death.

Homo sapiens went from being born free and living free, owing allegiance to no one, with total freedom to choose, to today being, at best, law-abiding and subject to the rule of law, or at worst, subject to racism, outright ethnic cleansing, state-directed persecution, economic and political hostility, death, or forced migration to survive.

This heist of Homo sapiens' freedoms included:

- From being one that called all the shots and owed allegiance to neither man, king, queen, nor any other ruler, to being a mere servant, peasant, slave, or military conscript;
- From once being free and able to choose survival as our forebears did to being an insignificant voice with little to no ability to influence this choice or to change outcomes;
- From the economic freedom and independence of hunting and gathering to economic dependency on a landlord-farmer as a landless peasant or slave;
- From the freedom to make a living within Earth's ecosystems to "working for the man" on a farm outside those ecosystems;
- From hunting and gathering to support family and group to farming to support kings, queens, priests, the elite, bureaucrats, craftsmen, and traders with the surplus grain produced;
- From little work and nearly unlimited leisure time to unending back-breaking work with little to no leisure or free time;
- From freedom to live virtually anywhere to essentially being chained exclusively to a particular patch of land;
- From an era of plenty and a free, healthy, and well-fed lifestyle to one of poverty with a constant risk of failed crops, starvation, malnutrition, and deficiency diseases;
- From a largely healthy and mostly disease-free lifestyle to one plagued by animal diseases, unhealthy living conditions, and improper, unsanitary disposal of human, food, and animal waste;
- From an organic, secure, reliable multisource food supply that enabled a healthy, nutritional diet, to inadequate nutrition from a limited number of agricultural crops.

And today, government of the people, for the people, and by the people has turned out to be the decline and fall of all the people, perpetrated by some of the people, upon the rest of the people.

This is not just an accident of history or an inadvertent transfer of these freedoms but a deliberate power grab, as deliberate as that of the exploitative European colonialists and today's multinationals that continue to prey upon countries and their peoples. This was outright theft that created such economic dependency and loss of economic freedom and independence that it remains highly doubtful whether "the rest" of twenty-first-century Homo sapiens, as individuals or as a species, still retain any freedom to choose survival over extinction.

Given the current existential threats Homo sapiens faces and the apparent lack of desire by "the few" to change course or implement appropriate and timely responses to avert them, what can or should the rest of Homo sapiens do? Is it still possible to choose to survive, and if so, can we regain enough of our freedoms to make that choice?

As noted in the beginning of this chapter, at some point during its evolution, an intelligent species will inevitably be confronted with at least one existential threat, and it must be free, willing, and able to choose to survive to avoid extinction. Today, the Homo sapiens species is confronted not by one but by numerous existential threats. As stated earlier, making this choice will not only be the most important conscious decision we as a species will ever have to make, but more important, it might be the sole reason for our existence in the first place. Failing to choose to survive won't be merely choosing to fail; it will be choosing to go extinct as a species.

The key issue arising from this chapter is, given the clear preference of "the few" in control of Spaceship Earth to stay the course, if it isn't already too late, what might the rest of humanity do to regain enough of those stolen freedoms so as to make that fateful choice in time to survive?

If Homo sapiens were to regain its freedom to choose, in the next chapter we explore how it might redirect its civilization to ensure survival over extinction.

16.

From Civilization to Ecolization

GOING BACK TO THE PAST TO SAVE THE FUTURE

We the People of the United States, in Order to form a more perfect Union, establish Justice, insure domestic Tranquility, provide for the common defense, promote the general Welfare, and secure the Blessings of Liberty to ourselves and our Posterity, do ordain and establish this Constitution for the United States of America.

—*Preamble to the United States Constitution*

IT IS NOT EVERY DAY ONE GETS TO DESIGN OR SPECIFY the laws that would govern a new nation, let alone an entire civilization. But in 1776 the American Founding Fathers tried their hand at defining what came to be known as the American Constitution, the preamble to which began as shown above.

It was signed on September 17, 1787, by delegates to the Constitutional Convention in Philadelphia. To date, there are twenty-seven constitutional amendments, and while it might still be regarded (though fast becoming less so) by many as being a better guide than most other governments', it is also becoming increasingly unclear whether, 240-plus years later, it can continue to serve as that guide for "We the people." America was not the first or only country or nation to come up with a set of laws and rules to guide itself. Yet, notwithstanding these laws and constitutions developed by countless empires, kingdoms, authoritarian governments, and dictators, our species still finds itself unable to create a set of rules or laws that can steer nations away from collapse and a civilization capable of avoiding self-extinction—rules and laws that if followed can ensure the survival of "We the people." Why is that so?

Why is that so! Have you heard of anyone who did not spend some time in a womb or did not need parents to come into the world? No one, not even the framers of the Constitution, gets to choose whether one needs parents, or who will be one's parents, or the town, city, state, or country where one is born. Yet parents always turn out to be humans, and that place of birth is never on another planet but always somewhere on Planet Earth. Why is that always so, too? And those are not the only constants that accompanied Homo sapiens' evolution that we take for granted. The day is always twenty-four hours, and the year 365 days, more or less. Because Earth revolves on its axis, we can count on the sun rising and setting like clockwork, just as we can count on the rotation of seasons because it orbits the sun. Similarly, we have come to expect numerous varieties of plants to grow and bear fruit whether we plant them or not, and animals, birds, fish, reptiles, and amphibians will be born, grow, and continue to supply the food chain. And thankfully this all happens without requiring an amendment to the Constitution, or edict of an emperor or king now long gone, or even the fiat of an almighty god. Why is this always so?

Clearly, embedded among the forces of the universe are the rules and patterns of evolution, repeating processes that work together to ensure the continuity of species, whose existence and survival depend upon these processes occurring without fail. There is a reason no one goes to bed worrying whether or not the sun will rise tomorrow, or whether spring will follow winter next year, but more and more of us have begun to seriously worry about how high the global temperature and sea level will rise by next summer. Again, why is that so? Something obviously changed. One or more of those planetary autonomic evolutionary processes, which functioned like clockwork for hundreds of millennia, have begun to behave differently—and we are the ones driving the change.

What else have we tinkered with? You guessed it—the survival-driven laws and patterns an intelligent species' civilization should follow if it's to avoid self-extinction and survive. Civilizations, like bacteria colonies and anthills, are not designed or created by edicts or constitutions—they evolve.

In almost every case, species, even intelligent species, are driven by the need to survive. They may end up taking on a characteristic shape or behavior pattern. They appear to grow and converge around familiar forms and patterns we have come to recognize and, in many cases, seem to follow repeating rules. So predictable are some rules and patterns that when imitated, as farmers often do, outcomes are similar to those that evolve naturally. In some cases, we have discovered patterns and deciphered rules like those that fashion a leaf, design a snowflake, or sketch an anthill, but we have yet, it seems, to find and imitate those that can guide an intelligent species civilization away from self-extinction. If the patterns and rules we follow to build our civilizations are laid down by evolution, the outcome will be survival. If, on the contrary, they are derived from our non-survival-driven constitutions and laws, they will continue to fail and lead to collapse and eventually to self-extinction.

By now it should have become clear that all of nature's flora and fauna, including ourselves, continue to survive, not as a result of the laws and constitutions our species has conjured up, but only because our very existence is itself a continuing automatic execution of evolution's survival rules and patterns, such as the rotation of the planet or its revolving around the sun. Apart from such planetary autonomic systems evolution appears to have set on automatic with their ability to ensure our survival, what other evolutionary rules and patterns might an intelligent species seek out and follow to evolve a civilization that leads to survival rather than extinction?

It's also clear most species have gone extinct—but none by self-extinction. This means that if species can avoid self-extinction, and in addition, can adapt, innovate, mature, and migrate to escape the universe's existential threats, as intelligent species have the ability to do, extinction can be avoided altogether. Surely Homo sapiens can follow these patterns and rules to avoid extinction. Yet doing so has obviously not been as easy as it sounds—and here is why.

All Earth-life species that avoided self-extinction were only able to do so because they remained in and followed the rules of the ecosystems in which they evolved. Homo sapiens, on the contrary, is the only Earth-life species that has abandoned the ecosystem and, perhaps as a consequence, has now ended up potentially engineering the means to its own extinction. Evolving and remaining in the ecosystem is then the only way for Earth-life species, including Homo sapiens, to survive.

As the human body is regulated by the internal biological processes of homeostasis, so energy flows and nutrient recycling regulate and enable species survival within an ecosystem. Similarly, as the rotation of the planet and its revolving around the sun supports Earth-life survival at the planetary level, so does the flow of energy and nutrient recycling within the Earth's ecosystem enable survival at the habitat and ecosystem levels. Indeed, almost everything in the universe, it would seem, needs to evolve and exist within some kind of cradle—an enclosure or framework wherein forces, energy, and matter can interact to enable evolution and survival.

Galaxies can evolve and exist only in a universe that provides the context, material, and forces that enable their formation and continued existence. Stars tend to form and thrive within stellar nurseries. Planets evolve from the dust and planetesimals that accrete to form an emerging solar system. Habitats capable of supporting the evolution of complex animal and intelligent species life are likely to form on planets in a solar system that orbit within the continuous habitable zone of their home star. Similarly, species can evolve within ecosystems that form within those habitats, as they did on Earth, and it is in these ecosystems the evolutionary forces that enable species evolution and survival converge. Earth's ecosystems vary by environment: forests, grasslands, deserts, and tundra, to name a few of land environments, and freshwater, marine, wetlands, mangroves, and coral reefs, to name a few of the aquatic environments.

As there needed to be a universe for galaxies to exist, galaxies for stars to exist, stars for solar systems and planets to exist, planets for habitats to exist, and habitats for ecosystems to exist, so there must be ecosystems for species to evolve and survive. One should expect it to be as problematic for species to survive outside ecosystems whence they evolved as it must be for life to evolve and survive on a planet revolving outside the continuously habitable zone (aka the Goldilocks zone) around its star. What the Goldilocks zone is to habitable planets, ecosystems are to species evolution and survival.

Not surprising, too, this pattern of life needing to evolve and survive in enclosures fit for purpose is also seen in the evolution of species, including the human female's womb—that's why we've all started there—and even down to cellular life as in eukaryotic cells (where the genetic code, the very machinery of life, is stored). And as physiological processes in the human body exist to operate, maintain, and regulate the relatively constant conditions of the body's properties, so too does the ecosystem for the species sharing its environment. Ecosystems, among other things, are a critical component of the biosphere and recycle the biologically important elements species need to survive. As homeostatic regulation within the human body turns out to be critical to its continued survival, so too staying within and following the rules of the ecosystem are indispensable to all Earth-life species survival—including Homo sapiens.

So, here are a few takeaways about species that live within and follow the laws of the ecosystem and habitats wherein they evolved:

(1) They find the sustenance for their lives from within the ecosystem in which they evolved;

(2) Their adaptations, innovations, maturing, and migrations are enabled and choreographed by the rules of and done within the ecosystem;

(3) They destroy neither ecosystems nor habitats in which they evolved;

(4) All their actions, everything they do, are survival-driven;

(5) Their method of living ensures a sharing of the resources of the ecosystem and habitat in a way that allows for the mutual survival of participating species;

(6) They drive neither themselves nor other species to extinction.

On the contrary, none of the above can be said for Homo sapiens, the lone surviving intelligent Earth-life species that has chosen to live outside Earth's ecosystems. Like Elvis, Homo sapiens has left the ecosystem, and the consequences are plain for all to see.

Built into every species except our own, it would seem, is this evolutionary restraint, a kind of built-in live-and-let-live instinct, of having to live within the Earth's ecosystems without destroying them, other species, and themselves. And that is exactly what all species have done since each evolved, including our hunter-gatherer forebears over the millions of years of their existence. Modern Homo sapiens, their descendants, became the first and so far only Earth-life species to change this evolutionary survival equation by choosing farming as its way to sustain its life, and thereby ended up doing so outside of the Earth's ecosystems. It was this seemingly simple change that was the root cause of, and led to, the kind of civilization our species has evolved, and the conditions with which twenty-first-century Homo sapiens now wrestles.

Today our species has begun, once more, to tweak that evolutionary equation, this time by rapidly driving up atmospheric CO_2 content, thereby increasing global warming, and yet again changing the conditions in Earth's ecosystems and habitats, rendering some incapable of sustaining life and driving many species to extinction.

Herein, then, lie the root causes of all the problems confronting twenty-first-century Homo sapiens civilization: First, the transition to farming as the predominant subsistence method, a choice to maintain their lives from *outside* of the Earth's ecosystems; and, second, the choice to extract and burn fossil fuels as a source of energy, essentially short-circuiting the long-term carbon cycle by prematurely returning to the atmosphere the CO_2 evolution had sequestered within the Earth over geologic time.

> *The choice to farm and the choice to use fossil fuels stand like towering bookends on either side of the last 10,000 years of our species' civilization. Any attempt to redirect twenty-first-century Homo sapiens civilization onto a path that ends in survival must begin at the farm and run all the way through to the fuel pump.*

Above we asked: Is there a way Homo sapiens might build a civilization that does not end up destroying Earth's species, habitats, and potentially itself? What this question really translates to is: How should our species change the way it adapts, innovates, matures, and migrates if it wants to avoid extinction? How can humans re-assimilate into local ecosystems, instead of living outside of them; begin to collaborate and integrate with other species, instead of driving them to extinction; and begin to sustain Earth's habitats so they might in turn sustain us?

Indeed, how? We already know that no human constitution, law, decree, or set of religious commandments and practices will suffice—none has so far.

Instead, each day that the sun rises, the seasons change, and the Earth continues to offer its bounty of plants and animals, we have proof that those evolutionary rules and patterns that energize Earth's habitats, ecosystems, and biosphere are always at work. Built into every evolutionary system and species is the set of rules it needs to follow to survive. And, like all species, Homo sapiens also has that built-in urge to survive. We simply need to reawaken, reprioritize, and re-elevate that survival urge so it becomes once again the overarching motive that drives all our actions as a species as it once did our hunter-gatherer forebears.

However, this survival urge needs to be pointed in the right direction, and in Chapter 13 we referred to this direction as our true north and identified a test that, like a compass, can point us in that survival-driven direction. We called it the Survivability and Habitat Sustainability Test. It's how as a species, our hunter-gatherer forebears found and stayed on the path to survival, and we can once again use it to evaluate and prioritize what's really important to our species' survival. Nothing gets done if it fails to drive species, habitat, and Homo sapiens survival—it's that simple.

> *When we follow our true north, we will steer clear of the pursuit of profits, wealth accumulation, and market growth, and instead get back on the road that leads to evolutionary sustainability and survival.*

Getting there is what the rest of this chapter explores. It is the essential question Homo sapiens must now ask and answer before it's too late.

How differently might civilization have turned out if, like their hunter-gatherer forebears, early humans had continued living *within* the Earth's ecosystems, preserving its habitats and countless species from extinction and itself from creating the means of its own extinction? We might never know. Instead, we chose to adapt.

Adaptation is the first of the evolutionary species survival patterns (*adapt, innovate, mature and migrate* to *survive,* or go *extinct)* discussed in earlier chapters. We have seen how, since choosing to live outside of the Earth's ecosystems, potentially the first Homo sapiens adaptation, each successive adaptation became the source of a new set of issues, and how each succeeding fix led to deeper and more complex adaptations and problems.

Agriculture eventually led to genetically modified crops. Fertilizers and simple manure morphed into chemical fertilizers and deadly pollutants such as herbicides, insecticides, and fungicides. Going even further, we have beaten our plowshares not only into swords but also into AR-15s and AK-47s; our pruning hooks not only into spears but also into nuclear-tipped warheads on intercontinental ballistic missiles, with North Korea the latest in the growing club of nations with the potential, to varying degrees, to end civilization as we know it.

What began as seemingly commonsense adaptations were followed by innumerable others across all aspects of an emerging civilization—economic, political, religious, social, cultural, technological, scientific, and more. These adaptations have today brought our species face-to-face with the reality that we have become the architects of the means of our own potential extinction. Could we have known that taking so seemingly simple a step as farming would eventually, 10,000 or so years later, land us here?

Next in these species' survival patterns came the trio of innovate, mature, and migrate, indispensable preambles to a successful outcome in the survival versus extinction decision each species must confront. And in the chapter on innovation, we saw that while Homo sapiens' innovations accelerated exponentially, the number of individuals necessary to drive our species to extinction had plummeted inversely. This occurs when innovation greatly outgrows maturity and exposes an unmistakable correlation between a rapid advance in innovation without a corresponding growth in an intelligent species' maturity.

We saw, too, how the Maturity/Self-Extinction Index in Chapter 13, "Evolutionary Maturity," shows the closest we got to the self-extinction indication (asymptotically approaching zero evolutionary maturity) was in 1945. On July 16 of that year, J. Robert Oppenheimer, head of the Manhattan Project, witnessed the first detonation of a nuclear weapon, which he had developed. He said later the experience brought to his mind the famous quote from the Bhagavad-Gita, "Now I am become Death, the destroyer of worlds."

> *On that day, Homo sapiens entered a new phase in its civilization: it had innovated the means of its own potential extinction.*

How is it that evolution's first and potentially only nascent intelligent technological species on Earth, perhaps in the galaxy, evolved with the innate intelligence to choose either survival or extinction but seems so unable to mature, and even less able to migrate to survive, while growing increasingly capable of and seemingly bent on self-extinction?

It was Jane Goodall, the primatologist most known for her long-term study of wild chimpanzees in Tanzania, who remarked, "The most intellectual creature to ever walk Earth is destroying its only home." What is it about our species that has driven us to the edge of this survival-extinction precipice, that has led us to develop a civilization that appears to be headed to self-extinction?

Isn't there another way an intelligent species can build a civilization that does not require destroying Earth's species, habitats, and potentially itself in the process? Is there a way to solve evolution's equation of life other than the way Homo sapiens seems to have chosen? There has to be another way, and indeed there is.

Homo sapiens already "know" of this way. It's the way our hunter-gatherer forebears lived for millions of years without driving themselves to this survival-extinction precipice. It's also the way all other species have followed. And since there is such a way, can we rediscover and follow it to ensure our survival as they did theirs? The answers are—yes, and maybe.

Yes, there is such a way, and we have already rediscovered it. This book, for instance, has identified and outlined that way, describing the evolutionary survival patterns and a Species Survival Maturity Model, which, like that yellow-brick road Dorothy followed, can guide us as it did our forebears, all the way past the point of self-extinction. And maybe—because we may not be prepared to abandon our current path to make the fundamental changes necessary to follow this new way.

To identify these necessary changes, one must first be able to describe two very different ways to survive. In this book they are referred to as civilization and ecolization. Ecolization is the way our ancestors lived as hunter-gatherers, sustaining their lives *inside,* rather than *outside,* Earth's ecosystems for millions of years before civilization began. Civilization came after and is what Homo sapiens called what followed the transition to agriculture and began only 10,000 years ago. And from everything we have seen so far in this book, the road to Homo sapiens' survival points not to more of the civilization we have had since that transition but might well be a return to ecolization.

Should that turn out to be the case, then Homo sapiens, like Marty McFly in *Back to the Future*, might have to return to its past to save its future, and itself, from self-extinction.

If it's reasonable to assume that given half a chance, Homo sapiens would choose survival over extinction, then our ultimate and overarching goal as an emerging intelligent species cannot be to go extinct, but rather to survive, and hence the choices we make must align with that goal. But as we have seen in Chapter 15, "Freedom to Choose," Homo sapiens' choices may no longer be our own to make, and our current choices are definitely not in alignment with a desire to survive. This is so because our ability to choose has been hijacked by a small subset of our species, "the few," whose choices are entirely orthogonal to the evolutionary requirements for survival. This means our species might well have lost the ability to even choose to alter its current civilization to survive.

More than likely, then, it won't be easy for Homo sapiens to make the needed civilizational changes to ensure survival, unless and until one of these things occurs: control is successfully wrested away from "the few" in time to make the needed changes, or civilization collapses. This collapse could be due to any number of causes, such as pandemics (COVID-19), a planet-wide calamity, or global warfare that brings an end to the control of "the few," enabling survivors, if any, to change course.

Some think there is also the possibility of persuading "the few" that it is in their own self-interest to relent and abandon their planet- and species-destroying mania for profits and economic growth. However, others pointing to their unwillingness to even explore capping fossil fuel use think that such a change, even if possible, is unlikely to occur before it's too late.

The way forward, then, remains unclear. But should Homo sapiens emerge free, willing, and able to once again make its own decisions, as our forebears did 70,000 years ago to leave Africa to survive, how might we go about rebuilding a civilization that won't lead to making the same mistakes?

We have all seen movies with actors caught in a time loop, the same scenes repeating again and again, each time ending in the same disastrous outcome until something happens to break this cycle. Unfortunately, like those repeating scenes, so too has hominins' evolution and Homo sapiens civilizations played out over geologic time, with modern Homo sapiens the "last man standing." We, too, could end up making the same mistakes that ended each time in extinction unless we can find a way to break free.

We must peer into the past to see more clearly what the future might hold. Throughout this book we have explored Homo sapiens' past to find the evolutionary survival patterns that choreographed our species' existence over geologic time. Hopefully, we can use those patterns to break free from this endless cycle of extinction. One way to use the past to better see the future is to compare the outcome of that past with present and potential near-future outcomes and see where each has led and might still lead. That is the approach taken in this concluding chapter.

While we've already looked at how making a living *outside* Earth's ecosystems has led us to the very edge of a survival-extinction precipice, we have not yet thought much about the kind of outcome that could emerge if Homo sapiens were to instead try the evolutionary way of sustaining life *within* the Earth's ecosystems, and *without* destroying them, or driving other species and potentially our own to extinction.

Keeping in mind evolution's survival-first goals, aligned with most Homo sapiens' desire for survival, we ask in this chapter what a Homo sapiens civilization would look like under the live-and-let-live evolutionary constraints of ecolization. What might be the outcome, and how would it differ from what we already have today?

While it might not be possible to spell out the specifics of such a civilization, there are clear indications of its guiding principles. One can compare the outcome of our forebears' millions of years of existence as hunter-gatherers sustaining their lives from *within* the Earth's ecosystems, or ecolization, with civilization—the outcome of Homo sapiens' existence *outside* of the Earth's ecosystems since the transition to agriculture.

One can compare various aspects of the two, such as their respective fundamental principles, values and motivations, guiding laws, life's focus and ambitions, individual and species goals, economic freedoms and independence, political freedoms and independence, adaptations and innovations, freedom and ability to migrate to survive, and freedom and ability to choose survival over extinction when faced with an existential threat. Having done so, one could draw conclusions about which of these is best suited for attaining evolution's and Homo sapiens' overarching goal of survival.

Still, you say, I am having difficulty seeing how such a comparison can help our species get to survival unless our species is able to regain control of its ability and freedom to choose, or, has to pick up and start over after a civilization collapse. Indeed, a mere comparison wouldn't do any of that. Yet, should either occur, the following thoughts may help jump-start re-building a Homo sapiens civilization around ecolization—the evolutionary way our forebears successfully sustained their lives for millions of years without the threat of self-extinction.

The following table then groups a few key life-impacting civilization categories for contrast and comparison. It looks at and asks, given these two very different approaches to sustaining life on Planet Earth that ecolization versus civilization requires, which is more or less likely to drive Homo sapiens survival versus extinction. It is not intended to be comprehensive but merely to point a direction and start the conversation. It begins with a definition of Ecolization and Civilization and compares these across the following categories: Fundamental and Guiding Principles and what motivates Homo sapiens; Values—how is nature viewed? Guiding Laws; Adaptation and Innovation Focus; Political Perspective; Economic Perspective; and Desired Civilization Outcome-Intelligent Species-Wide.

Ecolization vs. Civilization

Ecolization	Civilization
Definition: The process by which a species organizes to sustain itself ***within the Earth's ecosystem***s without destroying itself, Earth's habitats, and other species.	**Definition:** The process in which the species Homo sapiens organizes to sustain itself ***outside the Earth's ecosystem***s, which destroys Earth's habitats, drives other species to extinction, and puts its own survival at risk.

Planet-wide focus. One world. One species, Homo sapiens—the only surviving emergent intelligent species on Planet Earth.	Fragmented. Country and regional focus. East vs. West; aligned vs. nonaligned; democratic vs. nondemocratic; developed vs. Third World; global vs. regional; American vs. British vs. Chinese, etc. Fragmented by race, skin color, creed, wealth, and other distinguishing characteristics.

Understandings and Beliefs

Understands the Evolutionary Game of Life. Life is about survival and not about making money	**Understands the Money Game.** Life is about making money and not about survival.
Understands the evolutionary role and reason for intelligent species existence primarily is to evolve and survive beyond the point of self-extinction.	Most believe God created the universe and man is meant to rule the Earth. Some believe in evolution but have no idea why intelligent species evolved.
Allows that intelligent species survival instead of extinction might well have been the evolutionary reason for intelligent species evolution.	Thinks God is in control. Heaven or hell is the outcome, and evolution has nothing to do with Homo sapiens' survival or extinction.
Understands Earth's habitats and species are the life support systems of the Earth and are needed for intelligent species survival.	Thinks Earth's habitats and species are given to man by God to be used as man so chooses.
Aware of, understands, and follows the evolutionary survival patterns and their potential role in guiding intelligent species to survival.	Looks to religions, philosophies for guidance on how to live. Unaware of the possible existence of evolutionary survival patterns.
Deeply aware of the evolutionary role climate change, and its timing, plays in driving intelligent species to mature, migrate and survive, or go extinct.	A few are aware that climate change has shaped human evolution, but that understanding stops short of recognizing its role in driving intelligent species to mature and migrate to survive, or go extinct.

Deeply aware of and accepts the universal role migration plays in intelligent species survival. Would encourage and support Homo sapiens' terrestrial migration to survive. Deeply aware of and accepts the universal role migration plays in other species' survival. Understands that the next Homo sapiens species-wide survival-driven migration will have to be interplanetary and would begin preparation.	Mostly opposed to Homo sapiens' terrestrial migration and does not yet understand the role migration plays in intelligent species survival. Understands the role of migration in other species' survival but simultaneously pursues policies that block or limit migration. Only some have begun to grasp that the next Homo sapiens species-wide survival-driven migration will have to be interplanetary. No species-wide preparation as yet.

Fundamental and Guiding Principles
What motivates Homo sapiens?

Survival:	**Profits and economic growth:**
Of Homo sapiens—one's own species. Of all other species. Of all Earth's habitats and ecosystems.	For a few individuals—the 1%. For most corporations. For some countries.
Pursuit of evolutionary maturity and survival.	Pursuit of health, wealth, and happiness. Preoccupied with personal, corporate, and national wealth accumulation.
Understands the evolutionary yin and yang pattern that says habitat survival ensures species survival and species survival ensures habitat survival.	Fails to make the connection between species survival, habitat survival, and Homo sapiens survival.
Live and let live.	Winner takes all.
Sees itself as a part of a community of species and not elevated above them or here to rule over them.	Sees itself as being separate, apart and above other species and can do with them as it pleases.
A culture built on an overarching sense of mutual interdependence and care for all species, Earth's ecosystems and habitats, and one's self.	Cares for self and family mostly; close friends some of the time; much less so for utter strangers; and even less for non-pet species and Earth's ecosystems.

Values

How is nature viewed?

Species, ecosystems, and habitats are the life-support systems of the Earth and are to be preserved at all cost. Understands Homo sapiens would not survive if these go extinct.	Species, ecosystems, and habitats are commodities to be bought and sold for a profit. Fails to make the connection between other species' survival and its own survival.

Guiding Laws

The Survivability and Habitat Sustainability Test that ensures no activity that can result in the extinction of species, ecosystems, and the Homo sapiens species itself is undertaken.	Global and nation-state laws built mostly around economic development, national security and public safety, and minimally concerned about species, ecosystems, and Homo sapiens' survival.
Applicable to all species and all ecosystems and habitats.	Human rights laws. Species and ecosystem laws are afterthoughts.
Anything that can result in species, Homo sapiens, and habitat destruction is a nonstarter.	Mostly anything, regardless of species and habitat destruction, that can make a profit can proceed. Economic growth, market share growth, return on investment, shareholder dividends rule.
Nothing can proceed until and unless it satisfies this law.	Everything can proceed except where some non-human species and habitat protection rights exist and can't be easily ignored.

Adaptations and Innovations Focus

Survival-driven. No adaptations or innovations are possible unless the process and outcomes support species, habitat, and Homo sapiens survival.	**Profit- , growth-driven.** Scientific discovery-driven. Species, ecosystems, and Homo sapiens survival are not prerequisites. Environmental impact regulations routinely thwarted and/or rolled back.
Homo sapiens, all other species and habitat survival-driven.	Self, corporate, academic, or R&D lab and sometimes country-driven.

Will never under any circumstances and for any reason adapt or innovate capabilities that enable species-wide self-extinction. Could never develop WMDs.	Have some qualms about adapting or innovating capabilities that enable Homo sapiens' self-extinction but can still proceed as long as it can be shown to make a profit, increase market share, and if strategic for economic, political, military purposes and to achieve world dominance.
Interested solely in preserving life, habitat, and species survival and in adaptations and innovations that are survival-driven. Has no interest in profit for its own sake, market share, and world-dominance-driven innovations and adaptations.	Interested especially in profits, market share, and domestic and/or world market dominance-driven adaptations and innovations. Preserving species life and habitats is considered at best secondary and often viewed as overrated.

Political Perspective

Egalitarian. One level, and all are on the same level. No rulers, no elite, no poor, no wealthy, no haves and have-nots, no government, no constitution, no laws, no law enforcement.	Stratified, layered hierarchies of rulers, elite, rich and poor, haves and have-nots, governments, constitutions, endless laws and law enforcement with some more equal than others.

Economic Perspective

Aims for economic independence and freedom at individual and species level.	Built-in bias for economic dependence, limited and controlled freedom for individuals and species.
Cooperation focused	Competition focused
Coexistence focused	Domination focused
Mostly co-dependent	Mostly independent and separate
Sharing the Earth's bounty with all—one for all and all for one; no jobs, no homeless, no slaves, and no serfs.	Selfish and unequal— exclusive ownership, uncaring, at-will employment, homelessness, slaves and serfs.

Desired Civilization Outcome

Intelligent Species-Wide

Homo sapiens maturity and survival beyond the point of self-extinction.	Has no articulated species-wide outcome goals. Does not think or function as one global species but as a fragmented collection of competing nation-states with muddled mostly military, economic and financial and sometimes citizen goals. Except for threat of nation-state nuclear war, Homo sapiens' self-extinction is not taken seriously or seen as a real possibility. Little to no awareness of the prior cyclic nature of mass-extinctions.
Preservation of Earth's ecosystems and habitats and survival of all of Earth's species. To get Homo sapiens to evolutionary maturity—beyond the Survival Plateau.	Except for climate change, has no specific species survival goals. Frequently engages in activities that can lead to self-extinction.
Species-wide focus on the development of capabilities to escape from or defeat existential threats to survive.	Fragmented capabilities by nation-states and focused on economics, increasing profits, market share, and nation-state dominance. Marginal efforts to address the universe's existential threats.
Overarching goal is to avoid self-extinction, to survive existential threats, and to be able to migrate to other planets to survive if necessary.	Has no nation-state let alone species-wide survival plans in place and may or may not be able to survive as a species if and when confronted with species-wide existential threats.
Extraterrestrial migration and spreading mature intelligent species life across the galaxy.	Has no species-wide plan or capability to migrate to other planets or solar systems even to survive let alone spread mature intelligent species life across the galaxy.

The above skeletal comparison between civilization and ecolization is one way to begin to imagine how different things might be, but it's only a start, and leaves much room for elaboration and clarifications. Even when viewed from such a high level, though, it doesn't require a rocket scientist to envisage how vastly different ecolization outcomes might turn out if given 10,000 years of Homo sapiens survival-driven adaptations, survival-driven innovations, an evolutionary maturing species, and survival-driven migration.

Earlier in the chapter we asked how Homo sapiens civilization might have turned out had our species continued to sustain itself living *within* Earth's ecosystems, preserving its habitats and countless species from extinction. Would we have ended up where we are right now, confronting the realization that we have become the architects of our own potential extinction?

Think about it. How different will the world be when species, habitat, and Homo sapiens survival replaces profits, economic growth, and corporate and personal wealth accumulation as de facto guiding principles for all of what humans do? I encourage contemplating this new and different world. I can't wait to see how Adam Smith's economics would dissolve and what minds like Thomas Piketty's and Thomas Friedman's might make of an economic system driven, not by the economic laws Smith envisioned and others developed, but by evolution's Survivability and Habitat Sustainability Laws.

To stimulate the imagination, consider the following points of departure between these worlds:

- Many if not most of the species driven to extinction under civilization over the last 10,000 years might still be around today under ecolization;
- Burning fossil fuels as a source of energy would have been a nonstarter, and the resultant CO_2-driven global warming that threatens civilization today wouldn't have happened; similarly
- Most weapons of war and particularly WMDs would have been inconceivable;
- Most political, economic, social, religious, and cultural structures that emerged to support the money-driven economies and capitalism, as well as nation-state governments and their militaries, would probably not exist or be entirely different, performing entirely different functions

from the perspective of preserving species, habitats, and Homo sapiens survival.

Consider, also, these questions:

- Could power and economic resources across the globe have turned out to be as unevenly distributed as they are today, with very rich versus very poor countries?
- Could those eight, white, mostly American men own more wealth than 50% of the world's population?
- Could colonialism and the scourge of slavery have emerged in a civilization that prioritized Homo sapiens' freedom and survival?
- Would the West's liberal, economic, capitalist-driven world, lubricated by money and choreographed by multinationals, free trade, and international institutions, have emerged?
- Could hideous male domination of females, discrimination of people of nonbinary genders, and denial of reproductive freedom exist in such a world?
- Could political ideologies such as communism in China and Russia and democracy in Western countries, or the unending stream of strongmen and dictators who, under the guise of governing, suppress Homo sapiens, have even seen the light of day?

And where would all those religions be that today enslave the thinking of Homo sapiens the world over? Some of these we have to thank for the false belief in a god that supposedly endowed his followers with authority over all of Earth's species to do with as they please. Such false beliefs repositioned how Homo sapiens saw itself with respect to the rest of nature and may well have led to much of the devastation of the Earth's habitats and biodiversity in the last 200 years.

Much of the foundation of our current "civilization"—which began with the need to manage surplus grain after the transition to agriculture and morphed into today's profit-driven-economy-based entity fueled by greed and desire for personal and corporate wealth accumulation, a motive that is reflected by our innovations, adaptations, institutions, commerce, ideologies, customs, culture, beliefs, and more—most likely wouldn't have emerged in a world order that prioritized species, habitat, and Homo sapiens survival above all else.

It's entirely possible that what would have emerged in its place would seem strange, if not actually alien, to our "civilized" eyes. But however different it may have turned out, one can pretty much bet it would not have ended up 10,000 years later with Homo sapiens becoming the architects of its own destruction, left to contemplate, as it does today, its potential self-extinction through its own innovations.

Some might consider ecolization unrealistic. But as naïve, or unbelievable, as these proposed differences between ecolization and civilization might seem, one need only look back into our past to a time when capitalism was not even a word; when forms of government such as democracy and communism did not exist; when the sciences and innovations that led to the WMDs that today threaten self-extinction were as-yet unborn ideas; and when none of the religions that pervade the world today even existed in a shaman's imagination. If one were to go back even further, before farming and agriculture facilitated the hijacking of our species' freedoms by "the few," one would find a world order that evolved around the evolutionary principles of ecolization reflected in the way our forebears existed within Earth's ecosystems for millions of years before civilization emerged.

> *Ecolization has been habitat-survival-tested, species-survival-tested, and Homo sapiens-survival tested for millions of years and managed to steer our forebears clear of all the existential threats that civilization has unleashed just in the last 10,000 years. Far from being naïve, unbelievable, or even alien, then, ecolization offers a way to live that is consistent with the evolutionary survival equation and potentially the only path to Homo sapiens' survival.*

Living in a world order founded on the evolutionary principles that underpin ecolization does not condemn Homo sapiens to a life of hunting and gathering, as some might imagine (not that hunting and gathering was so terrible; on the contrary, it may have been a more affluent, politically free, and economically independent lifestyle than most Homo sapiens have today), but rather simply institutes the obligation and requirement to preserve and ensure, above all else, the survival of Earth's ecosystems and habitats, its flora and fauna, and the Homo sapiens species itself.

Similarly, such a life does not mean there would be no adaptations and innovations either, or that they would be fewer or dumbed down, but only that they be survival-driven adaptations, survival-driven innovations, and survival-driven scientific research. Indeed, as noted earlier, evolution's survival-driven adaptations and survival-driven innovations were all executed within Earth's ecosystems. Carried out by our nonintelligent microbe forebears over billions of years, these dwarf into insignificance those humans have so far attained by doing so outside Earth's ecosystems.

In fact, given a choice, who wouldn't prefer that those rare minerals used in our smartphones and the diamonds that sparkle in our rings and necklaces didn't depend on the suffering, death, and exploitation of peoples caught up in conflict? Wouldn't that box of exquisite chocolates not only taste much sweeter but sit much better in one's conscience if you knew that the real cost of a chocolate bar was not the loss of the rain forests in countries in West Africa; that cocoa-exporting countries like the Ivory Coast had not replaced 80% of their forests over the last fifty years mainly with cocoa plantations and farms; and that fair trade practices had been observed and no one had suffered, no locals had been defrauded by a multinational, to make that tasty treat affordable?

Wouldn't it be wonderful, too, to once again enjoy salmon and mackerel and other seafood without the risk of ingesting harmful mercury, and other chemicals? Would the oceans be choked with islands of plastic and garbage, or growing numbers of dead zones, if species and ecosystem survival were the preferred outcomes, rather than profit and economic growth? And think of all the harmful side effects modern pharmaceuticals cause, or the harm that comes from a healthcare and food industry driven by profits rather than concern for the health and survival of Earth's species, ecosystems, and the Homo sapiens species itself.

As Homo sapiens begins to apply ecolization's Suvivability and Habitat Sustainability-first approach, many of the ills that plague the modern world, whether scientific, technological, economic, political, medical, cultural, religious, or personal, will begin to recede. There would still no doubt be issues, if only because wherever humans are, there are issues. However, it is likely that the problems such a world would face will be orders of magnitude more manageable than the intractable ones that now bedevil and beset modern Homo sapiens civilization.

So if the path to Homo sapiens' survival requires a return to some form of ecolization, how can modern Homo sapiens begin while still ensnared in a civilization managed by capitalist principles and controlled by "the few"? How, who, or what could bring about such a change? Could mass protests, political action, big business, religion, innovations, or wars bring about this transformation? If not, is there anything one can do as an individual?

Could mass protests bring about this transformation—they mostly haven't so far but this time maybe? Let's look first at why they mostly haven't. Not to disdain or discourage any of the valiant attempts our species has made to regain its lost freedoms, but such an across-the-board, transformation more than likely won't occur as a result of *protests as we have known them*, and to bring it about by force clearly isn't a good idea.

> *Such a transformation will have to be species survival-driven, voluntary at both the individual and species-wide level to be effective.*

Protests as we have known them, whether they are peaceful or violent, for political, economic or humanitarian reasons, or by any other means or for any other purposes, historically tend to have no more success than the French accomplished in their decade-long revolution over 200 years ago. The Russians, too, got rid of their czar only to end up with communism and 100 years later with another czar in the person of Vladimir Putin. Similarly, the Chinese under Mao Zedong in 1976 embarked upon a Cultural Revolution, supposedly to get rid of bourgeois elements that might turn back to capitalism, closing schools and killing millions in the process, only to return to capitalism a dozen years later under Deng Xiaoping's special economic free market zones.

And will the 2019 Hong Kong anti-extradition law protest against Beijing be any more successful than the prior Hong Kong democracy protests of 2014, or the devastating Tiananmen Square protest of 1989? And how can one still retain hope in the promise of protests after seeing how millions of Syrians, driven from their farms by drought caused by climate change, were brutally slain by their government aided by Russia and others—now left to wander as refugees, their homes demolished, and their lives destroyed? Not even the American Civil War was able to rid the U.S. of slavery and racial bigotry: last count, America was home to more than 60,000 slaves. African-Americans and all involved in the American civil rights movement suffered to lift the legacy of slavery, only to find that decades later racism is still alive and well in America, perhaps even in the highest offices in the land.

And what about others of our species, during the not-so-distant past and innumerable others still over geologic time, who rose up only to be struck down by the powerful among their fellow Homo sapiens?

What the above *protests as we have known them* have in common that render them unlikely to succeed are that they are location and issues specific—gun violence in America by city or state; political oppression: millions of people are tired of despotic rulers and strongmen in Syria, Russia, Hungary, Venezuela, and various African countries; apathy, partiality and/or disunity driven by religion, politics, and numerous other forms of manipulation—there is no end of issues where it has proved impossible to get enough people to understand the facts and agree on a common problem and solution; obfuscation, fake news, science deniers by tobacco, climate-denying big governments, Big Pharma, auto, Big Tech, and on and on that lead people astray. And, because control of most or all of the levers of power are today consolidated in the hands of "the few," *protests as we have known them* are even less likely to succeed. It is probably now too late in our species' evolutionary journey for *protests as we have known them* to bring about such worldwide transformations.

However, there is a new form of protest occurring (let's call it Homo sapiens Survival-driven Protests) that has a chance of succeeding and is beginning to make a difference because it is—a worldwide species survival-driven protest.

Here's what different about this kind of protest:

It's humanity wide. It has the potential to become united around a protest engaged by all humanity—species-wide.

The last time our species rose up as one in pursuit of survival was 70,000 years ago when they migrated out of Africa. They did so because of failing habitats due to climate change brought on by the last ice age. We wouldn't be here had they hesitated, and our species might well go extinct if we fail to act as they did.

It represents a single issue. It is something everybody can understand—the threat to Homo sapiens survival. Survival from potential species-wide extinction due to climate change. Everyone gets it and can now root for the same goal—survival.

It is worldwide. It is not confined to a single region, country or geographic location, but, rather, it affects the entire planet thereby making it harder to play one country or region against another.

It impacts all social classes. The rich, middle class and poor are all affected, so derailing such protests by setting one class against another is unlikely to succeed.

It's driven by the younger generation and children. Indeed, out of the mouths of babes and sucklings—who would have thought?—and it includes all generations, including the unborn, because all will likewise perish unless urgent action is taken.

It has no ideological, political, economic, religious, or special-interest focus—just survival. As we have noted, all Homo sapiens want to survive—Monied sapiens, Techno sapiens and every other kind of Special Interest sapiens, even including, you guessed it, "the few" who run things now.

It might have succeeded at least once before. It resembles in some respects another successful protest—Martin Luther's protest against what he regarded as a dominant issue of his day, the sale of indulgences. Luther didn't like the fact people could buy indulgences, or reduced punishment, after death as practiced by the dominant power of the day—the Catholic Church. And like Luther's protest, it might even be accelerated by the emergence of new technologies, a free press aided by widespread access to social media and digital technologies, as his was enabled by the timely emergence of the technology of his day—the printing press.

Just as Luther's protest and the printing press enabled a choice to "know" the "Word" rather than to blindly believe what the Church claimed to be the "Word," so these climate change protests might well enable our world to become aware of the real threats to our survival, rather than continue to be deceived by flat-out lies and distortions by climate deniers, profit-driven multinationals, and most governments and politicians who conceal the truth and the existential risk climate change poses, as the Church did back then.

It was Luther's protests and the printing press that opened the eyes of "the many" and may well have sowed the seeds for the Reformation that may well, too, have eventually made possible the Industrial Revolution. So, it is not impossible then that these climate change protests in our time, enabled by a vibrant free press and prevailing digital technologies like social media, may well become the basis of yet-another reformation—this time though, a reformation of twenty-first-century civilization from being profit- and growth-driven to being habitat-, species-, and human-survival-driven.

Finally, this kind of change requires a paradigm shift. And, paradigm shifts throughout history didn't necessarily come from governments or technologies and science—these may have facilitated a shift but were not the root cause. Neither did they arise from economic thought and commerce or the multinationals driving trade, growth and profits. Paradigm shifts come from people who decide to make a change, and they tend to do so en masse when they perceive their survival has been put at risk—as climate change threatened our forebears with extinction 70,000 years ago and as it seems ready to try to do again today.

And within twenty-first-century Homo sapiens civilization today those who need to make this change happen to insure their survival are Homo sapiens, of course, not governments, institutions, ideologies, science, AI, or technologies—just us, hence protests.

These protests are just starting to gather steam and include numerous worldwide climate change protests by students. Various organizations are taking a stand in many countries, like Extinction Rebellion in the U.K. Their chances of success will depend on their becoming a worldwide survival-driven process focused on one thing and one thing only—Homo sapiens survival. Only time and participation will tell.

World-war-driven transformation. There are those who look to a third world war to break the stranglehold of capitalism and rid humanity of "the few," by wiping away much of what ails mankind (seen in many science-fiction movies, such as *Star Trek: First Contact*), as if such a global conflict will somehow take out only the rich and powerful, leaving only survivors from among "the rest" to remake the world. Homo sapiens will have learnt its lesson and will somehow develop a new civilization, free from capitalism and the pursuit of wealth, and will go on to develop the technologies for interplanetary migration.

It's not obvious why anyone would envision a world war as a path to such transformation. It sells well in the movies, no doubt at least partly because they never bother to show the third world war itself, picking up the story long after. It seems as if the people who write these screenplays haven't had much to do with the last two world wars and don't know much about what a global thermonuclear war will do to the planet, and the scant chance for sustainable ecosystems to survive one.

What about government and politics? If *protests as we have known them* and wars are unlikely to bring about this transformation, could political change do it? It's true that there has been a significant decline in warfare in the last century, due to the growth in a shared form of government (democracy), greater international collaboration, international trade agreements, and global institutions. But by the look of things, the political winds are starting to blow back toward fascism, strongmen, and a president-for-life style of government, with leaders who are intent on dismantling the very democratic institutions that brought them to power.

Far from bringing nations together around a common Homo sapiens family, America, the one remaining superpower, seems to be leading the way back to isolationism, dismantling long-standing global organizations, taking an axe to NATO, shredding or withdrawing from trade organizations, the Paris climate accord, and other key international agreements. Together, these agreements with "former" European allies whom America is now busy alienating helped establish the world order that led to the decline in war during the last century. One shouldn't rule it out, but does anyone see Russia, China, and America coming together anytime soon, if ever? If not in the short to intermediate term, what is the chance of a Homo sapiens-wide civilizational transformation in time to avert the worst of climate change?

The reduction in warfare during the last century is undoubtedly a huge achievement, but this success notwithstanding, political change would seem an unlikely way to bring about such a transformation.

Not even the strongest democracies have managed to escape the partisanship and political deadlock that render such states incapable of solving even grievous life and death issues, such as gun control in America and migration in Europe and Central America, even when a supermajority of citizens desire change. The rising tide of nationalism and strongman politics in these same democracies has begun to show signs of taking them apart. One shouldn't be surprised if the second half of the twenty-first century turns out to be politically more like the first half of the twentieth century than the second. So, yes, political change will occur, but don't expect it to usher in an ecolization-like transformation.

Could religion be the answer? It hasn't been so far. If one believes in any of a number of religions that promise deliverance by a messiah or prophet and an afterlife in a heaven or a restored Earth, well, good luck with that. One doubts that will work for those Homo sapiens who credit evolution rather than a god with their existence.

Can capitalism, free-market-driven commerce, entrepreneurial businesses, and multinational corporations bring about such a change? They are already guilty of creating the status quo. In fact, any venture driven by the profit motive is antithetical to a survivability and sustainability outcome. Most of the devastation wreaked on the planet and its species, including humans, has been profit-driven and carried out by these so-called legal persons. So, capitalism in its current form doesn't stand a chance. Nonprofits, however, can begin to resemble what such enterprises might look like under ecolization. Consider RideAustin, the company set up as a replacement after Uber and Lyft abandoned the city of Austin, Texas.

Unlike those it replaced, Ride Austin (*www.rideaustin.com/*) set out to drive down costs and increase accessibility by creating a ride-sharing service dedicated to serving the city and underserved areas. At the time of this writing, it was local and dedicated to its drivers, who kept tips and 100% of the fare. It had plans to contribute to the local labor market by hiring drivers as full-time employees with benefits. Operating more like a philanthropic organization, it gave back to the community rather than extracting, like the for-profit businesses it replaced. It was transparent and shared its data with the city to facilitate transportation planning. In other words, its focus was on the livelihood of its drivers and the transportation needs of the city, rather than profits.

In addition to such nonprofits, could a more humane form of capitalism work? Andrew Yang, for example, speaks of a kind of "human capitalism," as described in his book *The War on Normal People: The Truth About America's Disappearing Jobs and Why Universal Basic Income Is Our Future*. Yang envisions a different form of capitalism, one whose core tenets hold that humanity is more important than money (amen to that!), each person rather than each dollar is the unit of an economy (amen to that, too!), and markets exist to serve human goals and values, not the other way around (for this, nothing short of an alleluia amen will do!).

Indeed, such a change in capitalism would improve on the status quo by several orders of magnitude. Yet from an evolutionary species survival perspective, such a scheme doesn't go nearly far enough. What's missing? The fundamental change that needs to occur is a replacement of the motives that drive capitalism—personal and corporate profit, wealth accumulation, and economic and market growth—with a *return* to an economy that puts first, and is driven by, the survival of humans, habitats, and species. Adam Smith, however, wouldn't consider this capitalism.

Could science and innovation bring about that transformation? Many look to science and innovation as Homo sapiens' escape mechanism, but over the last 100 years, these innovations have done more to bring our species closer to extinction than to survival. Putting our hopes in science and technology may sound like a good strategy, but only until one realizes that it is precisely because of our science and technology, when paired with our evolutionary immaturity, that our species stands on the brink of a survival-extinction precipice. Perhaps we will be able to build spaceships that can take some (think "the few") to Mars, or some other planet to escape the troubles we are creating here on Earth. But not even the most hopeful scientists believe that will be possible for another 100 to 200 years. Meanwhile, our ability to adapt and innovate has already created much bigger problems than the ones we were hoping they would solve.

Can individuals bring about such a transformation? If transformation cannot be wrought by *protests as we have known them* or politics, religions or innovations, warfare, or new forms of capitalism, can individuals bring about such a transformation? Yes and maybe. Yes, because as an individual, one can take steps to bring about such a change at a personal level, and there is no lack of advice on how one can play one's part. Such personal transformation will have to be replicated across our entire species to become a species-wide preference, if not a species-wide choice. Such a transformation will have to be voluntary, both at the individual and the species-wide level, to be effective. Individuals must want this transformation first before it can become the desire of the entire species as a whole. It must transform from an individual to a worldwide species survival-driven protest. Yet, there is still the maybe aspect of this question.

Maybe, because even with all or most Homo sapiens wanting this transformation, control and power remain almost entirely in the hands of "the few." It will require more than likes and posts from me and you, and my neighbors and your neighbors, and all of their neighbors, on Facebook, Google, Pinterest, or Baidu to make it so. Indeed, neither "Old Power" nor "New Power" might be enough. Instead, modern Homo sapiens is going to need evolution's help as well, just as our forebears needed it and got it to migrate out of Africa. Let me remind you how it all went down.

It has been more than 70,000 years since that fateful event, and at least 10,000 years since early humans abandoned their evolutionary way of living *within* Earth's ecosystems and adopted instead agriculture and farming as their new and preferred way to maintain life on a planet that in the long term would require the former way of life to survive.

> *Now it's time to face up to the very obvious fact that the switch we made way back then has not worked and never did, even after 10,000 years of trying, and it probably never will. The plain and simple truth, if we could admit it, is that all Homo sapiens' troubles are the direct result of and are traceable back to this initial choice.*

Yet all is not lost. How so?

It is unlikely evolution would have taken the trouble to evolve in the 4 billion years of our planet an entire genus of intelligent species, Homo sapiens being the last so far, seemingly intending it to develop the capability to survive in a universe ensnared in an endless cycle of birth, life, and death, just to stand by and watch it all end in extinction in these final hours without a fight. Is Evolution about to lose to the universe this battle to successfully evolve a life-form capable of surviving its worst existential threats, a mature intelligent species that can go on to spread mature intelligent complex life across the galaxy? Probably not.

So, yes, help for Homo sapiens may yet arise from a corner almost no one—least of all, and in spite of, "the few"—would expect. But how might evolution pull this off?

Recall that it was plate tectonics 25 million years ago that began the separation of the African and Arabian plates, resulting in the formation of that land bridge. And 70,000 years ago, toward the end of the last ice age, it was climate change that locked up much of Earth's waters into ice caps just in time to make the Red Sea low enough to enable our forbears' migration out of Africa. And this wasn't the only time, it would seem, evolution intervened to shape our species' evolutionary journey.

Indeed, hominins' and early Homo sapiens' transformations have often been triggered by climate change. Hence, wherever there was climate-change-driven habitat loss, and thereby Homo sapiens transformations, there is a pretty good chance that would have been evolution at work.

Climate change, toward the end of the last ice age may well have rendered hunting and gathering in the Ethiopian Great Rift Valley untenable, forcing that survival-driven migration out of Africa. By the way things appear to be shaping up, it might again be climate change, already in progress, that might well undo yet another Homo sapiens civilization, one that will trigger yet another species transformation or extinction. This time, however, it's going to be modern Homo sapiens' civilization, *our* civilization, and hopefully this will be the transformation that will reopen that yellow-brick road and make possible a return to ecolization, and thus the path to Homo sapiens' survival.

How can one be so sure, you ask? One can't. But if one had to choose between Warren Buffett's investment advice paying off and evolution coming through once more, given the faithful repetition of evolution's survival patterns over Homo sapiens' history, my money will be on evolution. Because betting on evolution won't just be betting on "The Next Big Short," but rather, betting against "The Mother of all Shorts"—against the extinction of our species. If evolution went to such lengths to ensure our forebears' migration out of Africa and across the entire Earth, what are the chances it might just be getting ready once again to help their descendants to get across not just the Red Sea and the planet, but this time to mature past the point of self-extinction and migrate to the stars? Who or what would you bet on, again?

Yes, it may seem fanciful, and yes, it may seem unrealistic, even unbelievable, but the history of our species says it is possible to build an ecolization-based world order if only because it existed and survived for millions of years, across climate change events, before civilization and capitalism came along, and because it's the only answer evolution gave to solve the equation of life to survive—to live on Planet Earth without destroying its species, its habitats, and one's self.

And to avoid extinction, Homo sapiens may well have to return to its past to find its future, and the way forward to its survival.

GLOSSARY

Adaptation

Evolutionary species must adapt to changing conditions in their habitats or migrate to survive. Failure to adapt or migrate in the face of life- threatening conditions will result in extinction. Evolutionary survival-driven adaptation—Must be survival-driven to preserve habitats and biodiversity.

Anthropocene

Anthropocene is a proposed epoch that defines Earth's most recent geologic time period as being human-influenced, or anthropogenic, based on overwhelming global evidence that atmospheric, geologic, hydrologic, biospheric and other Earth system processes are now altered by humans.

www.anthropocene.info/

Biodiversity hotspots

A biogeographic region with significant levels of biodiversity that is threatened with destruction. There are thirty-five areas around the world that qualify as hotspots, and while they represent just 2.3% of the Earth's surface, they support more than half of the world's plant species that are found no place else and nearly 43% of bird, mammal, reptile, and amphibian species that are also found no place else on the planet.

Carbon cycles

The short-term carbon cycle is dominated by plant life and can increase atmospheric carbon dioxide by about 25%.

The long-term carbon cycle—If the carbon from plant life remains locked up in tissue or plant and gets buried without being consumed, it becomes part of a large organic carbon reservoir within the Earth's crust and is no longer available to the short-term carbon cycle. It then becomes part of the **long-term carbon cycle,** which involves very different kinds of transformations, the most important of which is the transfer of carbon from the rock record into the ocean or atmosphere and back again. The time scale of this transfer is measured in millions of years.

Civilization

The process and structure by which Homo sapiens organized themselves 10,000 years ago to sustain their lives ***outside of the Earth's ecosystems.*** That process so far has resulted in the destruction of numerous Earth habitats and the extinction of more than half the planet's species, as well as putting at risk the survival of the Homo sapiens species itself. Civilizations replaced nearly all hunter-gatherer communities and began by a transition to agriculture, which enclosed the land for farming, domesticated animals, and led to the practices of forced labor and slavery, creating the world's first poor population, depriving them of the native economic freedom and independence with which they evolved.

Climate change's role in hominin evolution
Climate change appears to be evolution's tool to prod species on to greater maturity or to extinction. It acts like the ultimate species survival chaperone, prompting what can turn out to be extinction or adaptation events and thereby a growth in maturity from the time a species evolves to the time it either goes extinct or reaches a higher level of adaptation and maturity and thereby survives.

Climate engine, Climate machinery
El Niño, La Niña, ENSO (El Niño-Southern Oscillation)

According to the United States Geological Survey, **El Niño** ("young boy" in Spanish) is an irregularly occurring and complex series of climatic changes affecting the equatorial Pacific region and beyond every few years. It is characterized by the appearance of unusually warm, nutrient-poor water off northern Peru and Ecuador, typically in late December.

El Niños often mean droughts, famines, forest fires, and floods, and the world has seen many devastating El Niños that resulted in millions of deaths and destruction.

El Niño and La Niña ("little girl" in Spanish) are opposite phases of what is known as the El Niño-Southern Oscillation (**ENSO**) cycle. The ENSO cycle is a scientific term that describes the fluctuations in temperature between the ocean and atmosphere in the east-central equatorial Pacific (approximately between the International Date Line and 120° west).

La Niña is sometimes referred to as the *cold phase* of ENSO, and El Niño as the *warm phase* of ENSO.

Cuban missile crisis
The event in October 1962 pitted U.S. President John Kennedy against Soviet statesman Nikita Khrushchev, America against Russia, with the fate of our entire species dangling around the finger of either man. For the first time, as far as is known, Homo sapiens' existence as a species might well have come down to the whim of a single man. Precisely at that point in

time (October 16 to 28, 1962) Homo sapiens' evolutionary maturity caught up and intersected with Homo sapiens' innovation and seemed inexorably headed in opposite directions. Homo sapiens' ability to innovate had finally surpassed its ability to mature and with devastating consequences.

Dead zones

Dead zones are oxygen-depleted areas in the oceans and are becoming commonplace across the Earth. In some places, the oxygen is getting so scarce that fish and other animals cannot survive. They can either leave the oxygen-free waters or die. The Louisiana Universities Marine Consortium reported that this year's dead zone in water along the Gulf of Mexico shoreline covers 7,722 square miles. There are now more than 400 coastal dead zones around the world.

Doomsday Clock

The Doomsday Clock was created by the board of the Bulletin of the Atomic Scientists in 1947 as a response to nuclear threats.

Drought

"Drought is the great enemy of human civilization. Drought deprives us of the two things necessary to sustain life—food and water. When the rains stop and the soil dries up, cities die and civilizations collapse, as people abandon lands no longer able to supply them with the food and water they need to live," writes Jeff Masters in a March 21, 2016, article titled "Ten Civilizations or Nations That Collapsed From Drought," published at *WunderBlog*, in which he provides a list of collapsed civilizations that includes many one might not have known were casualties of drought due to climate change.

https://www.wunderground.com/blog/JeffMasters/ten-civilizations-or-nations-that-collapsed-from-drought.html

Earth-life

All species (flora and fauna) that evolved and survive on Earth.

Ecocide

Ecocide, or ecocatastrophe, is the extensive damage to, destruction of or loss of ecosystem(s) of a given territory, whether by human agency or by other means, to such an extent that peaceful enjoyment by the inhabitants of that territory has been or will be severely diminished.

https://en.wikipedia.org/wiki/Ecocide

Ecolization

"Ecolization," as in "civilization," is a term used to describe how Homo sapiens organized themselves and lived for most of the 3 million years that preceded agriculture and civilization. It is the outcome of the process by which hunter-gatherers and all species organized to exist ***within the Earth's ecosystems*** without destroying itself, the Earth's habitats, and other species.

Endocrine Disrupting Chemicals (EDCs)

Endocrine disruptors are synthetic chemicals that interfere with human endocrine (or hormone) systems at certain doses. Any system in the body controlled by hormones can be derailed by hormone disruptors. They mimic or partly mimic naturally occurring hormones in the body like estrogens (the female sex hormone), androgens (the male sex hormone), and thyroid hormones, potentially producing overstimulation. They bind to a receptor within a cell and block the endogenous hormone from binding. These disruptions can cause cancerous tumors, birth defects, and other developmental disorders.

Equation of life

A species must be able to obtain all the energy it needs to survive from within the Earth's ecosystems and its habitat and must be able to do so without destroying or rendering these incapable of sustaining life, or driving other species and/or itself to extinction.

Evolution of the human mind

Homo sapiens' effort to master the capabilities and harness the powers of the mind for the good of self and all other species remains a work in progress.

Evolutionary maturity

Evolutionary maturity is a measure of a species' ability to survive. The maturity scale ranges from the zero or near zero maturity-level, indicating the near certainty of its self-extinction, to the fully matured level, the point at which it surpasses the ability to drive itself to extinction. When a species has matured beyond the likelihood, and perhaps even the possibility, of self-extinction, it is said to have reached a ***Survival Plateau***.

Evolutionary sustainability

Evolutionary sustainability is the care and feeding of its habitat by an intelligent species to ensure mutual survival, even during climate change cycles, and for as long as it takes to attain evolutionary maturity and migrate to survive.

Evolutionary survival patterns

A recurrent climate-change-driven set of survival and sustainability challenges present throughout hominins' history that appears to determine their survival or extinction and that repeats across geologic time.

All through hominins' history climate events appear to have been used to drive our species to (**A**) adapt, (**I**) innovate, (**M**) mature, (**M**) migrate, (**S**) survive, or (**E**) go extinct—(**AIMM–S/E**). These appear to constitute an unbreakable, sequential, and recurrent set of species survival patterns an

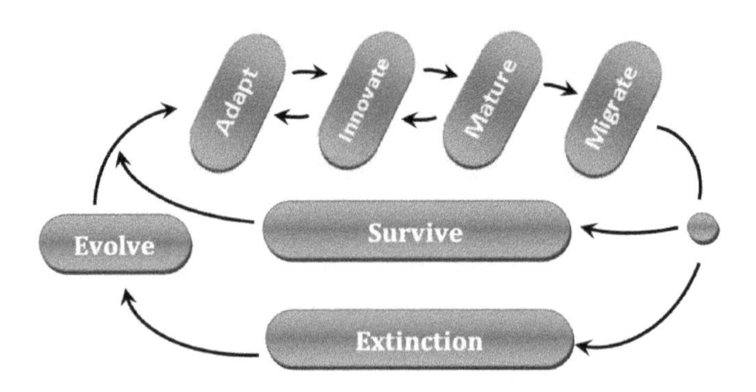

emergent intelligent species must successfully execute to survive or it will go extinct.

These evolutionary survival patterns work together in sets that enable maturity, migration, and survival on the one hand, or extinction on the other. These sets are the species survival set: **A**dopt, **I**nnovate, **M**ature, and **M**igrate to **S**urvive (**AIMM-S**) and the species extinction set: **A**dopt, **I**nnovate, **M**ature, or go **E**xtinct. (**AIMM-E**).

An intelligent species' survival is choreographed and circumscribed by this set of repeating evolutionary survival patterns driven by climate change. If an intelligent species were to discover and follow these patterns, its chances for survival would be significantly enhanced, if not ensured.

Extraterrestrial species evolution
Extraterrestrial species evolution is evolution that might occur on other planets or solar systems across the galaxy and/or universe.

Feudalism
From the fifth to the fourteenth century, social relations were based around agriculture and feudalism, a caste system that prevailed in Europe and later Japan. It was a way of structuring society around relationships derived from the holding of land in exchange for service or labor. It was a social and economic system centered on land worked by serfs as agriculture producers bound to the land. Historians describe feudalism as imposing an open political form of domination with the king and his vassals exerting authority over serfs who were attached to the land.

Freedoms—Economic freedom and independence

Economic freedom was obtained through hunting and gathering and is the evolutionary model for sustaining life from within the ecosystems of the Earth. Economic freedom and independence has been the native state of Homo sapiens for all but the last 10,000 years of its more than 3-million-year history. Hunter-gatherers were economically free and sufficient without the notion, need, or motive to pursue economic growth, profits, wealth, money, or to destroy Earth's species and ecosystems to survive. They hunted and gathered to meet all their needs and freely shared with, not sold to, members of their group.

Freedoms—Essential freedoms

The essential freedoms hunter-gatherers had prior to the emergence of agriculture and civilization that Homo sapiens today appear to have lost include the freedom to be free; economic freedom; freedom to choose to eat healthy foods; freedom from the need to own "stuff"; freedom from most diseases and Big Pharma; freedom from government, laws and politics; freedom from religions and belief in a god; freedom from profit-driven and extinction-capable innovations and R&D; freedom to use only renewable energy sources; freedom to choose to act to avoid climate change; freedom to migrate to survive; freedom to choose to exist within Earth's ecosystems without destroying Earth's species and habitats; freedom to choose survival over extinction.

Futurists

Futurists are Homo sapiens who are bullish about what the future holds for humanity.

Glacials, Interglacials

Major ice ages (glacials) and warm periods in between (interglacials) are associated with Earth's orbit around the sun (its eccentricity, which occurs once every 100,000 years).

Genus *Homo*

The genus *Homo* belongs to the Hominidae family, which includes the great apes and humans. Genus *Homo* and the hominins that followed, and thereby Homo sapiens, may well have comprised a nascent breed of intelligent species.

Great Filter theory

This theory holds that other intelligent species might have also emerged on other planets in other solar systems in the galaxy and may have gone extinct, perhaps of their own doing or as victims of one of the universe's mass extinction events.

Great Oxidation Event (Great Oxygenation Event)

Cyanobacteria, and then photosynthesis, generated most of Earth's oxygen and by so doing cyanobacteria, it is believed, triggered the Earth's first mass extinction in the so-called Great Oxidation Event and thereby may have transformed the planet forever.

Great Rift Valley

The Great Rift Valley is a contiguous geographic trench, approximately 6,000 kilometers in length, that runs from Lebanon's Beqaa Valley in Asia to Mozambique in Southeastern Africa. Believed to be the birthplace of early humans and the location from which the ancestors of current humans migrated.

Greenhouse mass extinctions

Greenhouse mass extinctions include the Devonian, Permo-Triassic (or Permian-Triassic), Triassic-Jurassic (T-J) extinctions and multiple times during the Jurassic and Cretaceous periods ending with the last known greenhouse extinction at the end of the Paleocene epoch some 60 million years ago.

These greenhouse extinctions occurred over a vast expanse of time, 400 million to 100 million years ago, and shared common characteristics of high temperatures, high CO_2 levels, and low oxygen levels. Research now shows that each occurred in a world of quickly rising CO_2 (and perhaps methane as well, based on yet-another line of evidence).

Hominidae family

The Hominidae family includes the great apes and humans. There are seven extant species in four genera: orangutans (two species in genus *Pongo*), gorillas (two species in genus *Gorilla*), chimpanzees (two species in genus *Pan*) and humans (genus *Homo*), according to the Encyclopedia of Life.

Hominins

Hominins are classified as a tribe (Hominini) of primates, a type of mammal. Hominins are part of the family, or larger group of primates, called hominids. Hominids include orangutans, gorillas, chimpanzees, and human beings. All hominins are hominids, but very few hominids are hominins.

According to paleontologist Lee Berger, hominins are primates that share characteristics such as "bipedalism, reduced canine size, and increasing brain size."

The only hominins alive today are human beings. There are many, many extinct hominins, a fraction of which are represented here. Fossilized remains of extinct hominin species have been found in parts of Africa, Europe, and Asia, many dating back millions of years. Learn more at National Geographic Hominin Story.

https://www.nationalgeographic.org/media/hominin-history/

Hunter-gatherers

Hunting and gathering was a highly stable, very long-lasting way of life for hundreds of thousands of years, until about 10,000 years ago, when this subsistence method began to change in a number of locations across the globe. It was the only way in which Homo sapiens were able to extract the necessary sustenance from the environment.

Intelligent species Survivability and Habitat Sustainability Test

This test is a survivability indicator and can be used to determine what can be added to or changed on the planet by an intelligent species while still remaining true to the overarching evolutionary goals of species survival and habitat sustainability.

To pass this test, every activity an intelligent species does must guarantee the survival of all species and sustainability of Earth's habitats. If an activity fails the test, it will have no place in an evolutionary mature intelligent species world.

If Homo sapiens wanted a blanket overarching guide as to what it can and cannot do to Earth's ecosystems, this would be it.

Innovations for Survival dashboard

This is a proposed website to track survival-driven innovations, what innovations should be focused upon immediately, in the short, intermediate, and long terms. The site lends focus and tracks progress as well as suggests potential areas for inventors, innovators, entrepreneurs, humanitarians, philanthropists, and anyone else who is concerned about survival.

Megafauna

Megafauna is a term used to describe animals with an adult body weight of over 44 kg (97 lbs). *Megafauna* can be found on every continent and in every country. For every living species of *megafauna*, there are a large number of extinct *megafaunas.* Pleistocene *megafauna* is the set of large animals that lived on Earth during the Pleistocene epoch and became extinct during the Quaternary extinction event.

https://en.wikipedia.org/wiki/Pleistocene_megafauna

Migration

Migration is a built-in evolutionary survival response, a survival pattern that appears to be independent and in spite of a species' innovative and technological prowess.

NASA's Mars Reconnaissance Orbiter

The multipurpose spacecraft found evidence of an ice age on Mars.

Neanderthals

Neanderthals (Homo neanderthalensis or Homo sapiens neanderthalensis) are an extinct species or subspecies of archaic humans in the genus *Homo*, who lived within Eurasia from circa 400,000 until 40,000 years ago.

https://en.wikipedia.org/wiki/Neanderthal

Metabolic rate of animals

Metabolic rate is the pace at which oxygen is used by an organism. As temperatures rise, metabolic rate goes up, and as metabolic rate rises, so, too, does the need for oxygen, as the chemical reactions of life are oxygen-dependent. Research shows that metabolic rate doubles to triples with each 10-degree rise in temperature. The consequences of this in a low-oxygen world would be major and life impacting.

Permian-Triassic extinction

It is believed as much as 95% of all species died during this event.

PFAS (Per- and polyfluoroalkyl substances)

PFAS, so-called forever chemicals, are a group of widely used synthetic chemicals that have been in use since the 1940s. These are the chemicals that make carpets stain resistant and fast food packaging able to repel grease and water. They're also used in fire-fighting foams and is what gives nonstick cookware it's non-stickines.

They're also known as "forever chemicals," because the molecular bonds that form them can take thousands of years to degrade, meaning that they accumulate both in the environment and in our bodies.

Phthalates

Phthalates are a group of chemicals used in hundreds of products such as toys, vinyl flooring and wall covering, detergents, lubricating oils, food packaging, pharmaceuticals, blood bags and tubing, and personal care products such as nail polish, hair sprays, aftershave lotions, soaps, shampoos, perfumes, and others.

Phytoplankton, Plankton

According to Dr Richard Lampitt in a discussion with *National Geographic*, phytoplankton are at the base of what scientists refer to as oceanic biological productivity, the ability of a body of water to support life such as plants, fish, and wildlife, and are the foundation of the oceanic food chain. About 50% to two-thirds of the planet's total atmospheric oxygen is produced by ocean phytoplankton, and therefore cessation would result in the depletion of atmospheric oxygen on a global scale. This would likely result in the mass mortality of animals and humans.

Quaternary glaciation

The "Ice Age," the most recent glacial period, occurred from around 110,000 to 11,700 years ago and is believed to be one event in a larger pattern of glacial and interglacial periods known as the Quaternary glaciation, which began about 2,588,000 years ago and continues to today.

Sapiens—Techno sapiens

Techno sapiens are Homo sapiens who are known for being particularly capable in the fields of inventions, technologies, and innovations and whose contributions enable them to be seen as leaders in the techno sphere.

Sapiens—Moneyed sapiens

Moneyed sapiens are Homo sapiens who have managed to amass enormous wealth and are known as being among the 1% that controls most of the wealth in the world.

Smithian stage

A million-year time interval around 247 million years ago that had the highest of all known temperatures since the time when animals first occurred.

Species-driven adaptations

Species-driven adaptations are forms of innovation that have been brought into being by an intelligent species (Homo sapiens in this case) and are above and beyond what evolution's natural selection has enabled.

Species-driven evolution

All life, including flora and fauna, is here because of evolution and is almost without exception biologically based. Species-driven evolution, however, refers to forms of evolution that can go beyond the biological and enable a species to develop capabilities unnatural to species, e.g., pacemakers, prosthetics, etc.

Species-driven innovation
Unlike evolutionary innovations, these are forms of innovation that are above and beyond what evolution and natural selection have enabled and that have been brought into being by a particular species.

Species-driven selection
Unlike evolutionary and natural adaptations, these are forms of adaptations that are above and beyond what evolution and natural selection have enabled and that have been brought into being by a particular species.

Species survival
The ability of a species to avoid all forms of extinction and remain alive.

Species Survival Maturity Model (SSMM)
A survival framework drawn from numerous evolutionary survival patterns that evolution appears to have left for intelligent species to follow to attain evolutionary maturity and thereby escape self-extinction and survive.

Species self-extinction
Intelligent species self-extinction occurs when such species engage in activities (e.g., global warfare) or enable innovations that can intentionally or accidentally bring an end to all extant members of the species.

Survival-driven adaptations
Most Homo sapiens' adaptations today are driven by market forces, focus on profits, and may or may not have anything to do with Homo sapiens' survival. Survival-driven adaptations, on the other hand, focus on species survival.

Survival-driven innovations
Most Homo sapiens' innovations are driven by market forces, focus on profits, and may or may not have anything to do with Homo sapiens' survival. Survival-driven innovations, on the other hand, focus on species survival.

Survival-driven Innovation Scorecard

A Survival-driven Innovation Scorecard measures Homo sapiens' innovations from the perspective of their focus on and ability to enable and enhance species survival and habitat sustainability.

Survival Plateau

The Survival Plateau, as indicated on the Species Survival Maturity Model, is the measure of the level of maturity an intelligent species has attained when it has matured beyond the likelihood, and perhaps even the possibility, of self-extinction.

Sustainable-capable habitat

A sustainable-capable habitat is one that self-resuscitates or can be revived notwithstanding multiple climate events. The Earth consists of multiple sustainable-capable habitats.

Survivability and Habitat Sustainability Test

The Survivability and Habitat Sustainability Test is an evolutionary test Homo sapiens might use to evaluate every activity so as to ensure the survival of all species and the sustainability of Earth's ecosystems and habitats. Any activity that results in the death of species or the destruction of Earth's habitats and ecosystems would not contribute to survival and hence would automatically fail the test.

"The few"

"The few" refers to that small fraction of Homo sapiens (rulers, governments, kings, queens, priests, the elite, bureaucrats, craftsmen, and traders) who managed to seize control starting some 10,000 years ago and have hung onto it, becoming the architects and overlords of the civilization we see today. Ever since farming drove hunter-gatherers off the land, "the few" in control began appropriating for themselves the produce, labor, and freedoms of "the rest" while relegating "the rest" to the basement of civilization.

The kings and queens, some still exist, mostly became nation-state governments, dictators and military strongmen; priests led to organized religions with offshoots that became institutions of learning like universities; the elite and bureaucrats soon became the wealthy; traders became the business class, companies and multinationals; and money changers are now our financial institutions and craftsmen the technology builders. What these all had and still have in common is a transfer of the freedom of the individual to institutions and organizations headed by "the few."

Universe's existential threats

These are species-extinction threats due to planetary, solar, galactic, and/or universal events. The asteroid that took out the dinosaurs posed an existential threat and did result in their extinction.

NOTES & REFERENCES

CHAPTER 1
EARTH

5 NASA. *Earthrise*, taken in space by astronaut William Anders and the crew during the Apollo 8 mission in 1968. "They also became the first to look back at their home planet and see the entire world in one glimpse." *http://www.nasa.gov/topics/history/features/apollo_8.html*

6 "*A Pale Blue Dot*," by Carl Sagan.

In the words of Carl Sagan from *Pale Blue Dot* (1994), when seen from 6 billion kilometers out (beyond Pluto), as photographed by Voyager 1.

8 Professor Brian Cox and Andrew Cohen: Human Universe. William Collins, Publisher. UK. 2016.

14 "It is clear that we are just an advanced breed of primates on a minor planet orbiting around a very average star, in the outer suburb of one among a hundred billion galaxies. BUT, ever since the dawn of civilization people have craved for an understanding of the underlying order of the world. There ought to be something

very special about the boundary conditions of the universe. And what can be more special than that there is no boundary? And there should be no boundary to human endeavor. We are all different. However bad life may seem, there is always something you can do, and succeed at. While there is life, there is hope." (Stephen Hawking)

CHAPTER 2
DOES THE UNIVERSE COME WITH EVOLUTIONARY SPECIES SURVIVAL MECHANISMS?

18 Sean Carroll, in his book *The Big Picture: On the Origins of Life, Meaning, and the Universe Itself*, says, "A law of physics is a ***pattern*** that nature obeys without exception."

18 According to quantum physics, says Steven Hawking in his book *The Grand Design*, "Nature does not dictate the outcome of any process or experiment, even in the simplest situations. Rather, it allows a number of different eventualities, each with a certain likelihood of being realized."

19 Sean Carroll, *The Serengeti Rules: The Quest to Discover How Life Works and Why It Matters*. Princeton Press. NJ. 2016.

Carroll, a cosmologist, reminds us that "every cell contains a society of molecules, every organ a society of cells, everybody a society of organs, and every ecosystem a society of organisms, and understanding the interactions within each of these societies has been the primary aims of molecular biology, physiology, and ecology."

Carroll continues: "We gathered fruits, nuts, and plants; hunted and fished for the animals that were available, and like the wildebeest and zebras on the Serengeti, we moved on when the resources ran low."

22 Yet today, only Homo sapiens remain. How did Homo sapiens manage to survive when all, even those closest to our species, have gone extinct? (BBC Earth). *http://www.bbc.com/earth/story/20150929-why-are-we-the-only-human-species-still-alive*

27 Professor Curtis Marean of the Institute of Human Origins at Arizona State University: (*https://www.theaustralian.com.au*)

"Shortly after Homo sapiens first evolved, the harsh climate conditions nearly extinguished our species." Some scientists even believe that the human race's population may have fallen to just a few hundred individuals who managed to survive in one location.

Professor Marean discovered ancient human artifacts in the isolated caves around an area known as Pinnacle Point, South Africa. Professor Marean said the caves contain archaeological remains going back at least 164,000 years.

Homo sapiens may well have survived, it would appear, not because of our big brain, capacity for innovation, and smarts but more than likely primarily because some of our species were lucky enough to have found a not-so-cold patch in Southern Africa and by becoming more intelligent and adaptable to exploit it during this period.

They may have been able to survive because of rich vegetation that was available in the area. Neanderthals, with whom our species shared the planet until just before the last glacial maximum, 20,000 years ago, may have struggled to survive as the rising and falling ice ate away at their habitat (although many other explanations for their extinction have been suggested) and, unlike Homo sapiens, who left Africa much later, may have failed to find sustainable habitats.

29 Research by Eelco Rohling of the University of Southampton in England suggests that we are now 2,000 to 2,500 years overdue for another ice age and that the reason it has not arrived yet has been the impact of Homo sapiens on the global climate.

CHAPTER 3
IS SPECIES SURVIVAL THE DEFAULT EVOLUTIONARY OUTCOME?

42 Walter Alvarez writes about the effect of plate tectonics in his book *A Most Improbable Journey—A Big History of Our Planet and Ourselves*. Published by W.W. Norton and Co. NY/London.

CHAPTER 4
AN INTRODUCTION TO INTELLIGENT SPECIES SURVIVAL PATTERNS

54 Peter D. Ward and Donald Brownlee, *Rare Earth: Why Complex Life Is Uncommon in the Univer*se (Univ. of Washington). Copernicus Books, NY, NY. 2003.

CHAPTER 6
THE ROLE OF CLIMATE CHANGE IN THE MAKING OF AN INTELLIGENT SPECIES

97 "Earth's climate has always been in a state of flux. Ever since our ancestors branched off the primate evolutionary tree millions of years ago, the planet has

faced drastic swings between moist and dry periods, as well as long-lived glacial freezes and thaws. It's clear that early humans were able to survive such changes—our existence confirms their success," so writes Brian Handwerk, in a *Smithsonian* article published September 30, 2014, titled "How Climate May Have Shaped Human Evolution" (https://www.smithsonianmag.com/science-nature/how-climate-change-may-have-shaped-human-evolution-180952885/) drawing on the works of Rick Potts, head of the Human Origins Program at the Smithsonian National Museum of Natural History.

98 "Scientists say, this past about our species' ability to survive climate change and adapt over millions of years can offer clues," writes Gayathri Vaidyanathan, in *ClimateWire* on April 13, 2010, in an article titled "Scientists attempt to understand how human ancestors adapted—or not—to previous periods of climate change" and publish*ed in Scientific Amer*ican.

98 "It is not true that the Sahara Desert has been a permanent feature for millions of years," said Peter deMenocal, a professor at the Lamont-Doherty Earth Observatory at Columbia University in a *Scientific American* article dated April 13, 2010. Africa oscillated between wet and dry every few thousand years, and each shift induced adaptation in the creatures that lived in the region. "Civilizations and populations can be very plastic that way," said deMenocal. "Climate change alters ecological landscapes, creates unnatural selection pressures, and promotes genetic selection to fit the pressures." Swings between wet and dry landscapes pushed some of our ancestors toward modern traits—and killed off others.

99 Rick Potts of the Human Origins Program at the Smithsonian National Museum of Natural History observed that "many organisms have habitat preferences, such as particular types of vegetation (grassland versus forests), or preferred temperature and precipitation ranges. When there's a change in an animal's preferred habitat, they can either move and track their favored habitat or adapt by genetic change to the new habitat. Otherwise, they become extinct."

Potts also said: "People think, we're such a successful species, nothing can happen to us." But, he pointed out, that most of our ancestors sooner or later went extinct. Homo erectus, the forerunner of modern humans, lived for 1.5 million years, he said. Homo sapiens, by comparison, have been around for only 200,000 years. Yet even they decreased in population size to between 600 and 10,000 breeding pairs when hit with megadroughts, heavy monsoonal rains, and the eruption of a volcano near Sumatra about 70,000 years ago.

101 Homo sapiens civilizations developed, thrived, and then collapsed. This repeated collapse of ancient Homo sapiens' civilizations many scientists now believe was driven primarily by repeated, extended, and severe bouts of climate change and reveals our species' struggle to understand and make sense of the Earth's climate machinery and its devastating cycles of monsoons, El Niños, La Niñas, and protracted multidecade droughts that wasted earlier civilizations.

102 In a March 21, 2016, article titled "Ten Civilizations or Nations That Collapsed From Drought," published at *WunderBlog*, Jeff Masters provides a list of collapsed civilizations or nations that includes many one might not have known were also casualties of drought due to climate change. *https://www.wundergrung.com/blog/JeffMasters/ten-civilizations-or-nations-that-collapsed-from-drought*

103 Brian Fagan, *Floods, Famines, and Emperors: El Niño and the Fate of Civilizations*. Published by Basic Books. 2009. Fagan lists a number of civilizations whose collapse coincided with huge climate events.

113 According to Franz Broswimmer in his book *Ecocide, A Short History of the Mass Extinction of Species* (published 2002 by Pluto Press, London, Michigan, USA), "Human history is replete with accounts of the early ecocidal activities of great empires such as Babylon, Egypt, Rome, ancient China, and Maya, all of which destroyed their forests and the fertility of their topsoil and killed off much of the original fauna through a combination of their linear thinking and their insatiable drive for material wealth."

120 According to an ongoing temperature analysis conducted by scientists at NASA's Goddard Institute for Space Studies (GISS), the average global temperature on Earth has increased by about 0.8° Celsius (1.4° Fahrenheit) since 1880; two-thirds of the warming has occurred since 1975, at a rate of roughly 0.15–0.20°C per decade. And if a picture is worth a thousand words, then Figs. 6-10 and 6-11, which show NASA's rendering of what the Earth's temperature profile might have looked like between 1885 and 1994 versus how it has changed between 2005 and 2014, tell the whole story. The year 2016 has gone down in history as the hottest year since our species has begun keeping global records. *http://earthobservatory.nasa.gov/Features/WorldOfChange-/decadaltemp.php*

126 According to Brian Fagan in his book *The Great Warming: Climate Change and the Rise and Fall of Cvilizations*, "We are entering an era when extreme aridity will affect a large portion of the world's now much larger population, where the challenges of adapting to water shortages and crop failures are infinitely more complex."

CHAPTER 7
HABITATS

135 Peter D. Ward and Donald Brownlee, *Rare Earth: Why Complex Life is Uncommon in the Universe* (Univ. of Washington). Publisher: Copernicus; 12/17/03 edition (December 17, 2003)

136 According to NASA, the galaxy has been around for about 13.6 billion years, almost as old as the universe itself, measures some 120,000–180,000 light-years in diameter, and is home to Planet Earth, the birthplace of humanity. *http://www.universetoday.com/22285/facts-about-the-milky-way/*

137 "The Best & Worst Places for Expats in 2016," *InterNations* (Internations.com). Taiwan came out on top, followed by an aspiring Malta, while Ecuador only just retained its place on the list. Kuwait, Greece, and Nigeria remain at the bottom of the pack.

145 In "A New Planetary Habitability Index" in *Popular Science*, a team of astrobiologists has now proposed a rubric that includes four groups of variables, each of which is weighted by its importance to sustaining life. *http://www.popsci.com/content/new-planetary-habitability-index*

158 According to Peter Ward in his book *A New History of Life*, oxygen and carbon dioxide levels (particularly oxygen) are the most important of all factors dictating animal survival, death, and diversity. Times of high oxygen give rise to an increase in species, whereas when low oxygen prevails, species die out.

159 Carbon dioxide sequestering may not be too bright an idea when viewed from the oxygen depletion perspective because it also sequesters oxygen in the process. *http://blog.hasslberger.com/2008/10/is_oxygen_depletion_more_worry.html*

160 In an article dated Sep 4, 2015, published in ACADEMIA, https://www.academia.edu/9373232/O2_dropping_faster_than_CO2_rising, titled "O_2 Dropping Faster than CO_2 Rising: Implications for Climate Change Policies," Dr. Mae-Wan Ho writes: "Within the past several years scientists have found that oxygen (O_2) in the atmosphere has been dropping, and at higher rates than just the amount that goes into the increase of CO_2 from burning fossil fuels, some 2 to 4 times as much, and accelerating since 2002-2003. Simultaneously, oxygen levels in the world's oceans have also been falling."

163 A study conducted by scientists from Britain's University of Leicester found that an increase of about 10.8°F (6°C) in the temperature of the world's oceans could prevent phytoplankton's oxygen production by disrupting photosynthesis. "It

would mean oxygen depletion not only in the water, but also in the air," said the research team. "Should it happen, it would obviously kill most of life on Earth." Phytoplankton may well turn out to be the canary in the ocean.

166 In a study published in the journal *Nature*, deforestation is said to be responsible for the removal of over 15 billion trees each year, and the number of trees has dropped 46% globally since the advent of human civilization, writes *The New Zealand Herald*.

167 Michael J. Benton writes in *When Life Nearly Died: The Greatest Mass Extinction of All Time* (Thames & Hudson, London): "The natural world is complex and consequences are often unpredictable. As tropical forests are cleared and reefs are poisoned, we are losing not only species, but also whole habitats. Destroying species and habitats piecemeal can lead to a runaway crisis as seemed to have happened in the past. Low levels of extinction can turn into high levels. It could be that removing one or two species from an ecosystem does little damage. But if another few species are picked off, then another few, and then a few more, a point may be reached when that ecosystem will collapse. And, once the world becomes locked into a spiral of downward decline, it is impossible to see how any human intervention could turn it back."

170 According to NASA, carbon dioxide levels reached 403.2 parts per million (ppm) in 2016., 405.12 ppm in November 2017 and 408.02 ppm in November 2018. By 2100, carbon dioxide levels in the atmosphere will have reached 935 parts per million, meaning the gas comprises nearly 0.1% of the atmosphere.

173 Both NASA and the National Oceanic and Atmospheric Administration (NOAA) recorded the highest average surface temperatures in 2016 since the agencies started tracking such data. NASA notes that since the late 1800s, average global temperatures have risen by 1.1°C. And from January to August of last year, NOAA says that each month broke its own warmest record. https://motherboard.vice.com/en_us/article/kb7pk9/july-hottest-noaa-nasa-climate-environment-global

173 NASA/Goddard Space Flight Center Scientific Visualization Studio, Jan. 18, 2017. Maximum temperatures in North Africa, the Middle East and northern India will exceed 45°C by 2100, as can be seen in the close-up of the region from the maps produced using NASA's new climate projection data set.

CHAPTER 8
SUSTAINABILITY

188 Serbian mathematician Milutin Milankovitch theorized that the ice ages occurred when orbital variations caused the Northern Hemisphere around the latitude of

the Hudson Bay and northern Europe to receive less sunshine in the summer. Milankovitch predicted that the ice ages would peak every 100,000 and 41,000 years, with additional "blips" every 19,000 to 23,000 years. The paleoclimate record shows peaks at exactly those intervals.

201 According to the 2017 Oxfam International's global income inequality report, just eight men own as much wealth as half the world. That's right. Eight men, mostly Americans, now control as much wealth as the world's poorest 3.6 billion people, according to the report. The men— Bill Gates, Warren Buffett, Carlos Slim, Jeff Bezos, Mark Zuckerberg, Amancio Ortega, Larry Ellison, and Michael Bloomberg—are collectively worth $426 billion. And in the U.S., the richest 1% control 42% of the wealth.

202 Using the World Bank definition of the global poverty line as $1.25/day, as of September 2013 (since raised to $1.90/day but poverty remains unchanged), roughly 1.3 billion people remain in extreme poverty. Nearly half live in India and China, with more than 85% living in just twenty countries.

203 According to the World Military Expenditure Report published by the Stockholm International Peace Research Institute (SIPRI), countries around the world spent a total of $1.68 trillion (1.56 trillion euros) on arms in 2016.

204 Immanuel Wallerstein comments on military spending data cited by Franz Broswimmer in his book *Ecocide: A Short History of Mass Extinction of Species*: "These inverted priorities are not the neutral decisions of a market, they are the priorities of powerful people in powerful nations, mostly men whose gender, race, and class interests drive the capitalists' political economic system and its worldwide system of accumulation and deprivation."

CHAPTER 9
ADAPTATION

230 Anthropologist Mark Cohen, in his book *The Food Crisis in Pre-History*, claimed megafauna extinction might well have been due to a food crisis resulting from the extermination of megafauna. In most places around the planet, the megafauna extinctions occurred shortly after the arrival of our forebears.

231 The words of Emily Dickinson come to mind as one contemplates our predicament: "Because I could not stop for Death, he kindly stopped for me."

233 Clive Ponting says in his book *A New Green History of the World* that human history over the last 8,000 years has been about the acquisition and distribution

of this surplus (from harvest), and the uses to which it has been put became the foundation of all later social and political change.

Ponting notes: The Christian Bible claims in Psalm 115, "The Heavens are the Lord's but the earth he has given to the sons of men" and in Psalm 8, "Thou has given man dominion over the work of thy hands." It should come as no surprise that early and medieval Christians accepted that their god had given humans the right to exploit plants, animals, and the whole world for their benefit. In this view, "Nature is not seen as sacred, but open to exploitation by humans without any moral qualms— indeed humans have the right to use nature in whatever way they think best."

Ponting notes that this belief system produced a highly anthropocentric view of the world and had an enduring impact on later European thought.

CHAPTER 10
INNOVATION

253 Brian Cox, in his book *Human Universe*, writes, "Two million years ago we were ape-men. Now we are spacemen."

254 Christopher Herd in an article titled "A Brief History of Humanity and the Future of Technology" published in *Peripheral Foresight*, October 11, 2016. https://medium.com/@ChrisHerd/a-brief-history-of-humanity-and-the-future-of-technology-8d40fe08176

258 Dr. Peter H. Diamandis and Steven Kotler, *Abundance: The Future Is Better Than You Think*. Free Press, New York. Mar 28, 2012.

273 Noam Chomsky once asked in a talk at MIT: "Is it better to be smart than to be stupid?"

281 The Doomsday Clock was created by the board of the Bulletin of the Atomic Scientists in 1947. *https://www.wired.co.uk/article/what-is-the-doomsday-clock*

283 IPCC Climate Change 2014 Synthesis Report Summary for Policymakers claims that human influence on the climate system is clear, and recent anthropogenic emissions of greenhouse gases are the highest in history.

284 In a *Climate Central* article by Andrew Freedman titled "Up to Five Billion Face 'Entirely New Climate' by 2050," (https://www.climatecentral.org/news/one-billion-people-face-entirely-new-climate-by-2050-study-16587), which is based on a study titled "The projected timing of climate departure from recent variability"

published Oct. 9, 2013, in the journal *Nature*, Freedman writes: "The mean annual climate of the average location on Earth will slip past the most extreme conditions experienced during the past 150 years and into new territory by between 2047 and 2069, depending on the amount of climate-warming greenhouse gases that are emitted during the next few decades."

284 *Climate Central* website article by Andrew Freedman based on a study titled "The projected timing of climate departure from recent variability." Published in *Nature*, October 2013. https://dimensions.altmetric.com/details/1814794/news

284 A recent study, published in the journal *Nature* (Oct. 9, 2013), used a new index to show for the first time when the climate—which has been warming during the past century in response to manmade pollution and natural variability—will be radically different from average conditions during the 1860-2005 period. The study shows that tropical areas, which contain the richest diversity of species on the planet as well as some of the poorest countries, will be among the first to see the climate exceed historical limits—in as little as a decade from now—which spells trouble for rain forest ecosystems and nations that have a limited capacity to adapt to rapid climate change.

293 According to the U.N. Food and Agriculture Organization, the human food chain is under continued threat from an alarming increase in the number of outbreaks of transboundary animal and plant pests and diseases (including aquatic and forest pests and diseases), as well as food safety and radiation events.

294 EPA's report titled "Persistent Organic Pollutants: A Global Issue, A Global Response," on persistent organic pollutants (POPs), toxic chemicals that adversely affect human health and the environment around the world. https://www.epa.gov/international-cooperation/persistent-organic-pollutants-global-issue-global-response

295 Endocrine disrupting chemicals (EDCs) article titled "Endocrine Disrupting Chemicals Ftound in Food" by Dr. Marcola. https://www.google.com/search?client=firefox-b-d&q=Endocrine+disrupting+chemicals+%28EDCs%29+article+by+Dr.+Marcola

299 Many wild bee species and other pollinators are in danger of extinction, warns a U.N. report titled "Pollination Assessment: IPBES," as reported by Seth Borenstein in an Associated Press article posted February 26, 2016. *https://www.pressherald.com/.../wild-bees-other-pollinators-in-danger-of-extinction-u-*. At least one-third of the world's agricultural crops depend upon pollination provided by insects and other animals.

299 Three-quarters of the world's crops—including fruits, grains, and nuts—depend on pollination, and the *insects* responsible are *disappearing*. "Could Disappearing Wild Insects Trigger a Global Crop Crisis?" February 28, 2013. https://www.smithsonianmag.com/.../could-disappearing-wild-insects-trigger-a-global-cr...

299 In a *HuffPost* article by Casey Williams titled "These Photos Capture the Startling Effect of Shrinking Bee Populations" published on April 20, 2016, one bee expert says, "Everything falls apart if you take pollinators out of the game. If we want to say we can feed the world in 2050, pollinators are going to be part of that."

301 Brian Fagan, *The Great Warming: Climate Change and the Rise and Fall of Civilizations*, Bloomsbury Press, New York. March 10, 2009. From the tenth to the fifteenth century the Earth endured a rise in surface temperatures, "The Great Warming," that changed climate worldwide.

303 According to a 2013 United Nations International Year of Water Cooperation report, 85% of the world population lives in the drier half of the planet; 783 million people do not have access to clean water; and almost 2.5 billion do not have access to adequate sanitation. Six to 8 million people die annually from the consequences of disasters and water-related diseases.

CHAPTER 11
MIGRATION

317 United Nations High Commission on Refugees (UNHCR) claims that "war and persecution have driven more people from their homes than at any time since records began, with over 65 million men, women, and children now displaced worldwide. This means one in every 113 people on Earth is either an asylum-seeker, internally displaced, or a refugee," according to the 2015 edition of UNHCR's Global Trends report.

322 According to Steve Connor, Science Editor, in an article published in *The Independent* on April 30, 2009, titled "World's most ancient race traced in DNA study," using DNA analysis scientists have identified what they refer to as the oldest race of people in Africa and on the planet. The San people of southern Africa, who have lived as hunter-gatherers for thousands of years, are likely to be the oldest population of humans on Earth, according to the biggest and most detailed analysis of African DNA. https://www.independent.co.uk/news/science/worlds-most-ancient-race-traced-in-dna-study-1677113.html)

CHAPTER 13
EVOLUTIONARY MATURITY

371 Paleontologist Dr. Niles Eldredge, curator-in-chief of the permanent exhibition "Hall of Biodiversity" at the American Museum of Natural History, in an ActionBioscience.org original article "The Sixth Extinction," says: "Agriculture represents the single most profound ecological change in the entire 3.5 billion-year history of life." And: "Homo sapiens became the first species to stop living inside local ecosystems."

CHAPTER 14
EXTINCTION

406 According to the World Wildlife Fund, Earth has lost half of its wildlife in the past forty years.

407 The Earth has seen at least five major mass extinctions and all without Homo sapiens' help. Moreover, at least two scientists, David M. Raup and J. John Sepkoski, Jr., calculated that these major extinction events of the past 250 million years occurred periodically at nearly constant intervals of 26 million years, as published in a paper titled "Periodicity of Extinctions in the Geologic Past," published October 11, 1993, in the *Proceedings of the National Academy of Sciences*.

410 Jan Zalasiewicz in his book *The Earth After Us: What Legacy Will Humans Leave in the Rocks*: All that would be left of our time on the planet, the innovations and civilizations we would have built notwithstanding, when seen 100 million years or so from the perspective of some future alien geologist, will be nothing more than a paper-thin layer in the rocks.

412 Jennifer Chu, *MIT News*, Feb. 10, 2014, in an article titled "Timeline of a Mass Extinction": "People have never known how long extinctions lasted," says Sam Bowring, the Robert R. Schrock Professor of Earth, Atmospheric and Planetary Sciences (EAPS) at MIT. Scientists have been updating their research on the end-Permian extinction, and an MIT-led team of researchers has been able to establish that the end-Permian extinction was extremely rapid, triggering massive die-outs both in the oceans and on land in less than 60,000 years, the blink of an eye in geologic time.

412 In an article titled "Recovering From a Mass Extinction" by the University of Bristol published in *ScienceDaily*, January 20, 2008, Sarda Sahney, one of the researchers, says, "Our research shows that after a major ecological crisis, recovery takes a very long time—at least 30 million years."

413 "The correlation of species extinction to Homo sapiens survival is becoming clearer as more and more species die off. The loss of one species can have unforeseen

consequences for many others, since ecosystems are connected by a complex web of interactions that we do not always fully understand," writes Sarah Kaplan in the *Independ*ent, dated October 8, 2015.

413 Peter D. Roopnarine, the curator of geology at the California Academy of Sciences, and paleobiologist Kenneth Angielczyk, an associate curator at Chicago's Field Museum, in an article published in *Science* titled "Community stability and selective extinction during the Permian-Triassic mass extinction."

414 Sean Carroll, *The Serengeti Rules: The Quest to Discover How Life Works and Why It Matt*ers, Princeton Press. NJ. 2016.

415 In his book *The Serengeti Rule*s, Sean Carroll shares useful insights about many food webs, the rules of the Serengeti and about nature as a whole. As per Carroll: "Without sunlight there would be no plants; without plants there would be no food for herbivores; without herbivores there would be no prey; without prey there would be no predators."

421 Ashley Dawson, *Extinction: A Radical Histor*y, OR Books, New York, London. July 22, 2016.

CHAPTER 15
FREEDOM TO CHOOSE SURVIVAL VS. EXTINCTION

439 "The original affluent society" is a view proposed by Marshall Sahlins back in 1966 and subsequently explored by James C. Scott in his book *Against the Grain: A Deep History of the Earliest States*. Yale University Press, 2017.

445 David C. Korten in his book *When Corporations Rule the World.* Berrett-Koehler, 2015.

464 Hervey C. Peoples and Frank W. Marlowe in "Subsistence and the Evolution of Religion." Published Sept. 23, 2013. *https://www.ncbi.nlm.nih.gov/pubmed/22837060*

466 Nick Lane in his book *Life Ascending: The Great Inventions of Evolution.* WW Norton Publishers. November 14, 2014.

467 Thomas Hornigold in an article titled: "The Enormous Promise and Peril of Bioengineering's Pandora's Box," published on *SingularityHub* on December 17, 2017.

468 A new study combining researchers from the U.S. and the U.K., published in *eLifeSciences,* titled: "Point of View: A transatlantic perspective on 20 emerging issues in biological engineering," by Bonnie C. Wintle et al; November 14, 2017.

LIST OF PHOTOS & CHARTS

Fig. 1-1	*Earthrise* photo from Apollo 8	4
Fig. 1-2	The pale blue dot	6
Fig. 1-3	Polar bears on iceberg	6
Fig. 1-4	Our neighborhood: The Solar System	7
Fig. 1-5	The asteroid approaches Earth	9
Fig. 1-6	Exoplanets discoveries through the years	11
Fig. 2-1	Hominin Timeline	23
Fig. 6-1	Climate change and Homo sapiens evolution	92
Fig. 6-2	Hominin species life-span	94
Fig. 6-3	The sun becomes a red giant	95
Fig. 6-4	Evolutionary change and adaptations over geological time	100
Fig. 6-5	Drought-driven peanut farm failure in Mozambique	105

Fig. 6-6	Drought-driven cattle die-off in Tanzania	106
Fig. 6-7	Locations of abandoned Mayan cities	108
Fig. 6-8	The Mayan city of Tikal	109
Fig. 6-9	Monks in Angkor's Ta Prohm temple	110
Fig. 6-10	Temperature profile of the Earth between 1885 and 1994	120
Fig. 6-11	This 2005-2014 profile shows how the temperature profile of the Earth has warmed since 1885	120
Fig. 7-1	The Milky Way Galaxy	136
Fig. 7-2	Potential habitats	139
Fig. 7-3	The New Planetary Habitability Index	145
Fig. 7-4	Earth changes over geological time	148
Fig. 7-5	Cyanobacteria (Phytoplankton)	152
Fig. 7-6	Historical Global Atmospheric Levels Graph	160
Fig. 7-7	A map showing levels of dissolved oxygen in the global oceans (a) and (b), how oxygen levels have declined or risen per decade since 1960.	161
Fig. 7-8	Dead zone spanning the Gulf 0f Mexico	162
Fig. 7-9	Cyanobacteria & phytoplankton	164
Fig. 7-10	Dynamic online forest alert system	165
Fig. 7-11	Global carbon dioxide emissions (1850–2000)	168
Fig. 7-12	800,000-year correlation between CO_2 and atmospheric temperature	169
Fig. 7-13	Visualization of peak temperatures in North Africa, the Middle East and northern India	173
Fig. 7-14	Ice-core data before 1958; Mauna Loa data after 1958	177
Fig. 8-1	The Survival Sustainability Cycle	184
Fig. 8-2	Biodiversity hotspots	196
Fig. 8-3	Views of Earth from space	200
Fig. 8-4	UN's Sustainable Development Goals (SDGs)	203
Fig. 9-1	Evolutionary survival patterns	216
Fig. 10-1	Technology growth	254

Fig. 10-2	Technology growth vs. time vs. population	255
Fig. 10-3	Sea Level Rise, Atmospheric CO2, Global Temperature-Differences and Arctic Sea Ice Minimum	256
Fig. 10-4	The Doomsday Clock	282
Fig. 10-5	Global trends in land and ocean temperatures 1850 - 2000	283
Fig. 10-6	Global temperature up to 2100	285
Fig. 10-7	Mean sperm count and global plastics production since 1970	297
Fig. 10-8	Map of global distribution of climatic zones	302
Fig. 11-1	Map of human migration	321
Fig. 12-1	Monarch butterfly on swamp milkweed in Michigan	340
Fig. 12-2	Species Survival Maturity Model	350
Fig. 13-1	Maturity/Self-Extinction Index	382
Fig. 15-1	Exploitation by colonialism	445
Fig. 15-2	African colonies after the Berlin Conference of 1884	446
Fig. 15-3	Share by country's population that is enslaved	450

LIST OF TABLES

Table 6-1 Longevity of some prior hominins 93

Table 7-1 A comparison of Earth with Mars, Venus, and the moons Titan, Ganymede, and Europa 141

Table 7-2 Gases comprising planetary atmospheres 146

Table 8-1 Ten most powerful countries' 2015 military spending 204

Table 10-1 Twenty-nine rules overturned 287

Table 10-2 Twenty-four rollbacks in progress 287

Table 13-1 Homo Sapiens Survival Scorecard 378

Table 13-2 An Evolutionary Maturity Model Scale for advanced civilizations 399

Table 15-1 Essential freedoms hunter-gatherers enjoyed compared to twenty-first-century Homo sapiens as a species and as individuals 475

INDEX

Adams, President John, 458
Adaptation, 213
adaptation schemes, downward spiral from, 240–242, 244–246, 250–251
agricultural, shift to, 218–219, 223, 224–229
anatomical adaptations, 83–84, 268
anthropocentric perspective and, 232–235
artificial habitat, development of, 227–229
artificial intelligence and, 85, 89, 237, 238, 240
civilizations, establishment of, 229–232
classical economic thought and, 232–233, 234
climate change-driven adaptations/innovations and, 117–119, 122, 225–226, 249
climate change-driven challenges and, 41–42, 188–190, 224
climate change, extinction/adaptation events and, 25–28, 41
complex animal/intelligent species, terraforming process and, 213
Earth's ecosystems, abandonment of, 218–219, 220, 226, 227
equation of life and, 217
evolutionary maturity and, 218, 248
fossil fuel consumption and, 223–224
Homo sapiens, superiority complex of, 218–219
hunter-gatherer populations and, 220–224
legacy of, 226–227, 239–242
limitations on, co-existence/mutual species survival and, 274
live-and-let-live reason for, 274, 275, 276, 277
megafauna, mass extinction of, 230–231, 240
mental modeling/belief systems, adaptations of, 232–235
migrations and, 26–27, 225–226
native economic freedom/independence, loss of, 218–219, 229–230
natural resources, exploitation of, 233
non-biological adaptations and, 274–275
photosynthesis-capable habitat and, 215

population explosion and, 225
prehistoric food crisis, megafauna extinction and, 230–232
religious beliefs/practices and, 232–234, 242–244
rules of survival-driven adaptation and, 211, 217, 218, 227, 241, 247–248, 251
scientific/technological progress and, 244–246, 277
self-extinction, potential for, 248, 277
species-driven adaptations and, 226–227
species survival-habitat survival-species survival loop and, 217
surplus harvests, social/political change and, 230
survival-driven adaptation and, 28–30, 46, 53, 215–220, 216 (figure), 227–229, 248, 251
survival-threatening events, adaptive change and, 41–42, 46, 218–220
sustainable-capable habitats and, 217–218
sustainable subsistence lifestyle, departure from, 218–219
writing, invention of, 242
See also Civilizations; Evolutionary maturity; Evolutionary survival patterns; Existential threats; Innovation
African-Americans, 322, 461, 505
Agriculture:
agribusinesses/corporate farming, destructive nature of, 206, 229, 238–239, 456
airborne organic chemicals and, 114
Ancestral Puebloan culture and, 111–112
animals, domestication of, 223, 225, 316, 460
artificial environments, development of, 228–229
bees/pollinators, loss of, 260, 299, 395
climate effects and, 101–103, 105–109, 110, 114
climate shift, flourishing small agriculture and, 105
crop failure and, 108, 126
crops, domestication of, 225
deforestation and, 107–108, 226, 228, 240
disease vectors and, 223
diversification in, 105
drought/flooding cycles and, 101–103, 105–109, 110
Earth's ecosystems, abandonment of, 218–219, 220, 226
enclosed lands and, 421, 440, 442
erosion/soil depletion and, 108
failing food production and, 220
fast-wood plantations and, 114
fertilizer use and, 227, 229, 240, 456
feudal caste system and, 235–237
flooding, Goldilocks condition and, 104
food chain, quality concerns and, 297–298
food production, shortfalls in, 229
genetically modified crops and, 240, 298
global temperatures, catastrophic rise in, 107
herbicides/insecticides/fungicides and, 206, 228, 240, 298, 395, 456, 489
intelligence, growing control by, 20–21
intensive farming techniques and, 225
irrigation systems and, 227, 228, 229, 240
methyl chloride dispersal and, 114
monsoon cycles, changes in, 104–107
profit motive and, 297–298
runoff, marine dead zones and, 228
shift to, 218–219, 223, 224–229, 231
slash-and-burn agriculture and, 226
soil erosion and, 227, 240
soil salinization and, 228, 229, 240
technological innovation and, 220
tools, development of, 225
transition to, 223, 224–229
water shortages and, 125–126
See also Chemical industry; Civilization; Modern humans
AIMM-E (adapt/innovate/mature/migrate-extinction) pattern, 310, 350 (figure), 356–357, 358, 436
AIMM-S (adapt/innovate/mature/migrate-survive) pattern, 310, 350 (figure), 356–357, 358, 436
AIMM-S/E (adapt/innovate/mature/migrate-survival/extinction) patterns, 310, 350 (figure), 356–357, 358, 436
Alvarez, Walter, 42
Amnesty International, 122
Amphibian species, 38, 197, 482, 526
Amsterdam Declaration, 194
Anders, Bill, 5, 137

Angielczyk, Kenneth D., 413
Animals:
amphibian species and, 38, 197, 482, 526
atmospheric oxygen, depletion of:
biological annihilation and, 406
Cambrian explosion and, 45, 69, 135, 151, 186
carbon dioxide, elevated levels of, 178
domestication of, 46, 223, 241, 2253
emergence of, 150
greenhouse mass extinctions and, 174–178
megafauna, extinction of, 230–231, 240, 277
metabolic rate of, 178
oceanic dead zones and, 162, 178
phytoplanktons/planktons and, 152, 157, 164–165, 164 (figure), 170
Smithian stage and, 177–178
Anthropocene, 341
Anthropocentric perspective, 232–235, 437
Anthropogenic effects. *See* Adaptation; Climate change; Climate machinery; Existential threats; Innovation
Artificial intelligence (AI), 85, 89, 237, 238, 240, 242, 254, 255, 257, 262, 269–270, 362, 453, 456
See also Existential threats; Innovation
Asimov, Isaac, 69, 290, 291, 292
Asteroid impact, 9 (figure), 10, 13, 24, 84, 200, 359, 407
Astronomical unit (AU), 7
Atmosphere:
aerosol particles and, 114, 115
anaerobic species, mass extinction of, 153–154
atmospheric gases, life-forms' influence on, 145, 312
bacterial ancestors/cyanobacteria, role of, 150, 152–154, 152 (figure)
carbon dioxide-based atmosphere, transformation to, 159–160, 160 (figure)
carbon dioxide in, 114, 153, 154, 155–157, 168, 170
carbon dioxide/oxygen levels, inverse relationship of, 158–159, 160
deforestation and, 114
dynamic disequilibrium and, 145, 157–158
early Earth atmosphere and, 153, 154–157
El Niño-Southern Oscillation cycle and, 116
global temperatures, radiated energy and, 121
greenhouse effect, fossil fuels consumption and, 29
greenhouse gases and, 29, 36, 96, 121, 159–160
greenhouse mass extinctions and, 174–179
Historic Global Atmospheric Oxygen Levels Graph and, 160, 160 (figure)
methane and, 115, 154, 175, 239
methyl chloride, dispersal of, 114
negative/positive feedback systems and, 167–168, 170
organic chemicals and, 114
oxygen and, 44–45, 153–154, 155–157, 159–160, 163, 167, 215
oxygen levels, change in, 157–158, 163, 286
photosynthesis, role of, 138, 151, 153, 157, 163, 167
phytoplanktons, atmospheric oxygen and, 157, 163, 164–165, 164 (figure), 170
planetary atmospheres, gaseous components of, 145–146, 146 (table)
Planetary Habitability Index and, 145, 145 (figure)
positive feedback cycle, switch to, 168–170, 168–169 (figures)
silicate rock, weathering of, 166
See also Carbon dioxide; Habitats; Oxygen
Atomic Energy Commission, 245
Avatar, 214, 389

Bacon, Francis, 234
Bacterial ancestors, 149–150, 152–154, 152 (figure), 213–214
Bees/pollinators,260, 299, 395
Benner, Steven, 312
Benton, Michael J., 167, 168, 412
Berger, Lee, 524
Berlin Conference of 1884, 446
Bicentennial Man, 64, 70, 88, 329, 403
The Bicentennial Man, 69
Big Pharma, 458, 459-460, 475, 504, 506
Bill and Melinda Gates Foundation, 460
Biodiversity, 194–196, 200–201, 205–208, 217,

372, 407
Biodiversity hotspots, 196–199, 196 (figure)
Biomimicry, 72, 73
Bionic life-forms, 85, 88
Bolsonaro, President Jair, 271
Bostrom, Nick, 257
Bowring, Sam, 412
Breakthrough Initiatives, 305
Brokaw, Tom, 220, 226
Broswimmer, Franz, 113
Brownlee, Donald, 54, 135
Bulletin of the Atomic Scientists, 281, 282, 293

Cambrian explosion, 45, 69, 135, 151, 186
Cameron, James, 214, 389
Capitalism:
- agricultural practices and, 220, 297–298
- alpha species status of, 237
- artificial intelligence and, 85, 89, 237, 238, 240, 242, 453
- benefits from, 236
- Big Pharma and, 458–460
- bio-capitalists and, 421
- capitalist system, abandonment of, 420–421
- commoditization trend and, 238–239, 387–388, 421
- consumption imperative and, 237
- corporate citizenship and, 272
- deforestation, arms deals/civil warfare and, 113
- Earth's biodiverse resources, exploitation of, 196–199, 196 (figure), 200–201, 205–208, 233
- Earth's ecosystem, destruction of, 48, 196–199, 196 (figure), 218–219, 227, 238–242
- economic development, foreign aid/ investment/trade and, 447
- evolutionarily-immature species and, 388–389
- existential threats and, 433
- feudal caste system and, 235–236
- free market forces and, 201, 219, 234–235, 237–238, 239, 240, 241–242, 388–389
- freedom from, 441–442, 443
- genetic material, patenting of, 421
- global economic institutions, autonomous nature of, 238–239
- global markets, ecosystem degradation and, 227, 471–472
- Global South nations, pillaged natural resources and, 445
- government, weakening of, 238
- *Homo sapiens* adaptations, failure of, 250–251
- *Homo sapiens*, demotion of, 237–238
- human extinction and, 420–421
- humanity's needs, reallocated funds for, 272
- hunting practices and, 231, 395
- income inequality and, 201–202, 258–259, 271
- industrial progress/growth and, 205–208
- institutionalized inequalities/racism and, 241
- invisible hand of economic power and, 238–239, 391, 421
- liberal economic thought and, 237, 241–242
- market-based authority, emergence of, 236
- mechanization trend and, 452, 453
- medical-industrial complex and, 458–460
- modern humans, market-driven adaptations and, 228–229, 232
- modern liberal economic thought and, 237
- multinational corporations and, 219, 227, 392, 433, 444, 445, 447, 448–449
- nationalized industries and, 238
- offshoring trend and, 452
- organisms, programmable manufacturing systems of, 421
- private enterprise and, 387
- privatization trend and, 238
- profit/greed motives and, 48, 81, 196–199, 200–201, 206, 207, 220, 231, 237, 251, 259, 263–264, 290, 297–298, 363, 376, 387, 421, 433
- rational self-interest/virtuous selfishness and, 201
- resource distribution and, 201
- Rust Belt cities and, 237
- shareholder profit, legal responsibility for, 459
- unemployment, pain of, 236–237
- unrestrained capitalism, regulatory efforts and, 240

wealth, pursuit of, 80, 81, 202, 206, 207, 220
workers, exploitation of, 376
See also Chemical industry; Economic systems; Energy industry; Essential freedoms; "The Few" power elites; Slavery

Carbon cycles:
atmospheric carbon dioxide and, 170
long-term carbon cycle, 169–170, 487
short-term carbon cycle, 168, 169

Carbon dioxide:
atmospheric carbon dioxide levels and, 114, 154, 155–157, 168–170, 168–169 (figures)
carbon dioxide-based atmosphere, transformation to, 159–160, 160 (figure), 164–165
Carboniferous period ice age and, 166
deforestation and, 114, 115, 165–167, 165 (figure)
early Earth atmosphere and, 153, 154, 155–157
ecosystem disturbances and, 114–115, 121
Eocene Epoch hothouse conditions and, 119
fossil fuel consumption and, 160, 168
global surface temperatures, rise in, 172–174, 173 (figure)
global warming trend and, 159, 168, 170, 172, 187
greenhouse mass extinction events and, 174–179
high-carbon dioxide world, viability of life and, 158–159, 165
long-term carbon cycle and, 169–170
oxygen/carbon dioxide levels, inverse relationship of, 158–159, 160, 174
photosynthesis, role of, 138, 151, 153, 157, 163, 167, 169
positive feedback cycle and, 168–170, 168–169 (figures)
short-term carbon cycle and, 168, 169
silicate rocks, weathering of, 166, 195
See also Atmosphere; Habitats; Oxygen

Carbon Dioxide Information Analysis Center, 168
Carlson, Richard, 312
Carroll, Sean, 18, 19, 20, 414
Caste systems, 235–236
Ceballos, Gerardo, 406
Center for American Progress on Climate Change, Migration and Conflict, 333
Centre for Evidence-Based Medicine, 459
Chapin, Harry, 290

Chemical industry:
agricultural applications and, 228
birth defects and, 296
chlorofluorocarbons and, 280
developmental disorders and, 296
emergence of, 297
endocrine disrupting chemicals and, 295–296
fertilizers and, 227, 229, 240, 294, 456
forever chemicals and, 274
glyphosates and, 298
herbicides/insecticides/fungicides and, 206, 208, 240, 294–295, 298, 395, 456, 489
per-/polyfluoroalkyl substances and, 274
phthalates and, 296–297, 297 (figure)
pollution issues and, 80, 294
sperm count reduction and, 296–297, 297 (figure)
toxic chemical pollutants and, 294–298, 297 (figure)
weaponized chemical agents and, 48, 245, 298
See also Agriculture; Energy industry; Existential threats; Modern humans

Chlorofluorocarbons (CFCs), 280
Choi, Charles, 104
Choice. *See* Freedom to choose
Chomsky, Noam, 273
Christianity. *See* Religion; Western Christian nations

Civilizations:
advanced technological civilizations, self-destruction of, 12–13, 255, 256 (figure), 257–262
agriculture, shift to, 229–232, 235
Akkadian Empire, 102
Ancestral Puebloan culture and, 102–103, 111–112
Angkor Wat/Khmer Empire, monsoon effects and, 110–111
climate shifts, exceptional historical events and, 101–115, 116

collapse of, 101–103, 105–112, 258, 300–301
deforestation, climatic effects of, 107–108, 113–114
development of, 21
Doomsday Clock and, 281–282
drought, civilization collapse and, 101–103, 105–112, 300–301
early civilizations, food distribution systems and, 230, 235
ecocidal activities and, 113–114
El Niño climatic pattern and, 105–107, 110
feudal caste system and, 235–236
food insecurity/starvation and, 106–107, 113
Harappan civilization, collapse of, 104–107
Inca Empire and, 102
innovation vs. evolutionary maturity and, 74–75, 247, 257–258, 260, 261–262
Khmer Empire and, 102, 110–111
Late Bronze Age civilization and, 102
Mayan civilization, 102, 107–109, 108–109 (figures), 244
Ming Dynasty and, 103
Moche civilization, drought/flooding and, 109–110
monsoon cycles, shifts in, 104–105, 110–111
native economic freedom/independence and, 218–219, 229–230
natural resources, exhaustion of, 235
slavery, role of, 276–277
Tang Dynasty and, 102
Tiwanaku Empire and, 102
urbanism, fading of, 105
See also Agriculture; Capitalism; Ecolization process; Economic systems; Essential freedoms; "The Few" power elites; *Homo sapiens*; Migrations; Modern humans
Climate change, 91
adaptations to, 98–99
agricultural activities and, 114
albedo effect, impact of, 114
anthropogenic global warming and, 125
atmospheric greenhouse effect and, 29, 36, 96
climate change-driven adaptations/ innovations and, 117–119, 122, 127–128
climate chaos and, 115–117
climate-refugee crisis and, 118, 190
climate shifts, exceptional historical events and, 101–115
current rate of, 96
cyclic climate change, evolutionary sustainability and, 188–189
deforestation, effects of, 107–108, 113–114
drought, civilization collapse and, 101–103, 105–112, 115
Earth's axis tilt, shifting of, 98
ecocidal activities and, 113–114
European cities, breaking point for, 260
evolutionary leaps and, 29
evolutionary maturity, movement toward, 91–92
evolutionary survival patterns and, 28–30, 92
extinction/adaptation events and, 25–30
extinction, threat of, 125–127
food insecurity/starvation and, 106–107, 109, 115
global issue of, 123
global response to, 122
global warming and, 74, 80, 81, 84, 96, 115, 119–122, 120 (figures), 125–127, 159
Goldilocks period and, 104
Great Drought and, 103
greenhouse gases and, 29, 36, 96, 121, 239, 260
habitable zones, destruction of, 94–95, 320
hominin species longevity and, 92–93, 92 (figure), 93 (table), 101
Homo sapiens evolution and, 92–93, 92 (figure), 96–101, 100 (figure)
hothouse conditions and, 29–30, 119
human precipitation of, 92, 96, 98
human vulnerability to, 20, 29–30, 96, 121–128
ice ages and, 23, 24, 25, 26–27, 28, 29
intelligent species evolution and, 92 (figure), 96–101, 100 (figure)
interplanetary migration and, 96
long-term climate change and, 119–122, 120 (figures)
mass extinction event, modern survivors

of, 127
monsoon cycles, shifts in, 104–105, 110–111, 117
Neanderthal extinction and, 23, 28
nuclear winters and, 26
out-of-Africa migrations and, 26–27, 320
population growth/density and, 123–124, 126
sedimentary evidence of, 27
self-extinction, potential for, 92, 94
short-term climate change and, 101, 105–107, 110, 115–117, 122, 125
supply chains, disruption of, 124
survival/sustainability challenges and, 29–30, 95–96
sustainability pressures and, 115–117
temperatures, catastrophic rise in, 107, 119–122, 120 (figures)
urban population density and, 124
water shortages and, 115, 125–126
See also Civilizations; Climate machinery; Evolutionary survival patterns; Existential threats; Global warming; Weather patterns
Climate Change 2014 Synthesis Report Summary for Policymakers, 283
Climate machinery:
anthropogenic warming/habitat destruction and, 125–126
civilization collapse and, 122–123
climate change-driven adaptations/innovations and, 117–119
deforestation and, 107–108
Earth's orbit-climate change connection, gravitational fluctuations and, 187–190
El Niño climatic pattern and, 101, 105–107, 110, 115–117
El Niño-Southern Oscillation and, 115, 116
food insecurity/starvation and, 117
global heat, buildup of, 121
greenhouse/icehouse states and, 119
La Niña climatic pattern, 101, 112, 115–117
long-term climate change and, 119–122, 120 (figures)
meta-stable periods and, 119
monsoon cycles and, 101, 104–105, 110–111, 117
Pacific wind patterns and, 116
positive feedback systems, activation of, 121
tectonic activity and, 97–98
trade winds, weakening of, 116
volcanic eruptions, nuclear winters and, 26, 97
See also Agriculture; Civilization; Climate change; Weather patterns
Clinton, President Bill, 164
Cohen, Mark, 230, 231
Cold War, 246, 282
Colonialism, 322, 421, 444–447, 445–446 (figures), 452
Common good, 201
Consciousness, 2
Continental drift, 42, 54, 315
Cook, Tim, 304
Cox, Brian, 8, 253
CRISPR gene-editing tool, 72, 73, 290–291, 467–468
Crowther, T. W., 166
Cuban missile crisis, 47, 255, 283, 381–382, 382 (figure), 383, 384, 392
Cyanobacteria, 149–150, 152–154, 152 (figure), 164, 187, 195, 215, 218, 312, 313
Cyber warfare, 86, 279, 283
Cybersecurity, 86

da Vinci, Leonardo, 245
Dawson, Ashley, 421
Dead zones:
Gulf of Mexico and, 162, 162 (figure), 228
ocean/coastal dead zones, 162, 239
Default outcome. *See* Evolutionary default outcome
Deforestation, 29, 107–108, 113–114, 165–167, 165 (figure), 179, 215, 228, 240
deMenocal, Peter, 98, 99
Denisovans, 22
Descartes, René, 234
Desertification, 189
Diamandis, Peter H., 258, 382
Dickens, Charles, 236
Dinosaur extinction, 10, 13, 42, 186, 313, 342, 411
Disease:
agriculture, early transition to, 459–460

chronic disease rates and, 297
civilizations, collapse of, 113
climate change pressures and, 126
community mobility and, 223
endocrine disrupting chemicals and, 295–296, 298
epidemics and, 223
food chain crisis and, 293–294
hunter-gatherer populations and, 222–223
infectious diseases and, 222
intestinal parasites and, 222
malnutrition/deficiency diseases and, 222
medical-industrial complex/Big Pharma and, 458–460
medical practice, harms from, 458–459
modern food production and, 297–298
nanobots and, 256
obesity and, 297
oxygen energy metabolism, failure of, 160–161
patented medicines and, 207
pesticide-induced cancers and, 298
water-related diseases, 303
Djoghlaf, Ahmed, 372
Doomsday Clock, 281–282
Doudna, Jennifer, 72
Drones, 258
Drought:
civilization collapse and, 101–103, 105–112
El Niño effects and, 117
existential threat of, 300–303, 302 (figure)
food shortages and, 202
Last Glacial Maximum and, 189
sustainability pressures and, 115, 202, 279
water shortages and, 115, 279
See also Civilizations; Climate change
Dylan, Bob, 210

Earthrise photo, 4, 5–6, 80
Earth's ecosystems, 137
anthropocentric perspective and, 232–234
destruction of, 48
Earth-sun geometry, sunlight exposure and, 188
Earth's evolution and, 148–149, 148 (figure)
Earth's orbit-climate change connection, gravitational fluctuations and, 187–190
Earth's orbit-driven climate change, solar/planetary phenomenon of, 188–190
ecocide/ecocatastrophe and, 113–114, 198–199, 200, 201, 286, 421
equation of life and, 217
habitability, requirements for, 138–141, 141–143 (table)
habitat destruction threat and, 42
Homo sapiens disregard for, 218–219
hunter-gatherer populations and, 220–221
long-term carbon cycle and, 169–170, 487
military expenditures, resource exploitation and, 202–204, 203 (figure), 204 (table)
preservation of, 199–201, 200 (figure)
progenitors of, 149–150
sunspot activity, mini ice ages and, 188
Survivability and Habitat Sustainability Test and, 293, 366, 374–375, 386, 469, 504
sustainable-capable habitats and, 128, 133–134, 135, 137
terraforming process and, 144, 149, 206, 213, 335
See also Atmosphere; Carbon dioxide; Earth's evolution; Ecolization process; Evolutionary sustainability; Great Oxidation/Oxygenation Event; Habitats; Oceans; Oxygen; Sustainability
Earth's evolution, 5, 6
asteroid impact and, 9 (figure), 10, 13, 24, 84, 200, 359, 407
bacterial ancestors/cyanobacteria, role of, 149–150, 152–154, 152 (figure), 213–214
complex animal/intelligent species life, emergence of, 14–15
early Earth habitat, characteristics of, 148–152, 148 (figure)
Earth-like exoplanets, habitable zone and, 10–12, 11 (figure), 36
Earth, location of, 9, 34, 36, 40–41
Earthrise photo, collective self-awareness and, 5–6, 80
Earth's existence, precarious nature of, 6
emergent technological species and, 9, 10–11
extraterrestrial civilizations, disappearance

of, 12–13
extraterrestrial intelligent life and, 8–9, 12
Great Filter theory and, 12
habitat destruction, threat of, 42, 44
Homo sapiens, long-term survival of, 13–14
intelligent life, evolution of, 12, 13, 14
Late Heavy Bombardment period and, 52, 148–149
Milky Way Galaxy and, 10–11
ozone layer, role of, 40, 72, 80, 280
photosynthesis, role of, 138, 163, 169, 178, 195, 214
phytoplanktons, role of, 152, 157, 163, 164–165, 164 (figure), 170, 215
plate tectonics/continental drift and, 42, 54, 97–98, 186, 315, 319
purposeful evolution and, 9–10
single-cell life, emergence of, 12
solar system, Earth's neighborhood and, 7–8, 7 (figure), 10
space exploration and, 7–8
supernovae explosions and, 40
technology, immersion in, 8
trees, energy/oxygen-supply grid and, 214–215
See also Earth's ecosystems; Intelligent species survival; Outer space; Species survival
Ecocidal activities, 113–114, 198–199, 200, 201, 286, 421
Ecolization process, 28–29, 274, 481–482
adaptation sequences, continuing problems and, 489
agriculture, transition to, 487, 491, 502, 512
American Constitution, non-survival-driven law and, 481–482
capitalist economic system, antithesis to survivability/sustainability and, 510–511
change strategies and, 504–514
civilization collapse and, 492–493, 494
definition of, 491
ecolization-based civilization, building of, 494, 500–501, 502–504, 514
ecolization vs. civilization comparison and, 494–499
ecosystem/habitat-based law, species living within, 486, 486–487, 491, 502–504
ecosystems, role of, 485–486
energy flow/nutrient recycling, habitat/ecosystem survival and, 484–485
evolution, role of, 512–514
evolutionary maturity, unrealized goal of, 490–491
"The Few", power of, 492, 502, 504, 512
fossil fuel consumption, shortened carbon cycle and, 487
freedom to choose, regaining of, 494
freedom to choose, usurpation of, 491–492
galaxies, existence/evolution of, 485
habitable zones and, 485
Homo sapiens, ecosystem/habitat-based law violations and, 486, 487, 493
Homo sapiens, survival vs. profits/economic growth motivations and, 496–498, 502
human capitalism, hope of, 511
individual efforts for transformation and, 512
innovation-maturity imbalance and, 489–490
innovations, exponential acceleration of, 489–490
intelligent species-wide civilization outcomes and, 499
live-and-let-live ethos and, 274, 275, 276, 277, 369, 370, 493–494
local ecosystems, *Homo sapiens* re-entry into, 487–488
nonprofit organizations and, 510
paradigm shifts and, 508
personal transformation, species-wide preference and, 512
political change, nationalism/strongman politics and, 509–510
protests as we have known them, failure of, 504–507, 509, 512
religion, Earth's devastation and, 501, 510
science/innovation-based transformation and, 511
self-extinction, potential for, 490, 491, 493, 500, 502
survival-based decision process and, 491–492
survival-driven patterns/rules and, 482–484
world war-driven transformation and,

508–509
world-wide species survival-driven protests and, 504, 506–508, 512
Economic systems:
biodiverse resources, exploitation of, 196–199, 196 (figure), 200–201, 205–208
classical economic thought and, 232–233, 234
climate change response and, 122
commoditization of species/natural resources and, 387–388
corporate citizenship and, 272
defense/military expenditures and, 202–204, 203 (figure), 204 (table)
deforestation, social/economic consequences and, 113–114
evolutionarily-immature species and, 388–389
exclusionary economic zones and, 271
existential threats, global economy and, 283
feudal caste system and, 235–236
feudalism and, 235–236, 444
"The Few" vs. "The Rest", power dynamics and, 201–202, 204–205, 207, 208
global economic institutions, autonomous nature of, 238–239
global income inequality and, 201–202, 258–259
government social programs and, 236
industrial economy, transition to, 236–237
invisible hand of economic power and, 238–239
liberal economic thought and, 237, 241–242
market-based authority and, 236
monetized biodiversity and, 196–199, 200–201, 219
money-based economy, shift to, 236
Moneyed sapiens and, 281, 304, 329, 377
native economic freedom/independence and, 218–219, 229–230
natural disasters, losses from, 118
politically based authority and, 236
rational self-interest/virtuous selfishness and, 201
See also Agriculture; Capitalism; Civilization; Essential freedoms
Ecosystems:
atmospheric gases, imbalance of, 114–115
deforestation and, 113–115
fast-wood plantations/commercial forests and, 114
natural forests and, 114
society of organisms, interactions within, 18, 19
tree planting efforts and, 114
See also Earth's ecosystems; Ecocidal activities; Habitats; Sustainability
Einstein, Albert, 82, 210, 340, 344, 347
Eisenhower. President Dwight D., 245
El Niño climatic pattern, 101, 105–107, 110, 115–117, 122, 190, 301
El Niño-Southern Oscillation (ENSO) cycle, 115, 116, 301
Eldredge, Niles, 371
Empathy, 47
Endangered species. *See* Extinctions; Human extinction
Endocrine disrupting chemicals (EDCs), 295–296, 298
Energy industry:
bio-based technologies and, 286
environmental degradation and, 207
exploitive/destructive nature of, 206–207
fossil fuel consumption and, 29, 114, 160, 163, 168, 207, 223–224, 271, 286, 297, 433, 470, 471, 472, 473, 487
fusion technologies and, 286
habitat destruction, choice in, 48
mining/drilling/logging operations and, 286, 472
renewable energy sources and, 206, 207, 246, 286, 469–470
See also Chemical industry; Existential threats
Equation of life, 217, 335, 487, 490, 514
Essential freedoms, 439
agriculture, transition to, 440, 441, 442, 444, 449, 450, 451, 459–460, 477
artificial intelligence revolution and, 453, 455–456
authoritarian despotic leadership and, 461–462, 463, 464, 472, 473
biodiversity, brinkmanship with, 466
bioengineering, peril of, 467–468
capitalist system and, 443–444, 445,

448–451, 452, 457–458, 465
civilization, cycles of economic dependencies and, 441, 442–443
climate change, citizens suing government and, 473
climate change-driven extinction, avoidance of, 471–474
colonialism, exploitative practices and, 444–448, 445–446 (figures), 452
descriptive assessment of, 474–476, 475 (table)
economic development, foreign aid/investment/trade and, 447–448
enclosure of lands/animals/people and, 44, 440, 442, 444, 449
exploitation/abuse and, 442–443, 445–448, 445–446 (figures), 452
feudalism, establishment of, 444
"The Few", power of, 442, 443, 445, 448, 451, 453, 462–464, 469, 472, 474, 476, 477, 479
free time, trading away of, 440
freedom/ability to migrate on Earth and into space, 474
freedom from Big Pharma/medical-industrial complex and, 458–460
freedom from politics/government/rule of law, 460–464, 472, 473
freedom from profit-driven/extinction-capable inventions and innovations, 465–469
freedom from religions/gods/superstitions, 464–465
freedom from stuff/big business and commerce, 457–458
freedom of economic independence/self-sufficiency, 441–451, 452
freedom to choose action against climate change-driven extinction, 471–474
freedom to choose healthy foods, 456–457
freedom to choose to exist freely, 439–441
freedom to choose/use renewable energy sources, 469–470
freedom to live within Earth's ecosystems protecting species/habitats, 451–456, 460, 466, 467
individual freedoms, transfer of, 434, 477–479
Intelligent Industrial Revolution and, 455
location of work, external determination of, 440
means of knowledge, preservation of, 458
mechanization trend and, 452, 453
medical-industrial complex/Big Pharma and, 458–460
migration patterns and, 440–441, 474
multinational corporate colonialism, pillaged natural resources and, 444, 445, 447
multinational corporations, exploitive practices and, 448–449
neocolonialism and, 447
self-extinction, potential for, 466, 467, 470, 471–474
slavery/forced labor and, 442, 449–450, 450 (figure), 452, 461
technology/innovation, twenty-first century religion of, 465–467
wage labor and, 444, 445, 447, 449, 452
work, evolution of, 454–456
work schedules, extension of, 440
working for a living and, 440, 441, 442, 444, 445, 447, 449, 450, 451–452, 453
See also Freedom to choose; Freedoms; Hunter-gatherer populations
Eukaryotes, 53, 69, 150
European Geophysical Union, 194
Evolution:
Cambrian explosion and, 15, 45, 69, 151, 186
change, evolutionary survival selection mechanism of, 29
chaos, order within, 2, 52, 53
complex animal/intelligent species life, incubation of, 14–15
Earth's environmental/climatic conditions and, 25–28
eukaryotes, plant/animal evolution and, 69
extraterrestrial species evolution and, 8–9, 12
hominin evolution, climate change and, 28–30
human mind, evolution of, 71–72, 75–79, 82–83
intelligent-species-driven evolution and, 57
intelligent species, environmental/climatic factors and, 25–28

intelligent species survival, evolutionary rules for, 129–130
intelligent technological species, evolution of, 12
last universal common ancestor (LUCA) and, 68–69
life/death and evolution, co-opetition between, 1–2
natural selection and, 27–28, 73, 82
progenitor species in, 149–150
purposeful evolution and, 2, 10
repeating universal pattern and, 2
single-cell prokaryotes, transition from, 52
species survival, default evolutionary outcome and, 1, 38, 39, 43, 48, 53, 185
survival-driven adaptation and, 28–30
survival, evolutionary factors in, 24
Universe's existential threats and, 2
See also Adaptation; Earth's evolution; Innovation; Intelligent species evolution; Intelligent species survival; Species-driven evolution; Species survival; Survival
Evolutionarily-immature species' world, 386
authoritarian despots and, 389–390, 392
capitalism, concept of, 388–389
commoditization of species/natural resources and, 387–388
common human views/practices and, 386–397
democratic governance and, 393
genocide and, 77, 210, 390, 391
habitat sustainability/species survivability, betrayal of, 394–395
income inequality and, 388
inequitable resource distribution and, 391
law enforcement/prison systems and, 391
migration, barriers to, 395
military-industrial complex and, 245–246, 258, 391
nuclear threats and, 390
ownership, concept of, 386–387
religious beliefs/practices and, 393–394
ruling class, need for, 391–393
sexes, inequality of, 396
sixth mass extinction and, 403–404
Survivability and Habitat Sustainability Test and, 386
want vs. need, greed/wealth accumulation and, 387
warfare, engagement in, 389–391, 392
worker replacement, artificial intelligence/robotics and, 391, 453
See also Evolutionary maturity; Modern humans
Evolutionary default outcome, 1, 38, 39, 43, 48, 53, 185
Evolutionary maturity, 367–368
advanced civilizations, classification systems for, 397–399, 399 (table)
characteristics of, 373–376
choice, capacity for, 368, 369–370, 373
clear model of, 58–59
climate change-driven extinction threat and, 372
definition/description of, 21, 91, 369–370, 373
edible energy sources, destruction of, 372–373
empathic awareness and, 47
energy production, photosynthesis and, 370–371, 372
Evolutionary Maturity Scale and, 369, 398–399, 399 (table)
evolutionary survival patterns and, 57, 58
"The Few", power of, 370, 376
fully matured level and, 354
habitat survival/health and, 373, 374
Homo sapiens Survival Scorecard and, 377–380, 378 (table)
human-caused extinctions and, 371–372, 372
human-driven destruction/exploitation and, 375–376, 381–383
human mind, evolution of, 71–72, 75
Industrial Revolution, declining evolutionary maturity and, 384
innovation-maturation imbalance and, 74–75, 247, 257–258, 260, 261, 383–384
innovation, survival intent and, 385
intelligent species, purposeful evolution of, 2, 10
intelligent species survival, prerequisites for, 368, 373, 374–375, 377
interplanetary migration and, 147–148, 311, 312–314, 370, 402–403
live-and-let-live ethos and, 274, 275, 276, 277, 369, 370

local ecosystems, living within, 370–373, 375–376, 380, 384
maturity scale and, 369, 398–399, 399 (table)
The Maturity/Self-Extinction Index and, 380–385, 382 (figure)
migration patterns and, 372
modern humans, stagnant evolutionary maturity and, 21, 205–208, 209–210, 317, 371–372, 377–380, 378 (table), 380–385, 382 (figure)
new evolutionarily mature civilizations, transformation to, 400–402
parent-offspring pattern and, 59–60
positive outcomes from, 47, 48
post-agricultural revolution humans and, 371–372, 383–384
self-extinction, potential for, 21, 46, 47, 57, 58, 311, 368, 370, 371, 380–385, 382 (figure)
species survival and, 46–47, 57, 58–59, 313–314, 370–374
Species Survival Maturity Model and, 13, 14, 15, 36, 39, 46, 47, 373, 377
species survival strategies, promotion of, 375–376
Survivability and Habitat Sustainability Test and, 374–375
survival-first culture and, 399
survival intent, innovation and, 385
Survival Plateau and, 92, 266, 368, 369, 370, 377
sustainable survival strategies and, 370–373
See also Climate change; Evolution; Evolutionarily-immature species' world; Evolutionary maturity; Existential threats; Innovation; Intelligent species evolution; Species-driven evolution; Species Survival Maturity Model (SSMM)
Evolutionary survival patterns, 190, 310, 366
AIMM-E pattern and, 310, 350 (figure), 356–357, 358
AIMM-S/E patterns and, 310, 350 (figure), 356–357, 358
AIMM-S pattern and, 310, 350 (figure), 356–357, 358
climate-change-driven challenges and, 25–28
See also Adaptation; Evolutionary maturity; Innovation; Intelligent species survival patterns; Migration; Sustainability
Evolutionary sustainability, 186–187, 192
biodiverse species/sustainable-capable habitats, synergistic function of, 194–196
biodiversity, symbiotic/interdependent relations and, 194
climate change cycles and, 185–186, 188–189
destructive impulses/habits and, 196–199, 200
ecocidal activities and, 113–114, 198–199
ecosystems, redundancy/interdependence and, 195
eukaryotic cell/complex animals, evolution of, 195
food webs/chains, development of, 195
global nature of, 192
habitat, care/feeding of, 185
interplanetary migration, unlikely event of, 193–194
migrations, role of, 189, 190
military expenditures and, 202–204, 203 (figure), 204 (table)
mutual survival, ensuring of, 274
photosynthetic life, free oxygen and, 195
self-regulating systems, habitability/species survival and, 194
species survival and, 183–184, 194, 195, 196
survival-driven design pattern and, 195
sustainable-capable habitats and, 186, 194–195
water, accumulation of, 195
See also Evolutionary survival patterns; Sustainability
Existential threats, 246, 277, 279–281
artificial intelligence and, 288–293
atmospheric oxygen, depletion of, 286
average climate conditions, radical departure from, 284–285, 285 (figure)
bees/pollinators, extinction of, 260, 299
carbon dioxide levels and, 285
chemical toxicity and, 294–298, 297 (figure)

climate change and, 283–288, 283 (figure), 285 (figure), 287 (tables), 301–302
cyber warfare and, 283
Doomsday Clock and, 281–282
drought conditions and, 300–303, 302 (figure)
ecocidal activities and, 113–114, 187–188, 200, 201, 286
endocrine disrupting chemicals and, 295–296, 298
environmental protections, rollback of, 286, 287 (tables)
evolutionary-driven survival, necessity of, 292
"The Few", intractable power of, 286, 288
flood events and, 283
food chain crises and, 293–299, 297 (figure)
food shortages and, 283
gene editing technology and, 290–291, 467–468
genetically modified organisms and, 298, 299
global economy, reinvention of, 283
global freedom, loss of, 432–434
global surface temperatures and, 283 (figure), 284–285, 285 (figure), 301–302
greenhouse gas emissions and, 283, 284, 285
Homo sapiens, threat of, 278–281, 289–290, 291, 354–355
Laws of Robotics and, 290, 291–292
natural disasters and, 284, 303
nuclear threats and, 282–283, 282 (figure), 288
ocean warming and, 279, 283 (figure), 285–286
persistent organic pollutants/toxic chemicals and, 294–298, 297 (figure)
phthalates and, 296–297, 297 (figure)
profit/greed motivations and, 290, 291, 298
refugee crises and, 283, 284
reproductive technologies, need for, 296
social media, privacy invasion and, 290
sperm count reduction and, 296–297, 297 (figure)
warming trend and, 284, 285–286
water pollution and, 294
water shortages and, 285, 300–303, 302 (figure)
See also Extinctions; Human extinction; Innovation; The Maturity/Self-Extinction Index
Extinction Rebellion, 508
Extinctions:
anaerobic species, extinction of, 153–154, 156
Anthropocene Epoch and, 44, 125
colonization, native population near-extinctions and, 322, 421
dinosaur extinction and, 10, 13, 42
emergent species, difference of, 128
extraterrestrial civilizations and, 12–13
global temperature rise and, 119–122, 120 (figures)
great filter theory and, 12
greenhouse mass extinctions and, 174–179
habitat destruction and, 135
home world, location of, 40–41
hominin species and, 126, 135
Homo sapiens-induced global extinction and, 263
ice ages, threat from, 26
inevitability of, 33–34, 43, 48, 53
mass extinction events and, 2, 10, 13, 42, 44, 107, 127, 148, 153, 156, 157, 164
maturity and, 46
megafauna and, 230–231, 240, 277
normal occurrence of, 21–22, 33–34
oxygen energy metabolism, failure of, 160–161
Permian-Triassic mass extinction and, 107, 127, 158, 160, 190, 286, 341
phytoplankton, role of, 164–165, 164 (figure)
Quaternary extinction event, 27
self-extinction, potential for, 21, 46, 47, 57, 74–75, 86–87
uncontrollable extinction-causing events and, 39–43
See also Climate change; Evolutionary survival patterns; Existential threats; Great Oxidation/Oxygenation Event; Human extinction; The Maturity/Self-Extinction Index

Extraterrestrial intelligent species evolution, 8–9, 12, 53, 54, 265, 267, 349, 359

Fagan, Brian, 103, 110, 115, 116, 124, 126, 301
Famine Early Warning Systems Network (FEWS NET), 107
Famines, 106–107, 109, 115, 117, 126, 459
Fermi, Enrico, 12, 56
Fermi Paradox, 12
Feudalism, 235–236, 444
"The Few" power elites, 201–202, 204–205, 207, 208, 259, 286, 288, 331, 370, 376, 388, 421, 422, 434, 435, 438, 442, 445, 462–464, 469, 472, 474, 476, 477, 479
Flora and fauna. *See* Agriculture; Animals; Plants
Food chain:
  carbon-to-carbon dioxide conversion and, 169
  chemical toxicity and, 294–298, 297 (Figure)
  climate-driven/environmental changes and, 41, 106–107, 115, 125–126
  cropland salinity and, 229
  drought and, 202
  ecosystems, interactions within, 18
  El Niño effects and, 117
  existential crisis in, 293–299, 297 (figure)
  food scarcity and, 189, 202, 223, 283, 297, 371, 451
  food waste and, 297
  food webs, species survival and, 413–415
  genetically modified organisms and, 298, 299
  Green Revolution and, 294, 298, 299
  modern food production, shortfalls in, 229
  multinational corporations and, 227
  nutritional content, reduction in, 298–299
  organic food and, 271
  ownership concept and, 387
  pesticides within, 294–295
  photosynthesis, role of, 215
  phytoplanktons/planktons, oceanic food chain and, 164–165, 164 (figure)
  poisoning of, 80, 274, 433, 456
  prehistoric food crisis and, 230–232
  quality concerns and, 297–299
  Stone Age foragers, freedom to migrate and, 124
  trans-boundary animal/plant diseases/ pests and, 293–294
  Tropic Levels in, 414, 415
  urban population density and, 124
  water, charging for, 207
  See also Agriculture; Existential threats
Fossil record, 22, 23, 81, 97, 98, 135, 157, 176, 211, 358, 360, 364, 365, 413
Frankenstein's monster, 257, 288
Free market. *See* Capitalism
Freedman, Andrew, 284
Freedom House report, 432–433, 462
Freedom to choose, 47–48, 68, 261, 263, 265, 272, 275, 277
  adaptation/innovation and, 436, 437
  AIMM-S/E pattern and, 436
  capitalist profit motive and, 433
  evolutionary maturity and, 436, 437
  existential threats, escape from, 433–434, 437
  "The Few", power of, 434, 435, 438
  galactic/universal-level survival game and, 437
  game of life, survival goal and, 435–438
  human extinction, choice and, 408–409, 418–420, 422, 425, 431–432, 434
  hunter-gatherers, essential freedoms of, 438–476, 475 (table)
  individual freedoms, transfer of, 434, 477–479
  intelligent species, evolution of, 437–438
  loss of, 432–433, 434, 438, 439–476, 475 (table)
  politicized decision process and, 433
  self-extinction, potential for, 434, 436, 437
  species-level survival game and, 437
  Species Survival Maturity Model and, 436
  survival, choice and, 432, 434, 435, 437–438
  survival-driven migration and, 432, 436, 437
  Survival Plateau and, 436
  wildebeest phenomenon and, 434
  See also Essential freedoms; Freedoms
Freedoms:
  destiny, control over, 21, 48, 59, 227, 262, 263, 265, 310, 369
  freedom to migrate, 124, 184, 309, 313, 315, 317, 318

individual freedoms, transfer of, 434, 477–479
loss of, 432–433, 439–476, 475 (table)
native economic freedom/independence and, 218–219, 229–230
See also Essential freedoms; "The Few" power elites; Freedom to choose
Fuller, Dorian, 105
Futurists, 81, 246, 253, 254, 256, 258, 263, 270, 271, 377

Galbraith, James, 419
Gamma ray bursts, 40–41, 52, 71, 139
Gates, Bill, 269, 288
Genetically modified organisms (GMOs), 298, 299
Geno-Graphic Project on Migratory Crossings, 318–319
Genocide, 77, 210, 390, 391
Genus *Homo*, 25, 100, 224, 248, 335, 361, 366, 368, 425, 435
Geologic time:
Anthropocene Epoch, 44
Cambrian explosion and, 45, 69, 135, 151, 186
Carboniferous period ice age and, 166
continental drift and, 42, 54
Cretaceous extinctions and, 174–175
Devonian period extinctions and, 174
dinosaur extinction and, 10, 13, 42
early Earth habitat, characteristics of, 148–152, 148 (figure)
Earth's evolution and, 148–149, 148 (figure)
Eocene Epoch hothouse conditions and, 119
evolutionary survival patterns and, 2
global climate change cycles and, 190–191
greenhouse mass extinctions and, 174–179
human evolution, environmental change and, 92 (figure), 97–101, 100 (figure)
intelligent species, evolutionary journey of, 57
Jurassic extinctions and, 174–175
Lucy, 98, 99
Paleocene Epoch extinction and, 175
Permian anoxic oceans and, 163
Permian-Triassic mass extinction and, 107, 127, 158, 174, 190, 286, 341, 372, 411–412
plate tectonics and, 42, 54, 97–98
Pleistocene Epoch, 25–27
Quaternary glaciation and, 27
Smithian stage and, 177–178
Species Survival Maturity Model and, 350 (figure)
Stone Age foragers, 124
temperature maximums and, 177–178
Triassic extinctions and, 174, 175, 178
Triassic-Jurassic extinction and, 174
See also Glaciations; Hominin species; Ice ages; Interglacials
Giosan, Liviu, 104, 105
Glaciations, 23, 24, 25, 26–27, 28, 41, 42, 54, 186, 188, 189, 190, 416, 471
See also Geologic time; Ice ages; Interglacials; Quaternary glaciation
Global forest Watch (GFW), 166
Global Policy Forum, 113
Global Trends report, 317
Global warfare, 44, 47
Global warming, 74, 80, 81, 84, 115, 119–122, 120 (figures), 125–127, 159, 163, 168, 170, 178–179, 187, 189, 354, 412, 416
See also Climate change; Climate machinery; Existential threats; Sustainability
Glyphosates, 298
Goddard Institute for Space Studies (GISS), 120
Goddard Space Flight Center, 173
Goldilocks conditions, 11, 104, 485
Goodall, Jane, 133, 490
Gøtzsche, Peter, 458
Great Drought, 103, 124
Great Dying, 127, 158, 286, 411
Great filter theory, 12
Great Oxidation/Oxygenation Event, 156, 174, 214, 218
See also Atmosphere; Earth's ecosystems; Extinction; Habitats; Oxygen
Great Rift Valley, 35, 52, 54, 70, 317, 319, 323, 325
Green Revolution, 294, 298, 299, 451
Greenhouse gases, 29, 36, 96, 121, 239, 260, 283
Greenhouse mass extinction events, 174–179, 372

Greenhouse state, 119
Gulf of Mexico, 162, 162 (figure), 228

Habitable zone, 10–12, 11 (figure), 36, 135, 138, 329, 351, 374, 485
Habitats, 133
- anaerobic species, mass extinction of, 153–154
- atmospheric carbon dioxide and, 114, 153, 154, 155–157, 159–160, 168–170, 168–169 (figures)
- atmospheric carbon dioxide/oxygen levels, inverse relationship of, 158–159, 160
- atmospheric oxygen, creation of, 44–45, 153–154, 155–157
- atmospheric oxygen levels, changes in, 157–158, 160
- bacterial ancestors/cyanobacteria, role of, 149–150, 152–153, 152 (figure)
- carbon dioxide-based atmosphere, transformation to, 159–160, 160 (figure)
- care/feeding of, 44, 147, 184–185
- civilization, impact of, 48
- complex animals, emergence of, 150
- deforestation and, 29, 107–108, 113–114, 165–167, 165 (figure)
- destruction, threat of, 42, 44, 48, 125, 164–165
- early Earth atmosphere, creation of, 150, 152–157, 152 (figure)
- early Earth habitat, characteristics of, 148–157, 148 (figure), 152 (figure)
- Earth, expiration date for, 94, 95, 96, 147, 311
- equation of life and, 44
- evolutionary sustainability and, 44
- fossil fuel consumption and, 160, 163, 168
- galactic neighborhood, familiarity with, 136–137
- global surface temperatures, rise in, 172–174, 173 (figure)
- greenhouse mass extinction events and, 174–179
- habitat destruction, extinction potential and, 135, 137, 147, 148, 174–179, 311
- hominin extinction and, 135
- interplanetary migration and, 137–141, 141–143 (table), 147–148
- life on Earth, history of, 133–136
- negative/positive feedback systems and, 167–168
- ocean oxygen level, decline in, 160, 161–162, 161 (figure)
- oxygen energy metabolism, failure of, 160–161
- photosynthesis, role of, 138, 151, 153, 157, 163, 167
- phytoplanktons, role of, 152, 157, 163, 164–165, 164 (figure), 170
- Planet B, nonexistence of, 146
- planetary atmospheres, gaseous components of, 145–146, 146 (table)
- Planetary Habitability Index and, 145, 145 (figure)
- replacement human habitats, search for, 137–141, 139 (figure), 141–143 (table), 146
- suicidal behavior, habitat destruction and, 180–181
- suitable habitat, characteristics of, 136, 139–141, 141–143 (table)
- Survivability and Habitat Sustainability Test and, 293, 366, 374–375, 386
- Survival Sustainability Cycle and, 184–186, 184 (figure)
- sustainability of, 18, 34–35, 37–38, 42, 44–45, 48, 135, 137, 183–184, 394–395
- sustainable-capable habitats and, 128, 133–134, 135, 137, 184, 186–187, 189, 217–218, 262, 352
- terraforming process and, 144, 149
- *See also* Atmosphere; Carbon dioxide; Climate change; Climate machinery; Earth's ecosystems; Global warming; Habitable zone; Oxygen; Sustainability; Temperature

Halpern, Daniel Noah, 296, 297
Handwerk, Brian, 97
Hawking, Stephen, 14, 18, 74, 75, 210, 269, 288, 309, 334, 418
He, Jinakui, 290–291
Helmholtz Center for Ocean Research, 161–162
Herd, Chris, 254
Historic Global Atmospheric Oxygen Levels Graph, 160, 160 (figure)
Ho, Mae-Wan, 160

Hobbits, 22
Hominidae family, 149
Hominin species, 9, 14
  archaic *Homo sapiens* and, 23
  art/language, development of, 24
  climate change, accommodation to, 96–101, 100 (figure)
  environmental/climatic factors and, 25–28
  evolution/extinction of, 20, 22–24, 23 (figure), 42, 52, 126, 135
  Homo sapiens, lone emergent hominin species and, 36
  lifespan of, 94 (figure)
  longevity of, 92–93, 92 (figure), 93 (table)
  Lucy and, 98, 99
  Neanderthal populations and, 23–24, 25, 26, 28, 41, 42
  out-of-Africa migration and, 15, 21, 23, 25, 26–27, 35, 42, 46
  supercontinents, breakup of, 42, 315
  See also Homo sapiens; Intelligent species evolution; Intelligent species survival; Intelligent species survival patterns; Species survival
Homo erectus, 26, 93, 99, 318, 362
Homo florensiensis, 22
Homo luzonensis, 22
Homo sapiens:
  agriculture, transition to, 20–21
  climate-change-driven challenges and, 25–28, 92–93, 92 (figure)
  emergence of, 12, 14, 20, 22
  evolution, control over, 43
  evolutionary journey of, 20, 22, 23 (figure), 24–25
  extinction, normal occurrence of, 21–22
  genus *Homo* and, 25, 100
  Great Rift Valley and, 35
  hominidae family and, 149
  innate abilities, survival and, 24–25
  native economic freedom/independence and, 218–219, 229–230
  purposeful evolution of, 2, 10
  subsistence method and, 20, 23, 29
  survival of, 13–14, 20, 24–25
  See also Agriculture; Civilization; Hominin species; Human extinction; Intelligent species evolution; Intelligent species survival; Intelligent species survival patterns; Modern humans; Neanderthals; Species-driven evolution; Species survival
Homo sapiens Survival Scorecard, 377–380, 378 (table)
Hornigold, Thomas, 467
Horton, Richard, 458
Hotspots. *See* Biodiversity hotspots
Human extinction, 405–406
  biodiversity, species survival and, 407, 413
  biological annihilation, foundations of civilization and, 406–407
  capitalist system, danger of, 420–421, 422
  carbon dioxide-rich atmosphere and, 411, 412
  choice, exercise of, 408–409, 418–420, 422, 425
  civilization collapse as liberation and, 421
  climate change threat and, 406, 416
  commoditization of life and, 421
  ecosystems, destruction of, 413–416, 420
  ecosystems, recovery time for, 412
  escape route from, 408–409, 411, 418, 419
  evolutionary maturity, species survival and, 423, 424–425
  food webs, species survival and, 413–415
  government/corporate leadership and, 419–420, 422
  interplanetary/inter-solar migration and, 411, 417, 424–425
  invisible hand of capitalism and, 238–239, 391, 421
  keystone species, loss of, 413
  mass extinctions, mechanisms of, 407–409
  mature intelligent species, evolutionary jump-start of, 417–418
  phytoplanktons, role of, 414–415
  post-agricultural revolution humans and, 371–372, 383–384, 412–413
  potential losses from, 409–410
  previous mass extinctions, guidance from, 411–412
  resources, exploitation of, 415
  signs/warnings of, 406, 412, 415
  sixth mass extinction, approach of, 406, 407, 411, 412, 416, 422, 425
  species extinction, interdependence and, 413–416
  survivable habitats and, 416

survival-driven migration and, 406, 412
survival-enabling patterns/rules and, 409
top predator status, native species extinction and, 412–413
See also Existential threats; Extinctions; Freedom to choose
Human Origins Program, 97
Human trafficking, 316, 317
Humanitarian relief efforts, 333
The Hunger Games, 356
Hunter-gatherer populations, 20
attrition of, 220
cooperative behaviors, development of, 28
diseases in, 222–223
economic insecurity, adaptation to, 229–230
egalitarian social structure of, 222
essential freedoms of, 439–476, 475 (table)
food sources for, 222
freedom to choose and, 438–439
glacial/interglacial periods, community survival and, 224
Greatest Generation status of, 220–221, 246–247, 439
infant mortality and, 222
life expectancy and, 222
mobility of, 223, 225
native economic freedom/independence and, 218–219, 229–230
Neanderthal populations and, 23–24, 25, 26, 28
out-of-Africa migrations and, 321–324, 321 (figure)
population levels of, 123
renewable energy use and, 223
settled communities, adaptation to, 229–232
subsistence method and, 20, 23, 29
tools, development of, 29, 36
See also Agriculture; Essential freedoms; *Homo sapiens*; Intelligent species evolution; Intelligent species survival

I, Robot, 367
Ice ages, 23, 24, 25, 26–27, 28, 29, 41, 42, 54, 121, 166, 186, 187, 188, 189, 190, 225
See also Geologic time; Glaciations; Interglacials
Icehouse state, 119
Income inequality, 201–202, 258–259, 271, 388
Indigenous peoples. *See* Native populations
Industrial Revolution, 45, 107, 297, 384, 451, 455, 507
Information Mastery Scale, 397, 398, 399
Innovation, 253
accelerating innovative rate, 253–256, 254–256 (figures), 260–261, 275
artificial intelligence and, 85, 89, 237, 238, 242, 254, 255, 257, 262, 269–270
bio-based technologies and, 286
climate change-driven innovations/adaptations and, 117–119, 122
climate-monitoring technology, forecasting strength and, 117–118
drones and, 258
entrepreneurs/inventors, role of, 259–260, 273, 278, 279, 281, 304–307
evolutionary innovation, species survival and, 262–267, 268
evolutionary maturity and, 74–75, 257, 263, 266–267, 270–273, 278, 281
extraterrestrial species evolution and, 265, 267–269
freedom to choose and, 261, 263, 265, 272, 275, 277
fusion technologies, 286
governments, ineffectiveness of, 118
greed, persistent problem of, 204, 208, 231, 251, 259, 271, 275, 290
Homo sapiens as threat to survival and, 278–281
innovation-maturation imbalance and, 74–75, 247, 257–258, 260, 261–262, 274–275
limitations on, co-existence/mutual species survival and, 274
live-and-let-live reason for, 274, 275, 276, 277
military-industrial complex and, 245–246, 258
modern scientific/technological progress and, 244–246
Moneyed sapiens and, 281
nanobiological intelligence and, 256
nanotechnology and, 254, 256
national evolutionary maturity and, 271
non-biological innovation and, 274–275

ocean-to-land migration and, 274
privacy, invasion of, 259
profit motive and, 259, 263–264, 270, 276
recorded knowledge, modern technology and, 254
reproductive technologies and, 296
robotics and, 254, 391, 453, 455, 458
Robotics Laws and, 290, 291–292
scarcity, functional myth of, 258, 259
sea water, conversion to potable water and, 303
self-extinction, potential for, 255, 256 (figure), 257–265, 269–270, 272, 274–275, 277
species-driven evolution and, 265–266, 268
species-driven innovation and, 268, 276–277
species survival/habitat sustainability goals and, 269–270
Survivability and Habitat Sustainability Test and, 293
survival-driven innovation and, 29–30, 46, 268, 270–273, 275–276, 305–306
survival-focused innovation and, 278, 293
Survival Plateau and, 266, 270
Techno sapiens and, 258–259, 260, 261, 269, 281
technologically-driven survival strategy and, 260–261
See also Adaptation; Existential threats
Innovations for Survival dashboard, 278
Institute of Human Origins, 27
Intelligent extraterrestrial species evolution, 8-9, 12, 53, 54, 265, 267, 349, 359
Intelligent Industrial Revolution, 455
Intelligent species evolution, 63–64
climate change pressures, adaptive innovation and, 64
er-engineered/enhanced *Homo sapiens* and, 64
progenitors of, 149–150
replacement intelligent species and, 96
species-driven evolution and, 64
survival-driven migration and, 64
See also Evolution; Evolutionary maturity; Hominin species; *Homo sapiens*; Modern humans; Species-driven evolution; Species survival; Survival
Intelligent species survival, 14–15, 17
agricultural societies, emergence of, 20–21
brain size and, 24–25, 28
change, evolutionary survival selection mechanism of, 29
civilizations, development of, 21
climate change pressures and, 26–28
cooperative behaviors, development of, 28
cultural technology, development of, 28–29
early existence, hunter-gatherer subsistence and, 20, 23
environmental challenges, adaptive response to, 28–30
environmental/climatic factors and, 25–30
evolutionary laws/patterns, waning knowledge of, 21
evolutionary rules for, 129–130
evolutionary survival-driven adaptation and, 28–30
extinction, normal occurrence of, 21–22, 33–34
hominin evolution/extinction and, 20, 22–24, 23 (figure)
Homo sapiens, evolutionary journey of, 20, 22, 23 (figure), 24–28
human destiny, control over, 21, 48, 59
innate capabilities of early *Homo* and, 24–25, 28
intelligence, growing control by, 20–21
interbreeding practices and, 24
laws of nature, recent understanding of, 20
natural selection pressures and, 27–28
nature/nurture, factors of, 24–25
out-of-Africa migration, climate change pressures and, 26–27
planetary/galaxy-wide patterns and, 19
rules/patterns, discovery/internalization of, 19–20
self-extinction, potential for, 21, 46
survival-impacting factors and, 30–31
tool development and, 29
Universe's laws/model, role of, 17–19, 20
See also Climate change; Earth's evolution; Freedom to choose; Hominin species; *Homo sapiens*; Intelligent species evolution; Intelligent species survival patterns; Modern humans; Species-driven evolution; Species survival;

Survival
Intelligent species survival patterns, 51–52
adaptive innovation, capability for, 28–30, 46, 53, 60
climate change pressures and, 55–56, 60
complex species evolution, singular process of, 53–54, 57, 60
evolution, default species survival and, 38, 39, 43, 48, 53, 56
evolution, guardian/companion roles of, 52, 53
evolutionary maturity, species survival and, 57, 58–59
extinction, inevitability question and, 33–34, 43, 48, 53
extraterrestrial intelligent life, evolution of, 8–9, 12, 53, 54, 56, 60
habitat destruction and, 55–56, 60
home world, location of, 34, 36, 40–41, 54
hominin species, multiple extinctions of, 52
innate survival instincts and, 54–55
intelligent-species-driven evolution and, 57, 58, 60
interplanetary migration and, 34, 35–36, 39, 44–45, 55, 56, 57–58
mature intelligent life, interplanetary dispersal of, 54, 56, 57, 60
out-of-Africa dispersal patterns and, 54–55, 56, 60
parent-offspring pattern and, 59–60
proto-intelligent-species seedling idea and, 56
reproductive ability, species continuity and, 38, 48, 53
self-extinction, potential for, 57, 58
solar/planetary events, survival threats and, 41, 52–53
survival-driven migration and, 26–27, 34–35, 45, 53–54, 56, 60
universal processes/evolutionary processes, counterbalancing of, 55
See also Intelligent species evolution; Intelligent species survival; Species-driven evolution; Species survival; Species Survival Maturity Model (SSMM); Survival
Intelligent Species Survival Scorecard, 377–380, 378 (table)
Intercontinental ballistic missile system, 282
Interglacials, 25, 29, 119, 188, 189, 190, 225, 471
See also Geologic time; Glaciations; Quaternary glaciations
Intergovernmental Panel on Climate Change (IPCC), 119, 283, 332, 470, 471–472
Intermediate-Range Nuclear Forces Treaty (INF), 282
International Monetary Fund (IMF), 304
International Rescue Committee (IRC), 122
International Space Station, 148
Interplanetary migration, 35-36, 39, 70, 88, 89, 185, 205, 268, 311, 313, 325, 326, 327, 329, 331, 334, 352, 353, 355, 359, 418, 496
Interplanetary settlements *See* Migration
Inter-solar migration, 70, 88-89, 147, 185, 311, 326, 331, 357, 378, 418, 423, 424
Interstellar, 331
Inter-Tropical Convergence Zone (ITCZ), 301
Ives, Burl, 240, 241

Jobs, Steve, 273, 363

Kaplan, Sarah, 413
Karate Kid, 153
Kardashev Scale, 397–398, 399
Karoo Basin fossils, 413
Kennedy, President John F., 47, 255, 382, 383, 384, 392, 433
Kepler Space Telescope, 11
Keystone species, 413
Khrushchev, Premier Nikita, 47, 255, 382, 383, 384, 392, 433
Kiel Declaration on Ocean Deoxygenation, 163
Kim, Jong-Un, 255, 389, 392, 433, 463
Kirschvink, Joe, 154, 158, 166, 174, 175, 178
Korten, David C., 445, 455
Kurzweil, Ray, 256, 263

La Niña climatic pattern, 101, 112, 115–117, 122, 190
Lampitt, Richard, 414
Lane, Nick, 466
Lang, Avis, 245
Last Glacial Maximum, 189, 190

Last universal common ancestor (LUCA), 68–69
Late Heavy Bombardment period, 52, 148–149, 341, 351
Laws of nature, 17–19, 30, 33, 36–37, 315, 343–344, 347
Lee, Kai-Fu, 453
Lennon, John, 204, 205, 270
London Interbank Offered Rate (LIBOR), 259
Long-term carbon cycle, 169–170
Lord of the Rings, 365, 401, 431
Louisiana Universities Marine Consortium, 162
Lovelock, James, 114, 194
LUCA (last universal common ancestor), 68–69
Lucy, 68, 69
Luther, Martin, 242–243, 507

Macron, President Emmanuel, 390
Malhotra, Aseem, 459
Manhattan Project, 490
Marean, Curtis, 27
Margulis, Lynn, 149
Marlowe, Frank W., 464
Mars rover Curiosity, 312
The Martian, 331
Marx, Karl, 242
Mass extinction events, 2, 10, 13, 42, 44, 107, 127, 148, 153, 156, 157, 164, 203–231, 341, 407–409
  See also Extinction; Greenhouse mass extinction events; Sixth mass extinction
Masters, Jeff, 102
Maturation. *See* Evolutionary maturity
Maturity scale, 369, 398–399, 399 (table)
  See also Evolutionary maturity; The Maturity/Self-Extinction Index; Species Survival Maturity Model (SSMM)
The Maturity/Self-Extinction Index, 380–385, 382 (figure), 466, 490
Megafauna, 230–231, 240, 277
Metabolic rates:
Methane, 115, 154, 175, 239
Microbial life, 149, 150, 151, 267, 289, 312, 313, 330, 351
Migrations, 1, 112, 309–310
  barriers to, 317–318, 324, 325, 395
  climate change-driven migration and, 26–27, 64, 112, 118–119, 190, 316, 332–333
  climate-refugee crisis and, 118–119, 190, 332–333
  cyanobacteria/phytoplanktons and, 312
  cyclic migrations and, 315–316
  early human migration routes and, 321, 321 (figure)
  early human migrations, conditions of, 322–324
  economic migrants and, 324
  evolutionary maturity, requirement of, 311, 313–314, 316, 318, 333–334
  evolutionary phase transitions, interplanetary migration and, 326–329
  "The Few", power of, 331
  freedom to migrate and, 124, 184, 309, 313, 315, 317, 318, 325, 333–334
  global cooperation/trust, interplanetary migration preparation and, 331
  government ineffectiveness/failures and, 118
  habitat destruction, extinction potential and, 311
  home planet expiration data and, 311
  human-caused triggers for, 316–317, 334
  interplanetary migration and, 34, 35–36, 39, 44–45, 55, 268, 311, 312–314, 320, 324–325, 329–330, 334
  Mars, life on/from, 312–313
  means to migrate and, 321
  meteorites, migrating life forms and, 312–313
  migration crises and, 260
  mobile lifestyle, unsuitability of, 316, 325
  natural disasters and, 333
  ocean-to-land migration and, 274
  open border policy and, 271
  out-of-Africa migration, 15, 21, 23, 25, 26–27, 35, 42, 46, 268–269, 310, 317, 318–319, 321–324, 321 (figure), 325
  planetary alignment and, 320
  population explosion and, 225
  refugee crises and, 39, 118–119, 122, 189, 258, 271, 283, 317, 318
  rhythm of the Universe and, 314–315

right-time window for, 319–320
slave trade and, 322
Species Survival Maturity Model, rungs on, 326
survival-driven migration and, 26–27, 34–35, 45, 53–54, 64, 269, 310–311, 315, 320, 325–326
war refugees and, 258, 317, 322
See also Species Survival Maturity Model (SSMM)
Military expenditures, 202–204, 203 (figure), 204 (table), 271, 272, 391
Military-industrial complex, 245–246, 258, 391
Milky Way Galaxy, 10–11, 13, 14, 36, 41, 43, 54, 136, 314, 315, 320
Milner, Yuri, 305
Mitochondrial DNA, 150, 279
Modern humans:
anthropocentric perspective and, 233–235, 437
atmospheric oxygen, depletion of, 178–179
belief systems/mental modeling of, 232–235
classical economic calculus and, 232–233
crimes against humanity and, 77
developmental wall, survival question and, 13
endocrine disrupting chemicals and, 295–296, 298
evolutionary maturity, stagnant development of, 21, 205–208
forever chemicals, PFAS and, 274
global income inequality and, 201–202, 258–259
greenhouse mass extinctions and, 178–179
habitat destruction, threat of, 42, 44, 48, 178–179
habitat sustainability and, 37–38
human mind, evolution of, 71–72, 75–79, 82–83
innovation vs. maturation, Cuban missile crisis and, 47
Native Americans, origins of, 26
religion, role of, 232–234, 242–244
scientific/technological progress and, 244–246
suicidal behavior, habitat destruction and, 180–181
See also Adaptation; Agriculture; Capitalism; Chemical industry; Climate change; Disease; Energy industry; Futurists; Human extinction; Innovation; Intelligent species evolution; Intelligent species survival; Species-driven evolution; Species Survival Maturity Model (SSMM); Sustainability
Moneyed sapiens, 281, 304, 329, 377
Montreal Protocol, 81
Morris, Ian, 450
Morse, Samuel, 8
Mount Toba eruption, 26, 100, 127
Movies:
Avatar, 214, 389
The Bicentennial Man, 69
Earth 2.0, interplanetary migration to, 331, 424
evolutionary mature human world, depiction of, 58-59
human evolutionary development, trajectory of, 69-70
The Hunger Games, 356
Interstellar, 331
I, Robot, 367
Karate Kid, 153
Lord of the Rings, 365, 401, 431
The Martian, 331
repeating scene cycles, disastrous outcomes and, 492
starship life-support systems and, 70-71
Star Trek, 69, 89, 201, 257, 267, 405, 419, 423, 424, 496, 508
Star Wars, 267
Titanic, 258, 406, 408,411,418,419
war-driven transformation and, 508-509
Multinational corporations. *See* Capitalism
Musk, Elon, 85, 269, 288, 304–305, 329
NASA (National Aeronautics and Space Administration), 7, 11, 70, 74, 120, 136, 142, 146, 165, 170, 173, 176, 189, 193, 285, 312, 320
National Climate Assessment of 2018, 473
National Defense Education Act (NDEA) of 1958, 245–246
National Integrated Drought Information System, 303
National Oceanic and Atmospheric Admin-

istration (NOAA), 173
Native populations, 26, 114, 233, 307, 322, 391, 421
Natural disasters, 41, 115–117, 118, 176–177, 202, 284, 303, 333
Natural law, 17–19, 30, 33, 36–37, 315, 343–344, 347
Nazi atrocities, 298, 389
Neanderthals, 9, 22, 23–24, 25, 26, 28, 41, 42, 54, 93, 224, 318, 362, 384, 471
Negative feedback systems, 167–168
Neocolonialism, 447
New Strategic Arms Reduction Treaty (New START), 282
Newton, Sir Isaac, 234, 344, 347
Nosbach, Marc, 106
Nuclear threats, 282–283, 282 (figure), 288, 381–382, 383, 384, 390, 490

O'Brien, Paul, 202
Oceans:
- bacterial ancestors and, 149–150
- biological productivity of, 414–415
- dead zones in, 162
- deforestation, effects of, 167–168
- El Niño/La Niña effects and, 115–117
- El Niño-Southern Oscillation and, 116
- food chain in, 164–165, 164 (figure)
- hunting practices and, 231, 395
- Kiel Declaration on Ocean Deoxygenation and, 163
- marine ecosystems, oxygen depletion and, 162, 178
- oxygen, historic fluctuations in, 163
- oxygen, low levels of, 160, 161–162, 161 (figure), 163, 412
- Permian anoxic oceans and, 163
- phytoplaktons/planktons, role of, 152, 157, 163, 164–165, 164 (figure), 170, 414–415
- Pliestocene Epoch glaciation and, 25–28
- sea level changes and, 26, 46, 121, 189, 229
- warming of, 121, 162, 163, 164–165, 279, 283 (Figure), 285–286
- *See also* Climate machinery; Habitats

Oppenheimer, J. Robert, 490
Oschlies, Andreas, 161
Outer space:
- advanced technological civilizations, self-destruction of, 12–13
- asteroid collisions and, 9 (figure), 10, 13, 24, 84
- Earth-like exoplanets, habitable zone and, 10–12, 11 (figure)
- exploration of, 18
- extraterrestrial intelligent life and, 8–9, 12, 53, 54, 60
- Fermi Paradox and, 12
- gamma ray bursts and, 40–41, 52, 139
- Goldilocks zone and, 11
- Great Filter theory and, 12
- human exploration of, 70
- interplanetary migration and, 34, 35–36, 39, 41, 44–45, 55, 70, 96
- interstellar catastrophe and, 13
- Kepler Space Telescope and, 11
- Late Heavy Bombardment period and, 52, 148–149, 341, 351
- liquid water, existence of, 11
- Mars exploration and, 8, 36, 39, 44–45
- mature intelligent life, interplanetary dispersal of, 54, 56, 57, 60
- Milky Way Galaxy and, 9–10, 13, 14, 36, 41, 43
- multi-planet species and, 34, 41
- Parker Solar Probe and, 7
- red giant sun and, 95 (figure)
- replacement human habitats, characteristics of, 137–141, 139 (figure), 141–143 (table)
- SETI program and, 8, 12
- sun, death of, 94–95
- sun, main sequence star classification of, 95
- Universe's laws/model and, 17
- *See also* Earth's evolution

Ownership concept, 386–387
Oxfam International, 201, 202, 271
Oxygen:
- atmospheric oxygen and, 44–45, 153–154, 155–158, 159–160, 163, 167, 215
- carbon dioxide-based atmosphere, transformation to, 159–160, 160 (figure)
- carbon dioxide/oxygen levels, inverse relationship of, 158–159, 160, 174
- deforestation and, 165–167, 165 (figure)
- depletion of, 158, 159–160, 163

Earth's atmospheric oxygen, generation of, 44–45, 153–154, 155–157, 164
Earth's atmospheric oxygen levels, change in, 157–158
global surface temperatures, rise in, 172–174, 173 (figure)
Great Oxidation/Oxygenation Event and, 174, 214
greenhouse mass extinction events and, 174–179
high-carbon dioxide world, viability of life and, 158–159
Historic Global Atmospheric Oxygen Levels Graph and, 160, 160 (figure)
ocean oxygen levels, decline in, 160, 161–162, 161 (figure), 163
ocean oxygen levels, historic fluctuations in, 163
oxygen energy metabolism, failure of, 160–161
photosynthesis, role of, 138, 151, 153, 157, 163, 167
phytoplaktons, oxygen production and, 157, 163, 164–165, 164 (figure), 415
See also Atmosphere; Carbon dioxide; Habitats
Ozone layer, 40, 72, 80, 280

Paris Climate Accord, 286, 470
Parker Solar Probe, 7
Payne, Jonathan, 411
Peoples, Hervey C., 464
Permian-Triassic extinction, 107, 127, 158, 160, 190, 286, 341, 372, 411–412
Persistent organic pollutants (POPs), 294
PFAS (per-/polyfluoroalkyl substances), 274
Pharmaceuticals, 207, 296, 458, 459-460, 475, 504, 506
See also Big Pharma; Chemical industry; Disease
Photosynthesis, 138, 163, 169, 178, 195, 214, 215, 239, 286, 370–371
Phthalates, 296
Phytoplanktons, 152, 157, 163, 164–165, 164 (figure), 170, 215, 239, 286, 312, 313, 414–415
Pinnacle Point, 27
Planetary Habitability Index, 145, 145 (figure)
Plankton. *See* Phytoplanktons
Plant species:
atmospheric oxygen, depletion of, 157–158, 163
bees/pollinators, loss of, 260, 299, 395
carbon dioxide levels, sensitivity to, 176
extinctions of, 372
greenhouse mass extinction events and, 174–176
phytoplanktons/planktons and, 152, 157, 164–165, 164 (figure), 170
short-term carbon cycle and, 168, 169
species-driven evolution and:
See also Photosynthesis
Plastic pollution, 239
Plate tectonics, 42, 54, 97–98, 186, 315, 319
Pleistocene Epoch, 25–27
Pliny the Elder, 259
Pollution. *See* Chemical industry; Energy industry; Existential threats
Ponting, Clive, 230, 233, 234, 469, 470
Poppick, Laura, 161
Popular culture. *See* Movies
Positive feedback systems, 168–170, 168–169 (figures)
Potts, Rick, 97, 98, 99, 360
Poverty:
climate change response and, 122
climate chaos and, 115–117, 122
climate-refugee crisis and, 118
drought/water shortages and, 115
evolutionary maturity and, 388
food insecurity/starvation and, 106–107, 109, 115
global income inequality and, 201–202, 258–259, 271, 388
global poverty, definition of, 202
native economic freedom/independence and, 218–219, 229–230
natural disasters, disproportionate suffering from, 118
sustained hunger and, 202
Power elites. *See* "The Few" power elites
Power industries. *See* Energy industry
Prokaryotes, 52, 53, 69, 150
Proxima B, 193
Proxima Centauri, 193, 320
Puebloan peoples. *See* Civilizations
Putin, Vladimir, 390, 392, 464

Quaternary extinction event, 27
Quaternary glaciation, 27

Rare Earth: Why Complex Life Is Uncommon in the Universe, 54, 135, 138
Raup, David M., 407
Reformation, 242–243, 507
Refugees, 39, 118–119, 122, 189, 258, 271, 283, 317, 322, 333
Religion, 232–234, 242–244, 347, 393–394, 464–465, 501, 510
"The Rest", 477, 479, 529
Ridley, Matt, 149
Robotics, 254, 391, 453, 455, 458
Rock record, 169
Rohling, Eelco, 29
Roopnarine, Peter D., 413, 414
Rutherford, Ernest, 255

Sagan, Carl, 260, 397, 398, 399
Sahlins, Marshall, 439
Sahney, Sarda, 412
Sapiens. *See Homo sapiens*; *Moneyed sapiens*; *Techno sapiens*
Scholz's Star, 320
Schrock, Robert R., 412
Schrödinger, E., 2
Scott, James C., 439
Self-extinction, 21, 46, 47, 57, 58, 74–75, 77, 86–87, 205, 206, 207–208, 248, 259–262, 263, 269–270, 288, 342, 348–349, 354–355
  See also Existential threats; Extinctions; Freedom to choose; Human extinction; The Maturity/Self-Extinction Index; Species Survival Maturity Model (SSMM)
Sepkoski, John, Jr., 407
SETI (Search for Extraterrestrial Intelligence) program, 8, 12, 56
Shelley, Mary, 257
Short-term carbon cycle, 168, 169
Sierra Club, 286
Silverberg, Robert, 69
Simmon, Robert, 168
Single-cell life. *See* Procaryotes
Sixth mass extinction, 44, 128, 171, 226, 234, 329, 341, 372, 403–404, 406, 407, 411, 412, 416, 422
  See also Human extinction
Slavery, 230, 276–277, 317, 322, 376, 421, 442, 449–450, 450 (figure), 452, 461
Smithian stage, 177–178
Smithsonian National Museum of Natural History, 97
Snowball Earth, 134, 214, 341, 355, 360, 416
Social Progress Index (SPI), 397, 398, 399
Solar system, 7–8, 7 (figure), 188–189
Space. *See* Outer space
Spaceship Earth, 80, 193, 196, 419, 420
  See also Evolutionary sustainability
Species-driven adaptation, 226–227
Species-driven evolution, 57, 58, 60, 67–68, 265
  adaptive/innovative capabilities and, 68, 69, 72, 73, 79–80, 82–83
  anatomical adaptations and, 83–84, 85, 88–89
  artificial intelligence, emergence of, 85, 89
  artificial life-support systems/sustainability features and, 70–71
  biomimicry and, 72, 73
  Cambrian explosion and, 69
  choice, freedom of, 68
  code of life, re-engineering of, 71–72
  complexity, evolution toward, 68–69, 73
  crimes against humanity and, 77
  CRISPR gene-editing tool and, 72, 73, 290
  definition of, 71, 265–266
  directed evolution revolution and, 71–72, 73
  eukaryotes, plant/animal evolution and, 69
  evolutionary chain, potential culmination of, 68, 69
  evolutionary maturity, goal of, 71, 73, 75, 77, 85–86, 87, 89
  evolutionary process, complexity/extinction phases and, 68–69
  genome manipulation, mature intelligent species and, 73
  habitat changes and, 68, 69
  habitat/species, destruction of, 68
  hominin evolution and, 69
  Homo sapiens survival, prerequisite for, 70, 88
  Homo sapiens, transformation of, 69, 70, 71, 73
  human mind, evolution of, 71–72, 75–79,

82–83, 84, 88
intent/goal of, 73–74
interplanetary migration and, 70, 73, 88, 89
last universal common ancestor/LUCA and, 68–69
mature intelligent species and, 73, 86–87
maturity-innovation imbalance and, 74–75, 76, 77, 79, 80
meaning, search for, 79, 80
mental constructs, creation of, 78–79
missteps, understanding/correction of, 79–81
natural selection, responsive nature of, 82
nature, control over, 74–75
self-extinction, potential for, 74–75, 77, 86–87
species-guided evolution and, 82–83
super intelligent life-forms, transformation into, 73, 88, 89
survival, goal of, 68, 70, 79–80, 81, 84–85, 87
technological innovation, imperiled humanity and, 81–82
wealth/profit, pursuit of, 81
See also Intelligent species evolution; Intelligent species survival; Intelligent species survival patterns; Species Survival Maturity Model (SSMM)
Species-driven innovation, 268, 276–277
Species-guided selection, 82–83
Species self-extinction. *See* Self-extinction
Species survival, 33
adaptation/innovation impulse and, 46
capitalist profit motive and, 48
chemical reactions, basic life processes and, 37
choice, freedom of, 47–48
climate-driven environmental changes, adaptative response to, 41–42
cultural survival narrative, development/integration of, 34
destiny, control over, 48, 57, 60
empathic awareness and, 47
evolution, default survival outcome and, 1, 38, 39, 43, 48
evolutionary-driven survival and, 292
extinction-causing events, avoidance of, 44
extinction, inevitability question and, 33–34, 43, 48
gamma ray bursts and, 40–41
habitat destruction threat and, 42, 44, 48
habitat sustainability and, 34–35, 37–38, 42, 44–45, 48
home world, location of, 34, 36, 40–41, 54
Homo sapiens, lone emergent hominin species and, 36
individual's survival instinct and, 45–46
innate extinction-defying abilities and, 38
interplanetary migration and, 34, 35–36, 39, 41, 44–45
laws of nature, predictability and, 36–37
Mars exploration and, 8, 36, 39, 44–45
maturity, positive outcomes from, 46–47, 48
migration, evolutionary survival response of, 26–27, 45
natural disasters and, 41
out-of-Africa migrations and, 15, 21, 23, 25, 26–27, 35, 42, 46
planetary environment/climate and, 34–35
plate tectonics, role of, 42
reproductive ability, species continuity and, 38, 48, 53
self-extinction, potential for, 21, 46, 47, 57
solar/planetary events, survival threats and, 41, 52–53
subsistence patterns, rapid changes in, 41
supernovae explosions and, 40
Survivability and Habitat Sustainability Test and, 293
survival-driven adaptation/innovation and, 46
survival-driven migrations and, 26–27, 34–35, 37
timing/luck factors and, 41, 53
tool development and, 29, 36, 46
uncontrollable extinction-causing events and, 39–43
See also Evolution; Extinction; Intelligent species survival; Intelligent species survival patterns; Species Survival Maturity Model (SSMM); Survival
Species Survival Maturity Model (SSMM), 13, 14, 15, 36, 39, 46, 47, 51, 336, 373, 377, 491
adaptation/innovation/invention capability and, 352, 354, 356, 360, 361

AIMM-S/E sequence and, 356–357, 358
changing circumstances, evolutionary survival selection mechanism of, 361
climate change-driven maturity and, 354, 355, 359–360, 361
diagram of, 349–350, 350 (figure)
essential initial survival conditions and, 351–353
evolutionary maturity, pursuit of, 87, 91, 349–350, 353, 354, 356–357, 360–361, 362–363
evolutionary survival patters/strategies and, 366
evolution's clock and, 326
extinction-enabling decisions and, 355–356, 362–363
extinction vs. survival, potential for, 355–356, 359–360, 362–363, 364–365, 366
extraterrestrial intelligent species evolution and, 349, 359, 363–364
fully matured level and, 354
habitat changes, geologic/climate events and, 354, 355
habitat unsustainability and, 359–360, 361
habitat viability and, 352, 365
hominin extinctions and, 310
Homo sapiens, evolutionary journey of, 13, 19, 310, 349–350, 350 (figure)
human-caused existential threats and, 354–355, 357, 362–363
intelligent species, evolution of, 365–366
interplanetary/inter-solar migration, need for, 352–353, 355, 359, 364
living/extant species and, 351
location-dependent survival and, 351
migration patterns and, 352–353, 355, 358, 362
planetary climate-driven events and, 354–355, 359–360, 365
profit/greed/power aspirations and, 363
reproductive success and, 351–352
self-extinction, potential for, 348–349, 356, 362
self-protective actions, lost ability/freedom for, 128
summary assertions of, 357–364
Survival Plateau and, 91, 92, 96, 336, 349–350, 350 (figure), 355, 356, 357, 359, 362, 363, 365
sustainable-capable habitats and, 352
timing/luck and, 351
See also Species survival; Survival
Star Trek, 69, 89, 201, 257, 267, 405, 419, 423, 424, 496, 508
Star Wars, 267
Starvation, 107, 109, 220, 222, 223, 451, 457, 459, 478
Sternberg, Samuel, 72
Stockholm International Peace Research Institute (SIPRI), 203, 204
Stone Age foragers, 124
Stromberg, Joseph, 108
Subsistence patterns, 20, 23, 29, 41, 218–219, 221
Sunspot activity, 188
Supercontinents, 42, 315
Supernovae explosions, 40, 139, 345
Survivability and Habitat Sustainability Test, 293, 366, 374–375, 386, 469, 504
Survival, 1, 339
chaos, order within, 2, 341, 347
choice in evolutionary outcome and, 342–344, 348
evolution, survival preference of, 1, 339, 340, 341–342
evolutionary factors and, 24, 28–30
extinction, inevitability question and, 33–34, 342
habitat survival and, 340
Homo sapiens, inept survival strategies and, 347–348
individual's instinct for, 45–46
intelligent species, requirements for survival and, 345–346
laws of nature and, 343–344, 347
life/death and evolution, co-opetition between, 1–2
migration, role of, 1, 26–27, 346
monarch butterfly example and, 339–340, 340 (figure)
scientific laws and, 347
self-extinction, potential for, 348–349
single intelligent species, survival of, 9
survival-impacting factors and, 30–31, 33–34
Universe's birth/life/death cycles and, 1, 340–344, 345, 346

See also Earth's evolution; Evolution; Evolutionary survival patterns; Extinction; Intelligent species survival; Self-extinction; Species survival; Species Survival Maturity Model (SSMM)
Survival-driven adaptation, 28–30, 46, 53, 215–220, 216 (figure), 227–229, 248, 251, 352, 356
Survival-driven innovation, 29–30, 46, 268, 270–273, 275–276, 305–306, 352, 356
Survival Plateau, 91, 92, 96, 266, 270, 336, 349–350, 350 (figure)
See also Species Survival Maturity Model (SSMM)
Survival Sustainability Cycle and, 184–186, 184 (figure), 292
Sustainability, 183
adaptive/innovative practices and, 189
altruism/common good, trampling of, 201
biodiversity hotspots and, 196–199, 196 (figure)
Cambrian explosion and, 186
carbon dioxide-driven global warming and, 187
climate events, sustainability pressures and, 115–117, 185–186, 187, 188, 192
Earth's biodiverse resources, exploitation of, 196–199, 196 (figure), 200–201, 205–208
Earth's ecosystems and, 185, 186–187, 194–196
Earth's ecosystems, preservation of, 199–201, 200 (figure)
Earth's orbit-climate change connections, gravitational fluctuation and, 187–190
Earth's orbit-driven climate change, solar/planetary phenomenon of, 188–190
ecocidal activities and, 113–114, 198–199, 200, 201
economic freedom and, 218–219, 229–230
economic gain, exploited biodiversity and, 196
evolutionary maturity, necessity of, 185, 186, 192, 209–211
evolutionary survival patterns and, 190–192
evolutionary sustainability and, 186–187, 192–196
extinctions, recurrence of, 186–187
"The Few" vs. "The Rest", power dynamics and, 201–202, 204–205, 207, 208
food scarcity and, 189
freedom to migrate and, 124, 184, 185
habitat destruction threat and, 42, 134
habitat sustainability and, 18, 34–35, 37–38, 42, 128, 183–184, 185, 186
industrial progress/growth, capitalist imperative and, 205–208
interplanetary migration and, 189, 190, 207
migration patterns and, 118, 189, 190, 191
military expenditures, resource exploitation and, 202–204, 203 (figure), 204 (table)
planetary magnetic fields, alterations in, 188–189
self-extinction potential and, 205, 206, 207–208, 210–211
solar system, heating of, 188–189
species survival, default evolutionary outcome and, 185, 205, 206
species survival, necessary priority of, 200, 201, 205
Survival Sustainability Cycle and, 184–186, 184 (figure)
sustainability planning, global perspective and, 192
sustainable-capable habitats and, 128, 133–134, 184, 186–187, 189, 217–218, 262
water shortages and, 115, 125–126, 189
See also Adaptation; Evolutionary survival patterns; Evolutionary sustainability; Innovation; Migration; Modern humans
Sustainable habitats, 18, 34-35, 37-38, 42, 44-45, 48, 128, 135-137, 183-184, 185, 186, 394-395
Syrian civil war, 298, 322, 390

Techno sapiens, 258–259, 260, 261, 269, 281, 304, 329, 377, 392
Technology. *See* Adaptation; Artificial intelligence (AI); Innovation
Temperature:
atmospheric temperatures, 168, 169

(figure)
carbon dioxide-driven global warming and, 158, 170–172
Carboniferous period ice age and, 166
drought conditions and, 300–303, 302 (figure)
El Niño-Southern Oscillation cycle and, 116
Eocene Epoch hothouse conditions and, 119
glacier melt and, 121
global heat, indicators of, 121
global surface temperatures, rise in, 107, 119–122, 120 (figures), 168, 172–174, 173 (figure), 176–178, 283 (figure)
global temperature/global heat, difference between, 121
global warming, mitigation of, 74
greenhouse mass extinctions and, 176–179, 177 (figure)
high temperatures, life span of, 177–178
ocean temperatures, 121, 162, 163, 279, 283 (figure), 285–286
Smithian stage and, 177–178
solar system, heating of, 188–189
See also Climate change; Climate machinery; Global warming
Terraforming process, 144, 149, 206, 213, 335
Terrorism, 113, 258, 260, 281, 391
Tibetan Plateau, 97
Titanic, 258, 406, 408, 411, 418, 419
Tools:
agricultural tools, 225
climate change, influence of, 101
CRISPR gene-editing tool, 72, 73, 290
hand axes and, 98
hunting, proficiency in, 29, 36
survival-driven adaptation and, 29, 36, 46
Trump, Donald, 255, 271, 281, 286, 392, 433, 463, 472, 473
Twins Study, 70
Tyson, Neil deGrasse, 245

United Nations Climate Change Conference (UNCCC), 304
United Nations Food and Agriculture Organization (UNFAO), 115, 293
United Nations Framework Convention on Climate Change (UNFCCC), 107
United Nations High Commission on Refugees (UNHCR), 317
United Nations International Year of Water Cooperation, 303
United Nations Sustainable Development Goals (UNSDGs), 203–204, 203 (figure)
United States Agency for International Development (USAID), 122
United States Department of Agriculture (USDA), 228
United States Department of Defense (DOD), 333
United States Geological Survey (USGS), 105, 116
Universe's laws/model, 17–18, 30, 33, 36–37

Vaidyanathan, Gayathri, 98
Volcanic activity, 26, 100, 127, 186

Wallerstein, Immanuel, 204
Wang, Sherry, 350
Ward, Peter, 54, 135, 148, 150, 154, 158, 166, 174, 175, 178
Warfare, 44, 47, 77, 113, 230, 243, 245–246, 260, 279, 283, 316, 317, 381–382, 389–391, 509
Water pollution, 294, 303
Water shortages, 115, 125–126, 189, 229, 279, 285, 297, 300–303, 302 (figure)
Wealth inequality, 201–202, 258–259
Weapons of mass destruction (WMDs), 245, 246, 255, 258, 260, 279, 282, 288, 298, 362, 376, 382, 383, 385, 390, 433, 463, 467
Weapons trafficking, 113
Western Christian nations, 232, 233, 234, 243
See also Religion
Weather patterns:
El Niño/La Niña, devastating effects of, 115–117
forecasting technology, improvements in, 74, 117–118
human vulnerability to, 20, 115
weather anomalies, tracking of, 118
See also Climate change; Climate machinery
Westheimer Institute for Science and Technology, 312

Woods Hole Oceanographic Institution, 104
World Bank, 122, 202, 260
World Food Program (WFP), 106
World Health Organization (WHO), 122
World Military Expenditure Report, 203
World War, I, 77, 381, 382, 382 (figure), 390, 509
World War II, 77, 353, 381, 382, 382 (figure), 383, 446, 509
World Wildlife Foundation (WWF), 406, 414
World's Dry Areas/Dryland Systems facts, 300
Wright brothers' flights, 8, 255

Xi, Jinping, 392

Y Combinator, 305
Yang, Andrew, 511

Zalasiewicz, Jan, 410, 422
Zuckerberg, Marc, 290

Enjoyed this book?

THANKS FOR JOINING ME IN THIS exploration of these Evolutionary Rules for Intelligent Species Survival. If you enjoyed the book, a review would be much appreciated as it will help other readers discover these Survival Patterns.

There is much more to this story of what drives Homo sapiens survival versus extinction, and if you would like to know when the next book in The Survival Series is released you can sign-up for my reader's list here:

http://samuellayne.com/sign-up/

Lightning Source UK Ltd.
Milton Keynes UK
UKHW012326211220
375668UK00004B/75/J